# Data-Centric Systems and Applications

More information about this series at
http://www.springer.com/series/5258

Seppo Sippu • Eljas Soisalon-Soininen

# Transaction Processing

## Management of the Logical Database and its Underlying Physical Structure

 Springer

Seppo Sippu
University of Helsinki
Finland
and
Aalto University
Finland

Eljas Soisalon-Soininen
Department of Computer Science
and Engineering
Aalto University School of Science
Finland

ISBN 978-3-319-35357-9      ISBN 978-3-319-12292-2 (eBook)
DOI 10.1007/978-3-319-12292-2
Springer Cham Heidelberg New York Dordrecht London

# Preface

## The Concept

The processing of transactions on databases is a well-established area, and many of its foundations were already laid in the late 1970s and early 1980s, most notably in a series of articles by the architects of the System R relational database management system. Most of the textbooks on database management contain chapters on transaction processing, concurrency control, and recovery from failures. Textbooks devoted solely to transaction processing include the classic works by Bernstein et al. [1987] and Weikum and Vossen [2002] on the theoretical side and that by Gray and Reuter [1993] on the implementation side.

Transactions are a concept related to the logical database as seen from the point of view of a database application programmer. A transaction is a sequence of actions on the logical database that, if declared as committed, should have its effects recorded permanently in the database and otherwise have no permanent effect on the logical database. The database system must be able to ensure that updates by committed transactions persist and that updates by uncommitted transactions are rolled back. System R also introduced the important database-programming paradigm of partial rollbacks, with which a transaction can roll back only the most recently executed actions, back to a preset savepoint, and then continue as a forward-rolling transaction.

The actions of transactions are executed by concurrently running server-process threads on pages in the underlying physical database, permanently stored on disk but parts of it buffered in main memory during access. A delicate interplay is needed between the management of transactions at the logical database level and the management of page accesses at the physical database level. In many cases the keys of the data items to be accessed by a transaction are determined only after accessing and latching the pages that hold them, after which appropriate logical key-range locks can be acquired to protect the logical action against forbidden accesses by other concurrently running transactions.

Index structures such as B-trees are used to accelerate key-based access to data pages that store the data items of the logical database. The consistency and balance of the indexes must be maintained under insertions and deletions by concurrent multi-action transactions as well as in the event of process failures and system crashes. The management of recoverable index structure modifications, although documented in the research literature, is a topic usually ignored or treated inadequately in database textbooks.

## The Focus

Our aim in this textbook is to bridge the gap between the theory of transactions on the logical database and the implementation of the actions on the underlying physical database. In our model, the logical database is composed of a dynamically changing set of data items (tuples) with unique keys, and the underlying physical database is a set of fixed-size data and index pages on disk, where the data pages hold the data items. We include total and partial rollbacks explicitly in our transaction model and assume that the database actions of transactions are tuple insertions, tuple deletions, tuple updates, and key-range scans on a relational database (or any database conforming to our model) and that server-process threads execute these actions on the data pages.

For efficient access to the tuples in the database, we usually assume that the logical database is stored in the leaf pages of a sparse primary B-tree index. Fine-grained (tuple-level) concurrency control is used to synchronize actions of transactions and short-duration latching for synchronizing page accesses, so that a database page can contain uncommitted updates by several concurrently running active transactions at the same time and uncommitted updates by an active transaction can migrate from one page to another due to structure modifications (page splits and merges) caused by concurrent transactions.

The B-tree is the most widely used index structure, because of its ability to accelerate key-range scans and to retain its balance under insertions and deletions. In addition to the traditional B-tree, we also consider index structures optimized for write-intensive workloads. Such workloads are more and more common in modern web services, and for such situations the basic update-in-place strategy of B-trees may not be efficient enough.

Unlike the abstract read-write model of transactions used in many textbooks, our logical database model more closely represents actions generated from SQL statements. This model allows for a rigorous treatment of isolation anomalies, such as unrepeatable reads (especially phantoms), and of fine-grained concurrency control protocols, such as key-range locking, that prevent such anomalies. As the basis of presentation of transactional isolation, we adopt the isolation-anomaly-based approach of Berenson et al. [1995], rather than the classical serializability theory, thus embracing also the cases in which transactions are run at isolation levels lower than serializability. The use of lower isolation levels such as "read committed" (which permits unrepeatable reads) may be essential for the performance of very-high-rate transaction processing when full isolation (serializability) is not absolutely necessary.

The treatment of transaction processing in this book builds on the "do-redo-undo" recovery paradigm used in the ARIES family of algorithms designed by Mohan et al. [1992a] and nowadays employed by virtually all database engines. With this paradigm, correct and efficient recovery from a system crash is achieved by repeating first the recent transaction history by redoing physically the missing updates by all transactions from the log records saved on disk and only then undoing (physically or logically) the updates by the transactions that were active at the time of the crash.

The standard physiological write-ahead logging (WAL) protocol and the steal-and-no-force buffering policy are assumed. The forward-rolling insert, delete, and update actions of transactions are logged with redo-undo log records and their undo actions with redo-only log records, both containing the page identifier of the updated page besides the logical tuple update to make possible physical redo and physical or logical undo. In this setting, with fine-grained concurrency control, correct recovery could not be achieved using the opposite "do-undo-redo" recovery paradigm in which the active transactions are first rolled back followed by the redoing of missing updates by committed transactions.

All the methods and algorithms presented in this book are designed carefully to be compatible with do-redo-undo recovery, write-ahead logging, steal-and-no-force buffering, and fine-grained concurrency control, even though in some cases we adopt an approach different from those previously proposed for ARIES-based algorithms. Most notably, in order to make the interplay between the logical and physical database levels as simple and rigorous as possible, we prefer redo-only structure modifications over redo-undo ones: we define each structure modification such as a page split or merge as a small atomic action that retains the structural consistency and balance of an initially consistent and balanced index.

## The Topics

Chapters 1–6 constitute the minimum of topics needed to fully appreciate transaction processing on a centralized database system within the context of our transaction model: ACID properties, fixing and latching of pages, database integrity, physiological write-ahead logging, LSNs (log sequence numbers), the commit protocol, steal-and-no-force buffering, partial and total rollbacks, checkpoints, the analysis, redo and undo passes of ARIES recovery, performing and recovering structure modifications, isolation anomalies, SQL isolation levels, concurrency control by key-range locking, and prevention of deadlocks caused by interplay of logical locks and physical latches. In these chapters, we do not yet give algorithms for index management but only explain the principle of atomically executed redo-only structure modifications that commit independently of the triggering transactions.

Chapters 7 and 8 include detailed deadlock-free algorithms for reading, inserting, and deleting tuples in a relation indexed by a sparse B-tree, under the key-range locking protocol, and for performing structure modifications such as page splits and merges triggered by tuple insertions and deletions. The B-tree variant considered is the ordinary one without sideways linkings. Saved paths are utilized so as to avoid repeated traversals. Together with the preceding six chapters, these chapters minimally cover the topic of the subtitle of the book: management of the logical database and the underlying physical structure.

The remaining chapters cover additional material that we consider important and is in line with the topics of the preceding chapters. In Chap. 9 we extend the tuple-wise key-range locking protocol to more general hierarchies of different granularities (multi-granular locking) and consider methods for avoiding deadlocks arising from lock upgrades (update-mode locking) and for avoiding the acquisition of many read locks. In Chap. 10 we extend our transaction model with bulk actions that read, insert, delete, or update a larger set of tuples and show how these actions are performed efficiently on a B-tree. In Chap. 11 we consider an application of bulk insertions: constructing a secondary index for a large relation while allowing the relation to be updated by transactions at the same time.

Several database management systems of today use transient versioning of data items in order to avoid read locks on items to be read. In Chap. 12 we review issues related to versioning, such as snapshot isolation and versioned B-trees. In Chap. 13 we consider the management of transactions that access data partitioned or replicated at several database servers in a distributed system. We also briefly discuss the requirements that have led to the design of systems called "key-value stores" or "NoSQL data stores" used to implement large-scale web applications. In Chap. 14 we consider a database system model, called the page server, in which transaction processing occurs at client machines, on cached pages fetched from the server.

Chapter 15 is devoted to issues related to the management of write-intensive transactions. In many Internet applications and embedded systems, relatively short updating transactions arrive in high frequency; in such systems, the database may reside in main memory, but log records are still written to disk for durability. Group commit is used to accelerate logging, and random disk writes are avoided by collecting updates into large bulks to be written with single large sequential writes onto a new location on disk.

## The Audience

This book is primarily intended as a text for advanced undergraduate or graduate courses on database management. A half-semester (6-week) course on transaction processing can be taught from Chaps. 1 to 6, possibly with selected topics from Chaps. 9 and 12 (such as basics of multi-granular locking and snapshot isolation), and the course can be extended to a full-semester (12-week) course by including all of Chaps. 7–15. The students are assumed to have basic knowledge of data structures and algorithms and of relational databases and SQL.

## Acknowledgements

This textbook has grown out from lecture notes on courses on database structures and algorithms, transaction processing, and distributed databases given by the authors for many years since the late 1990s through the 2010s at the Department of Computer Science, University of Helsinki, and at the Department of Computer Science and Engineering, Aalto University School of Science (formerly Helsinki University of Technology).

The lecture notes were originally written in Finnish. Earlier versions of the notes were commented and translated into English by Riku Saikkonen; much of his text still remains in this book. The LATEX figures were prepared by Aino Lindh. We also borrow somewhat from our articles written together with Tuukka Haapasalo, Ibrahim Jaluta, Jonas Lehtonen, Timo Lilja, and Riku Saikkonen. The contributions of Aino, Ibrahim, Jonas, Riku, Timo, and Tuukka are gratefully acknowledged. We also thank Satu Eloranta, Sami El-Mahgary, and Otto Nurmi for comments on the lecture notes.

The work was financially supported by the Academy of Finland.

Helsinki, Finland        Seppo Sippu
Espoo, Finland        Eljas Soisalon-Soininen
August 2014

# Contents

# Chapter 1
# Transactions on the Logical Database

The database as seen by an application programmer is called the *logical database*. In most cases the logical database is a relational database, so that application programs operate on tuples in relations through an SQL interface. A *transaction* is a sequence of read and update actions on the logical database, performed on the database upon a sequence of requests from an application. The action sequence constituting a transaction is atomic in the sense that either the effects of all the actions are recorded permanently in the database (in which case the transaction is committed) or none are (in which case the transaction is aborted and rolled back).

Integrity constraints specified by the database designer on the logical database determine which database states (i.e., database contents) are legal. Each transaction should be programmed in such a way that it retains the integrity of an initially integral logical database. Upon requests from applications, the actions included in a transaction are performed one by one on the database by the *database server* whose responsibility is to ensure that the logical database as well as the underlying physical database retain integrity under many concurrently executing transactions.

In this chapter we present the basic concepts related to transactions on a relational database. We model a transaction as an action sequence consisting of a forward-rolling phase possibly followed by a backward-rolling phase of undo actions, where the forward-rolling phase can also contain partial rollbacks. For ease of treatment, we define a primitive transaction model called the *key-range transaction model*, which we shall use throughout this book when discussing different issues of transaction processing.

## 1.1 Transaction Server

The database server of a typical relational database management system functions as a *transaction server*, also called a *query server* (or a *function server*). Such a server offers an interface through which client application processes can send requests to

© Springer International Publishing Switzerland 2014
S. Sippu, E. Soisalon-Soininen, *Transaction Processing*, Data-Centric Systems
and Applications, DOI 10.1007/978-3-319-12292-2_1

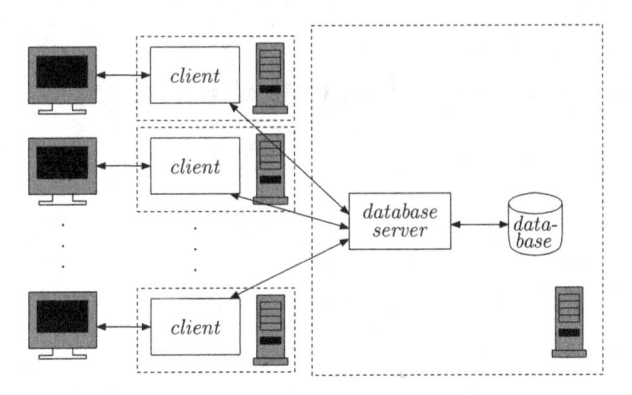

**Fig. 1.1** Typical two-tier transaction server

execute database actions (queries and updates) on the database stored at the server. The server processes the requests coming from the clients and returns the results to the clients. The requests are specified by SQL queries and update statements embedded in an application program or by a special database programming interface consisting of a collection of procedures and functions for communicating between an application program and the database server. Java Database Connectivity (JDBC) and Open Database Connectivity (ODBC) are examples of such interfaces.

The mode of operation of a transaction server is called *transaction shipping* or *query shipping* (or *function shipping*): a transaction (or a query or a function) is "shipped" from a client to the server, to be executed on the data stored at the server. Figure 1.1 shows a typical two-tier organization of many client machines connected to a database machine. The two "tiers" are the client tier and the database tier. For scalability, a large system with thousands of clients may be organized with additional tiers between the client and database tiers, such that a machine at a lower tier services only requests from a subset of machines at the next higher tier.

There are one or more *server processes* running at the server. Such a process is usually *multi-threaded*, that is, has several subprocesses or *threads* running concurrently in the same virtual address space. A typical server process services a number of clients, so that for each client application process there is a dedicated server-process thread.

A connection between a client application process and a server-process thread is created by the embedded SQL statement **connect to** *s*, where *s* is an identifier (network address) of the server. The statement creates a new server-process thread to service the application process or assigns an existing idle thread for the purpose.

## 1.2  Logical Database

A database application programmer sees the database as a *logical database*, a collection of data items defined and manipulated by the data definition and manipulation language offered by the database management system. In most cases the logical database conforms to the relational data model extended with features from the object data model, with SQL as the data definition and manipulation language.

The logical databases that we consider in this book are assumed to be purely relational, if not explicitly stated otherwise. Thus, the logical database is a collection of *relations* or *tables*, which are multisets of *tuples* or *rows*. A *multiset* or *bag* is an unordered set where the same element can occur multiple times. All tuples of a relation conform to the same schema, that is, are of the same type. A *relation schema* $r(Z)$ consists of the name $r$ of the relation and the names and types $Z$ of its attributes, along with possible integrity constraints defined on the relation (uniqueness of keys, referential integrity, etc.).

The operations of the relational model take relations as arguments and return a relation as a result. For example, the SQL statement

**delete from** $r$ **where** $A = a$

means the same as the assignment

$$r \leftarrow r \setminus \sigma_{A=a}(r),$$

where "$\setminus$" denotes the relational difference operation and "$\sigma$" denotes the relational selection operation, that is, $\sigma_{A=a}(r) = \{t \in r \mid t[A] = a\}$.

However, for the purpose of transaction management, this view of operations is too coarse grained. For one thing, as we will see shortly, we have to see a database transaction as a sequence of actions where each update action can be rolled back, that is, undone, if so desired. This requirement imposes a strong restriction on the complexity of actions: any single update action must be simple enough so that its undo action (inverse action) is easily defined and implemented. For another thing, modern database management requires that as much concurrency as possible be allowed among transactions active at the same time. This means that the granularity of data items used to synchronize transaction execution should be as small as possible. For example, we must allow several transactions to update a relation simultaneously, as long as the updates affect different tuples.

The usual solution (and the solution we adopt here) is to view the actions on the logical database as *tuple-wise actions*, so that each action always reads or updates (inserts or deletes) a single tuple only. Accordingly, in our view, the above **delete** statement is regarded as a sequence of tuple-wise actions:

delete $t_1$ from $r$; delete $t_2$ from $r$; ... ; delete $t_n$ from $r$,

where $\{t_1, t_2, \ldots, t_n\}$ is the bag of tuples with $A = a$. For the deletion of a single tuple $t$, it is easy to define its undo action: it is the insertion of the same tuple.

Thus, for the purpose of transaction management, we assume that the collection of logical database actions consists of tuple-wise actions such as the following:

1. Reading a tuple $t$ of relation $r$. The SQL query **select** $*$ **from** $r$ **where** $C$ generates a sequence of such read actions, one for each tuple satisfying the selection condition $C$.

2. Inserting a new tuple $t$ into relation $r$. Such an insert action is generated by the SQL statement **insert into** $r$ **values** $(t)$.
3. Deleting a tuple $t$ from relation $r$. The SQL statement **delete from** $r$ **where** $C$ generates a sequence of such delete actions, one for each tuple satisfying the selection condition $C$.
4. Updating the value of an attribute $A$ of a tuple $t$ in relation $r$. The SQL statement **update** $r$ **set** $A = e$ **where** $C$ generates a sequence of such update actions (if $e$ is a constant) or a sequence of pairs of a read action followed by an update action (if $e$ contains attributes whose values thus are read in computing the value of $e$), one for each tuple satisfying condition $C$.
5. Creating a new relation $r(Z)$ in the database. The input to the action is a relation schema, and the action is possible when there is no relation named $r$ already in the database. The action creates a new, empty relation on which other actions can subsequently be performed. This action is generated by the SQL statement **create table** $r(Z)$, where $Z$ is the SQL definition of the attributes and integrity constraints of the schema.
6. Deleting an empty relation $r$ from the database. After this action, the only action that can be applied to $r$ is action (5). The SQL statement **drop table** $r$ generates a sequence of tuple-deletion actions (3), one for each tuple in $r$, followed by the deletion of $r$.

Logical database actions also include certain *transaction-control actions* needed for managing transactions: beginning a new transaction, committing a transaction, aborting a transaction (i.e., starting rollback), and completing a rollback. The exact set of logical actions used in this book is fixed later, when the transaction model (the key-range model) is defined.

## 1.3  Integrity of the Logical Database

As is customary with treatments of relational database management, we may use the terms "database" and "relation" with two meanings: on the one hand, a logical database (resp. a relation) may mean a *database value* (resp. *relation value*), that is, some fixed contents of the database (resp. relation) as a bag of tuples, and on the other hand, it may mean a *database variable* (resp. *relation variable*), that is, a variable that can take a database (resp. relation) as its value. The context should make it clear which of the two meanings is assumed. Sometimes, when we wish to emphasize the distinction between the two meanings, we may talk about *database states* when referring to database values.

For each logical database and for each of its relations, considered with the latter meaning (as a variable), there is a set of associated *integrity constraints*, which have been specified at the time of creating the database or relation (with the SQL **create table** statement) or added afterwards (with the SQL **alter table** statement). An integrity constraint may be internal to a single relation such as a *key constraint* (**primary key, unique**) or it may span over two relations such as a *referential integrity constraint* (**foreign key**).

Integrity constraints restrict the values that the database and relation variables are allowed to take. A logical database (i.e., its state) is *integral* or *consistent* if it fulfills the integrity constraints specified for the database.

An attempt to perform an update action that violates an integrity constraint specified by SQL on the database either returns with an error indication, or triggers a sequence of corrective actions so as to make the violated constraint to hold again if such corrective actions have been specified. In addition to constraints that can be specified by SQL **create table** and **alter table** statements, the logical database usually satisfies some application-specific constraints. Applications using the database must check for these and keep them satisfied.

As a running example of a relation, we use

$$r(\underline{X}, V),$$

where $X$ and $V$ are sequences of attributes. The underlining of $X$ means that we have specified a key constraint on the relation: a relation that satisfies this constraint cannot contain two distinct tuples $t$ and $u$ with $t[X] = u[X]$. The attribute sequence $X$ is called the *primary key*, or *key* for short, of the relation.


## 1.4  Transactions

When a database application process is running, it generates a sequence of requests to perform SQL queries and update operations on the database stored at the server. The server-process thread allocated to service the application process parses, optimizes, and executes the queries and update statements in the order in which they arrive from the application. From the view of the logical database, the execution of a query or an update statement is a sequence of tuple-wise actions on the relations in the database.

The sequence of SQL requests coming from the application process is divided into subsequences by issuing occasionally an SQL **commit** or **rollback** request. The sequence of tuple-wise actions on the logical database resulting from one such subsequence of requests, that is, one that extends from the action immediately following a **commit/rollback** request (or from the first action of the application) to the next **commit/rollback** request, is called a transaction.

A transaction is an action sequence that the database application programmer wants to see as forming an *atomic* (i.e., indivisible) unit of work: either (1) all changes to the logical database produced by the transaction are done or (2) none of the changes appear in the logical database. Requirement (1) must hold in the case of a *committed transaction*, obtained by terminating the transaction with a **commit** request, while requirement (2) must hold in the case of an *aborted transaction*, obtained by terminating the transaction with a **rollback** request. Requirement (2) is also the only option in the case in which the database management system is unable to commit a transaction due to a process failure or a system crash occurring before or during servicing a **commit** request.

*Example 1.1* Assume that in the relation $r(\underline{X}, V)$, $V$ is a single numeric-valued attribute. The following fragment of an application program written in embedded SQL generates a transaction that doubles the $V$ value of all tuples in $r$:

> **exec sql update** $r$ **set** $V = 2 * V$;
> **exec sql select sum**($V$) **into** :new_sum **from** $r$;
> **exec sql commit**.

The application process sends to the server three requests, one by one, waiting for one request to be serviced before sending the next:

1. A request to execute the **update** statement
2. A request to execute the **select** query
3. A request to commit the transaction

Assuming that the contents of relation $r$ are initially the set of tuples $\{(x_1, v_1), \ldots, (x_n, v_n)\}$, request 1 results in performing the following action sequence at the server, where $B$ denotes the action of beginning a new transaction, $R[x, v]$ the action of reading tuple $(x, v)$, and $W[x, u, v]$ the action of changing tuple $(x, u)$ to $(x, v)$:

$$BR[x_1, v_1]W[x_1, v_1, 2v_1]R[x_2, v_2]W[x_2, v_2, 2v_2] \ldots R[x_n, v_n]W[x_n, v_n, 2v_n].$$

Because the semantics of the **update** statement does not specify the order in which $r$'s tuples are processed, the system selects the most efficient order, probably the order in which the tuples are physically stored in $r$'s file. After performing the update, the server returns to the application an indication of successful completion. Then request 2 is sent, resulting in the following action sequence:

$$R[x_1, 2v_1]R[x_2, 2v_2] \ldots R[x_n, 2v_n].$$

The computed sum is returned to the application, which assigns it to the program variable new_sum. Finally, request 3 is sent, resulting in the action

$$C,$$

denoting the commit of the transaction, assuming that no failures occur.     □

In addition to atomicity, *durability* is required for all committed transactions. This means that the changes produced by a committed transaction need to actually happen and stay in effect in the logical database, even in the presence of process failures and system failures occurring after the transaction has successfully committed. The only way to undo updates produced by a committed transaction is to program a new transaction (or several new transactions) for effectively compensating for the updates.

In the above example, if the "commit the transaction" action is finished successfully, so that the transaction actually commits, all of the updates in the $V$ attribute of

$r$'s tuples need to stay in effect. If, however, the transaction does not commit (e.g., because of a system failure), then all of the $V$ attributes have to retain their original values, so that none of them must be multiplied by two. To accomplish this, the database management system must be able to undo any updates that have already been done before the abort of the transaction.

## 1.5 Transaction States

Formally, we define a *transaction* as a pair $(T, \alpha)$, where $T$ is the identifier and $\alpha$ is the state of the transaction. The *transaction identifier* is a serial number or timestamp that uniquely identifies a transaction over a long period of time when the database system has been in use. More specifically, as will be evident later, transaction identifiers have to be unique over the set of transactions that either are currently active (i.e., have not yet committed or rolled back) or have committed or rolled back but still have traces (i.e., log records) retained in the available log files.

The *transaction state*, $\alpha$, is the sequence of actions performed for the transaction thus far. Each action in $\alpha$ includes the values for the input arguments with which the action was performed on the logical database and the values of the output arguments returned. For example, the transaction state

$BR[x_1, v_1]W[x_1, v_1, 2v_1]$

represents the result of performing the first three actions of the transaction given in Example 1.1. The arguments $x_1$ and $v_1$ are constants; $R[x_1, v_1]$ represents a read action that, when given input $x_1$, retrieved from the database the tuple $(x_1, v_1)$l; and $W[x_1, v_1, 2v_1]$ represents an update action that, when given input $x_1$, returned the value $v_1$ of the tuple with key $x_1$ and replaced the value in the tuple by the value $2v_1$.

We use the symbol $T$ (or subscripted, $T_i$) to denote both the transaction identifier and the transaction $(T, \alpha)$ as a whole; accordingly, we may say that transaction $T$ is at state $\alpha$. Also, when the state of the transaction is the interesting part, we may even talk about "transaction $\alpha$."

We distinguish between four different types of transaction states: (1) forward-rolling, (2) committed, (3) backward-rolling, and (4) rolled back. Every transaction starts as a *forward-rolling transaction*; in general, the state of such a transaction is an action sequence of the form

$$B\alpha, \tag{1.1}$$

where $B$ is the *begin-transaction* action and $\alpha$ is a sequence of read actions and normal (forward-rolling) update actions. The action sequence $\alpha$ forms the *forward-rolling phase* of the transaction.

The begin-transaction action can be thought of as an action that is needed to introduce a new transaction into the system, including the generation of a new transaction identifier. We assume that with database interfaces such as (embedded)

SQL that have no explicit request for starting a transaction, the first read or update action triggers the begin-transaction action as the first action of the transaction.

A forward-rolling transaction can be continued with read and update actions until a commit or an abort action is performed. The state of a *committed transaction* is an action sequence of the form

$$B\alpha C, \tag{1.2}$$

where $B$ and $\alpha$ are as in a forward-rolling transaction and $C$ is the *commit-transaction* action, the result of a successfully processed **commit** request. A committed transaction cannot be continued with any more actions.

The state of a *backward-rolling transaction* is an action sequence of the form

$$B\alpha\beta A\beta^{-1}, \tag{1.3}$$

where $B\alpha\beta$ is the state of a forward-rolling transaction, $A$ is the *abort-transaction* action, and $\beta^{-1}$ is a sequence of undo actions for the suffix $\beta$ of the forward-rolling action sequence $\alpha\beta$. The action sequence $\beta^{-1}$ forms the *backward-rolling phase* of the transaction. Such a transaction has rolled back the forward-rolling update actions in $\beta$; the update actions in $\alpha$ are still to be undone.

The abort action can be seen as marking the start of the service of an SQL **rollback** request or the start of the rollback of a transaction aborted due to an outside event such as a system failure. The service of the **rollback** request also includes performing the backward-rolling phase of the transaction back to the undo action for the first forward-rolling update action, after which it marks the transactions as rolled back.

The *undo sequence* for a forward-rolling action sequence $\beta$, denoted $\beta^{-1}$ or $undo(\beta)$, consists of undo actions for the sequence of update actions $o_1 o_2 \ldots o_n$ contained in $\beta$, in reverse order:

$$undo(\beta) = \beta^{-1} = o_n^{-1} o_{n-1}^{-1} \ldots o_1^{-1} = undo(o_n)undo(o_{n-1}) \ldots undo(o_1).$$

The *undo action* or *inverse action* for an update action $o$ is defined separately for each action:

1. $undo$(insert tuple $t$ into $r$) = delete $t$ from $r$.
2. $undo$(delete tuple $t$ from $r$) = insert $t$ into $r$.
3. $undo$(change the value of attribute $A$ in tuple $t$ of $r$ from $u$ to $v$) = restore the value of attribute $A$ in tuple $t$ of $r$ to $u$.
4. $undo$(create relation $r(Z)$ in the database) = drop empty relation $r(Z)$ from the database.
5. $undo$(drop empty relation $r(Z)$ from the database) = create relation $r(Z)$ in the database.

Unlike a forward-rolling transaction, a backward-rolling transaction cannot be continued arbitrarily. The next action to be performed is always uniquely defined:

the undo action for the last forward-rolling update action still undone is the one to be performed next. Thus, any backward-rolling transaction will eventually enter in a state in which all its forward-rolling updates have been undone; then the transaction is recorded to have rolled back.

The state of a *rolled back transaction* is an action sequence of the form

$$B\alpha A\alpha^{-1}C, \tag{1.4}$$

where $B\alpha A\alpha^{-1}$ is the state of a backward-rolling transaction and the $C$ action marks the transaction as being rolled back. Thus, a rolled back transaction has, as a result of its backward-rolling phase $\alpha^{-1}$, rolled back all of its forward-rolling updates.

The definition of a rolled back transaction implies that such a transaction has no permanent effect on the state of the logical database: whatever updates it does in its forward-rolling phase are all undone in the backward-rolling phase. As will be seen later, the physical database that stores the logical database, however, is not necessarily restored into its original state by the undo actions. Note, for example, that undoing a tuple deletion need not bring the tuple back to its original data page but may insert it to some other page allocated to the same relation.

We use the same action name, $C$, to mark the end of both committed and rolled back transactions. This is because from the point of view of transaction processing, the same procedure is performed in both cases, only the result indication to be returned to the application being different.

A committed or rolled back transaction is said to be *terminated*. A terminated transaction cannot be further advanced with any action. A forward-rolling or backward-rolling transaction is *active*. A backward-rolling or rolled back transaction is *aborted*.

*Example 1.2* Let us change the embedded SQL fragment of Example 1.1 so that instead of committing the transaction, it is aborted and rolled back:

**exec sql update** $r$ **set** $V = 2 * V$;
**exec sql select sum**($V$) **into** :new_sum **from** $r$;
**exec sql rollback**.

The following action sequence is generated:

$BR[x_1, v_1]W[x_1, v_1, 2v_1]R[x_2, v_2]W[x_2, v_2, 2v_2]\ldots R[x_n, v_n]W[x_n, v_n, 2v_n]$
$R[x_1, 2v_1]R[x_2, 2v_2]\ldots R[x_n, 2v_n]$
$AW^{-1}[x_n, v_n, 2v_n]\ldots W^{-1}[x_2, v_2, 2v_2]W^{-1}[x_1, v_1, 2v_1]C.$

Here an undo action $W^{-1}[x_i, v_i, 2v_i]$ restores the previous $V$ value $v_i$ of the tuple with key $x_i$.

Should instead a system failure occur during performing the **update** statement, the following transaction is generated, assuming that the update on tuple $(x_i, v_i)$ is the last that is found recorded (in the log saved on disk) at the time the system recovers from the failure:

$BR[x_1, v_1]W[x_1, v_1, 2v_1]R[x_2, v_2]W[x_2, v_2, 2v_2]\ldots R[x_i, v_i]W[x_i, v_i, 2v_i]$
$AW^{-1}[x_i, v_i, 2v_i]\ldots W^{-1}[x_2, v_2, 2v_2]W^{-1}[x_1, v_1, 2v_1]C.$

We conclude that in both cases, the logical database is restored to the state in which it was before the transaction started.

We leave as an exercise to figure out what should be done if a system failure occurs during performing the **rollback** statement.                                     □


## 1.6   ACID Properties of Transactions

The implementor of a database application must take care that each committed transaction $T$ produced by the application process is *logically consistent*: when run alone without outside disturbances or failures on a logically consistent database, $T$ keeps the database consistent.

Naturally, it has to be possible to program the transactions without taking care of the physical structure of the database, other concurrent transactions, or system failures. The function of the database management system is to ensure that both logical and physical consistencies are preserved when there are multiple logically consistent transactions running at the same time and system failures can occur.

A rolled back transaction is always trivially logically consistent, regardless of what it does in its forward-rolling phase and what the integrity constraints of the database are: in any case, the backward-rolling phase undoes any updates done in the forward-rolling phase.

A forward-rolling (and thus uncommitted) transaction does not need to be logically consistent. In some cases it may be difficult or even impossible to retain a referential integrity constraint in effect between two update actions; in such cases it is meaningful to require integrity checking to be enforced only at transaction commit.

In general, transactions are required to satisfy four named properties, called the ACID *properties*, of which we have already defined the following three:

1. *Atomicity*: all the updates performed by a committed transaction appear in the database, and all the updates performed by an aborted transaction are rolled back.
2. *Consistency*: committed transactions retain the consistency of an initially consistent logical database.
3. *Durability*: all the updates performed by a committed transaction remain permanently in the database.

The fourth ACID property is:

4. *Isolation*: each transaction has—more or less—the impression that it is running alone, without any other transactions interfering. More specifically, isolation means restricting how early updates by a transaction become visible to other transactions and how early values read by a transaction can be overwritten by other transactions. When transactions are run in full isolation, an update by a transaction becomes visible to, or a value read by a transaction is overwritten by, another transaction only when the first transaction has committed; then it seems

that all the transactions are executed serially, one transaction at a time, although in reality some transactions are running concurrently.

Two of the above properties, namely, atomicity and durability, are maintained solely by the database management system, according to the transaction boundaries, that is, the placement of the **commit** or **rollback** statements, as designated by the application programmer. The other two properties are maintained partly by the application programmer and partly by the database management system.

For consistency, the application programmer must ensure that each of her committed transactions, when run alone in a failure-free environment on a consistent logical database, retains the consistency of the database, while the system must ensure that the consistency of the logical database is also preserved in the presence of multiple concurrently running transactions, to the extent possible under the isolation levels set for the transactions. If all the updating transactions are set to run at the highest isolation level (i.e., serializable), the system is expected to provide full preservation of the consistency of the logical database.

For isolation, the application programmer must ensure that each of her transactions is designated to run at an isolation level high enough so that database consistency is preserved to a sufficient extent, while the system must ensure that the designated isolation level indeed is achieved in the presence of concurrent transactions.

*Example 1.3* Assuming the primary keys in $r(\underline{X}, V)$ are integers, the following embedded SQL program fragment reads the maximum key $x$ from $r$ and inserts into $r$ a new tuple with key $x + 1$:

**exec sql select max($X$) into :x from $r$;**
**exec sql insert into $r$ values(: x $+ 1, v'$);**
**exec sql commit.**

When run alone without interference from other transactions and in the absence of failures, the program fragment generates a committed transaction of the form

$$B \dots R[x, v] \dots I[x + 1, v']C,$$

where the action $R[x, v]$ retrieves the tuple $(x, v)$ with the maximum key $x$ from $r$ and the action $I[x + 1, v']$ inserts the tuple $(x + 1, v')$. (To determine the maximum key $x$, also some other read actions may be needed.)

The above transaction is obviously logically consistent: when run alone, it retains the consistency of an initially consistent logical database. The designer of the transaction (the application programmer) is allowed to assume that no other transaction can change the maximum key between the executions of the read action $R[x, v]$ and the insert action $I[x + 1, v']$.

A violation of the primary-key constraint would occur if some other transaction inserted a tuple with key $x + 1$. This other transaction could, for example, be generated from another instance of the same application, and both transactions may read the same maximum key $x$ and hence try to insert a tuple with the key $x + 1$.

It is the responsibility of the application programmer to designate the transaction to be run at a level of isolation that disallows the occurrence of such a violation of database integrity, while the concurrency-control mechanism of the database management system has to ensure that such an isolation level is indeed achieved.

The system has also to ensure that if the transaction indeed commits, the insertion of the new tuple is recorded permanently in the database and that otherwise if the transaction after all does not commit, the transaction is aborted and rolled back, undoing the insertion (if needed).                                                            □

Allowing transactions to be run at an isolation level lower than full isolation usually means more efficient transaction processing, because of the increased concurrency between transactions. However, with lower than full isolation, the execution of transactions may not correspond to any of their serial executions, being thus incorrect with respect to the usual notion of correctness. In many cases this is acceptable (see Sects. 5.5 and 9.5). Additionally, however, as shown in Chap. 5, if a transaction is run at a low isolation level, it may not be possible for the system to maintain logical consistency. For example, it would be highly risky to run the transaction of Example 1.3 at any isolation level lower than full isolation (serializable).

The program steps between two transaction boundaries in a database application program can be viewed as a mapping from a vector of input values to a vector of output values, where the input values are relations, constants, and values of program variables given as input to database actions, and the output values are results returned by those actions. The output values include both relations updated in update actions and the values returned by read actions.

It is in line with the above definition of the ACID property "C" and the responsibility laid on the programmer in this respect to regard such a mapping as *correctly programmed* so that an input vector is correctly mapped to an output vector when the program is run alone on a consistent logical database in the absence of failures. Again, serial executions are guaranteed to retain correctness, while correctness may be lost in nonserial executions.

*Example 1.4* Assuming the $V$ values in $r(\underline{X}, V)$ are numeric, the following embedded SQL program fragment generates a transaction, $T_1$, that scans the relation $r$ and computes the sum of the $V$ values of all tuples in $r$:

**exec sql select sum($V$) into** :s **from** $r$;
**exec sql commit**.

The mapping defined by this program fragment maps the set of $V$ values retrieved from $r$ to the sum stored into the program variable s.

Another transaction, $T_2$, generated from the following program fragment is run concurrently with $T_1$:

**exec sql insert into** $r$ **values** $(x, u)$;
**exec sql insert into** $r$ **values** $(y, v)$;
**exec sql commit**.

The mapping defined by this program fragment maps the constants $x$, $u$, $y$, $v$ and the current $r$ to the relation obtained from $r$ by inserting the tuples $(x, u)$ and $(y, v)$.

In serial execution, the sum computed by $T_1$ includes both of the values $u$ and $v$ if $T_2$ is executed first and includes neither of them otherwise. Both executions must be regarded as correct.

However, if $T_1$ is run at an isolation level lower than full isolation (serializable), the sum computed by $T_1$ may include only one of the values $u$ and $v$. For example, if $x < y$ and $T_1$ scans $r$ in ascending key order, it may happen that $T_2$ inserts $(x, u)$ after $T_1$ has already scanned a key greater than $x$, but inserts $(y, v)$ and commits before $T_1$ has scanned the greatest key less than $y$. In this case $v$ is included in the sum, but $u$ is not.

This phenomenon, called the *phantom phenomenon*, is possible if $T_1$ is run at an isolation level that permits an isolation anomaly called "unrepeatable read." It is up to the application programmer to decide whether or not that is acceptable. □

The ACID properties stated above only pertain to the logical database. The database management system is solely responsible for maintaining *physical consistency*, that is, integrity of the underlying physical database, regardless of whether or not logical consistency is maintained. This means, for instance, that a B-tree index structure for a relation is maintained in a consistent and balanced state even in the presence of logically inconsistent transactions and process failures and system crashes.

## 1.7 The Read-Write Model

In order to study transaction management in more detail, we need to define a database and transaction model that specifies the actions that are used in transactions. For the sake of simplicity, we assume that our logical database consists of only one relation with the scheme $r(\underline{X}, V)$. The tuples of the relation are pairs $(x, v)$, where $x$ is the unique *key* of the record and $v$ is the *value* of the tuple (i.e., the values of the other attributes in the tuple). Transactions always operate on $r$'s tuples using the key $x$.

In the simplest transaction model, called the *read-write model*, a transaction on the database $r$ can contain, besides the begin-transaction action $B$, the abort-transaction action $A$, and the commit-transaction (or complete-rollback) action $C$, the following two types of forward-rolling database actions:

1. *Read actions* of the form

$$R[x, v] \tag{1.5}$$

for reading the tuple $(x, v)$ with key $x$. The key $x$ is an input parameter for the action. The action fetches the unique tuple $(x, v)$ with key $x$ from $r$. If the tuple is not found, the action fails. A shorthand notation for the action is $R[x]$.

2. *Write actions* of the form

$$W[x, u, v] \tag{1.6}$$

for updating the value of the tuple with key $x$. The key $x$ and the new value $v$ are input parameters, and the old value $u$ is an output parameter. The action replaces the value of the tuple with key $x$ in the relation; the former value $u$ will be replaced by $v$. If the tuple does not exist, the action fails. Shorthand notations are $W[x, v]$ and $W[x]$.

*Example 1.5* The action sequence

$$BR[x, u] R[y, v] W[z, w, u + v]$$

is a forward-rolling transaction that reads the values $u$ and $v$ of the tuples with keys $x$ and $y$. Then, in the tuple with key $z$, the transaction replaces the previous value $w$ by the sum $u + v$. When augmented with the commit action $C$, this sequence becomes a committed transaction:

$$BR[x, u] R[y, v] W[z, w, u + v] C.$$

$\square$

In the backward-rolling phase of an aborted transaction or in a partial rollback to a savepoint, the *undo actions* of the forward-rolling write actions are performed in reverse order.

3. For the write action $W[x, u, v]$, the *undo-write action*

$$W^{-1}[x, u, v] = undo\text{-}W[x, u, v] \tag{1.7}$$

restores the value $u$, that is, its effect on the logical database is that of $W[x, v, u]$.

*Example 1.6* The action sequence

$$BR[x, u] R[y, v] W[z, w, u + v] A W^{-1}[z, w, u + v]$$

is a backward-rolling aborted transaction, which has undone its only write action. When augmented with the action $C$, this sequence becomes a rolled-back aborted transaction:

$$BR[x, u] R[y, v] W[z, w, u + v] A W^{-1}[z, w, u + v] C.$$

$\square$

Insertions and deletions of tuples cannot be represented in the read-write model. When transaction management based on the read-write model is examined in the literature, the targets of the actions are not tuples but abstract uninterpreted "data items," which are not precisely defined. If the sets of tuples of the logical database, for instance, the relations of a relational database, are selected as the data items, it is possible to model insertions and deletions of records by write actions on the set (relation). Similarly, if the pages of the physical database where the tuples are stored are selected as data items, insertions and deletions of tuples can be modeled as write actions on a page. In both cases, transaction management is rather coarse grained: the log must record changes to a whole relation or page, and the unit of synchronization for managing concurrency (the lockable unit) is a whole relation or page.

In modern database management systems, individual tuples are used in log records and in locking (as the most fine-grained units). In this case, when a transaction $T_1$ has updated a page and is still active, another transaction $T_2$ can update the same page. It is also allowed that, due to a structure modification (such as a page split in a B-tree) caused by $T_2$, the tuple that $T_1$ updated on page $p$ will be moved to another page $p'$ while $T_1$ is still active.

## 1.8   The Key-Range Model

In the transaction model used throughout this book and termed the *key-range model*, read actions return the least key in a given *key range*, and the update actions are insertions and deletions of tuples with a given key.

The set of key values for the tuples is assumed to be totally ordered. The ordering is denoted $\leq$ and its inverse relation $\geq$; the respective irreflexive ordering relations are referred to in the familiar way as $<$ and $>$. The least possible key value is $-\infty$, and the largest is $\infty$; we assume that these do not appear in any tuple in the actual database.

The key-range model is sufficient to model the most important principles that are used in physiological log-based recovery and tuple-level concurrency control (key-range locking of a unique key). The model can also be used to describe, in a natural manner, isolation anomalies (see Sect. 5.3) coming from concurrent key-range reads and insertions and deletions of individual tuples (for instance, the phantom phenomenon; see Sects. 1.6 and 5.3 and Example 1.9 below).

In the key-range model, a transaction on the database $r$ can contain, besides the begin-transaction action $B$, the abort-transaction action $A$ and the commit-transaction (or complete-rollback) action $C$, the following four types of forward-rolling database actions:

1. *Read-first actions* of the form

$$R[x, \geq z, v] \tag{1.8}$$

for reading the tuple $(x, v)$ with the first key value $x$ that satisfies the condition $x \geq z$. The key value $z$ is an input parameter, and the key $x$ and the value $v$ are output parameters. The action fetches the tuple $(x, v)$ whose key is the least key satisfying $x \geq z$ and $(x, v) \in r$. If no such tuple exists, $(\infty, 0)$ is returned. Shorthand notations for this action are $R[x, \geq z]$, $R[x, v]$, or $R[x]$.

2. *Read-next actions* of the form

$$R[x, >z, v] \tag{1.9}$$

for reading the tuple $(x, v)$ with the key value $x$ next to $z$. The key $z$, $-\infty \leq z <$ $\infty$, is an input parameter, and the key $x$ and the value $v$ are output parameters. The action fetches the tuple $(x, v)$ whose key is the least key satisfying $x > z$ and $(x, v) \in r$. If no such tuple exists, $(\infty, 0)$ is returned. Shorthand notations for this action are $R[x, >z]$, $R[x, v]$, or $R[x]$.

3. *Insert actions* of the form

$$I[x, v] \tag{1.10}$$

for inserting the tuple $(x, v)$ into $r$. Input parameters are the key $x$ and the value $v$. The action inserts the tuple $(x, v)$ into relation $r$. If $r$ already contains a tuple with key $x$, the action fails. A shorthand notation is $I[x]$.

4. *Delete actions* of the form

$$D[x, v] \tag{1.11}$$

for deleting the tuple $(x, v)$ with key $x$ from $r$. The key $x$ is an input parameter, and the value $v$ is an output parameter. The action deletes the tuple $(x, v)$ with key $x$ from relation $r$. If the tuple is not found, the action fails. A shorthand notation is $D[x]$.

For an insert or delete action $o[x]$, we define the *undo action* $o^{-1}[x]$ or *undo-o[x]* as follows:

5. The *undo-insert action*

$$I^{-1}[x, v] = undo\text{-}I[x, v] \tag{1.12}$$

undoes the action $I[x, v]$ by deleting the tuple $(x, v)$ from $r$.

6. The *undo-delete action*

$$D^{-1}[x, v] = undo\text{-}D[x, v] \tag{1.13}$$

undoes the action $D[x, v]$ by inserting the tuple $(x, v)$ into $r$.

*Example 1.7* The forward-rolling transaction

$$BR[x_1, \geq x', v_1] R[x_2, >x_1, v_2] R[x_3, >x_2, v_3] D[x, v] I[x, v_1 + v_2 + v_3]$$

reads three tuples with consecutive keys and replaces the value in the tuple with key $x$ by the sum of the values in the three tuples. This transaction is rolled back by first performing the abort action $A$ and then undoing the two updates and finally completing the rollback:

$BR[x_1, \geq x', v_1] R[x_2, >x_1, v_2] R[x_3, >x_2, v_3] D[x, v] I[x, v_1 + v_2 + v_3]$
$AI^{-1}[x, v_1 + v_2 + v_3] D^{-1}[x, v] C.$

The original value $v$ of the tuple with key $x$ is thus restored.                                □

Let $r$ be the logical database (a relation) and $o[\bar{x}]$ an action, $o \in \{B, R, W, I, D, C, A\}$ and $\bar{x}$ a sequence of constant arguments. We define when the action $o[\bar{x}]$ *can be run* on $r$ and what is the database (relation) $r'$ *produced* by the action—this is denoted $(r, r') \models o[\bar{x}]$.

1. $(r, r) \models o$, when $o \in \{B, C, A\}$.
2. $(r, r) \models R[x, \theta z, v]$, if $(x, v) \in r$ and $x$ is the least key in $r$ that satisfies $x \theta z$. Here $\theta$ is the operator $\geq$ or $>$.
3. $(r, r) \models R[\infty, \theta z, 0]$, if $r$ contains no tuple with key $x$ satisfying $x \theta z$.
4. $(r, r') \models W[x, u, v]$, if $r' = (r \setminus \{(x, u)\}) \cup \{(x, v)\}$.
5. $(r, r') \models I[x, v]$, if the key $x$ does not appear in $r$ and $r' = r \cup \{(x, v)\}$.
6. $(r, r') \models D[x, v]$, if $(x, v) \in r$ and $r' = r \setminus \{(x, v)\}$.
7. $(r, r') \models W^{-1}[x, u, v]$, if $(r', r) \models W[x, u, v]$.
8. $(r, r') \models I^{-1}[x, v]$, if $(r', r) \models I[x, v]$.
9. $(r, r') \models D^{-1}[x, v]$, if $(r', r) \models D[x, v]$.

To simulate an *exact-match read action*, we may write $R[x, \geq x, v]$. By (2), $(r, r) \models R[x, \geq x, v]$, if $(x, v) \in r$.

The action sequence $\alpha$ *can be run* on database $r$ and *produces* the database $r'$, denoted $(r, r') \models \alpha$, if either (1) $\alpha = \epsilon$ and $r' = r$ or (2) $\alpha$ is of the form $\beta o$, where the action sequence $\beta$ can be run on $r$ and produces a database $r''$ and action $o$ can be run on $r''$ and produces $r'$.

*Example 1.8* The transaction of Example 1.7 can be run on every database that contains the tuples $(x_1, v_1)$, $(x_2, v_2)$, $(x_3, v_3)$, and $(x, v)$ with $x' \leq x_1 < x_2 < x_3$ but no other tuples with keys in the range $[x', x_3]$. The forward-rolling portion of the transaction produces a database where the tuple $(x, v)$ has been changed to $(x, v_1 + v_2 + v_3)$ and the entire rolled back transaction restores the original database.
                                                                                                        □

*Example 1.9* The transactions of Example 1.4 can be modeled in the key-range model as follows:

$T_1 = BR[x_1, >-\infty, v_1] R[x_2, >x_1, v_2] \ldots R[x_n, >x_{n-1}, v_n] R[\infty, >x_n, 0] C,$
$T_2 = BI[x, u] I[y, v] C,$

where $\{(x_1, v_1), \ldots, (x_n, v_n)\}$ is the set of tuples in the relation $r(\underline{X}, V)$ and $x$ and $y$ are keys not found in $r$. The phantom phenomenon occurs if

$$x_i < x < x_{i+1} < y = x_j$$

for some $i$ and $j$, and the actions are executed in the following order (from left to right):

$T_1: \quad BR[x_1, v_1] \ldots R[x_{j-1}, v_{j-1}] \qquad\qquad\qquad R[x_j, v_j] \ldots R[\infty, 0] C$
$T_2: \qquad\qquad\qquad\qquad\qquad\qquad BI[x, u] I[y, v] C$

Thus, the first tuple, $(x, u)$, inserted by $T_2$ is not among the tuples read by $T_1$, while the second tuple, $(y, v) = (x_j, v_j)$, is.

We also note that the phantom phenomenon is not prevented by the simple locking protocol in which transactions obtain commit-duration shared locks (read locks) on the keys of tuples read and commit-duration exclusive locks (write locks) on the keys of tuples inserted, deleted, or updated.                                    □

## 1.9  Savepoints and Partial Rollbacks

The transaction model of SQL allows for *partial rollbacks* of transactions: a subsequence of the update actions performed by a forward-rolling transaction is rolled back without aborting and rolling back the entire transaction. After performing a partial rollback, the transaction remains in the forward-rolling phase and can thus perform any new forward-rolling actions.

The actions to be rolled back in a partial rollback constitute a sequence of actions from the latest update action by the transaction back to preset *savepoint*. Savepoints are set in the application program using the SQL statement

**set savepoint** $P$

where $P$ is a unique name for the savepoint. The SQL statement

**rollback to savepoint** $P$

executes a partial rollback to savepoint $P$: all forward-rolling update actions performed by the transaction after setting $P$ that are not yet undone are undone.

*Example 1.10*  Partial rollbacks can be nested (Fig. 1.2).

> **insert into** $r$ **values** $(x_1, v_1)$;
> **set savepoint** $P_1$;
> **insert into** $r$ **values** $(x_2, v_2)$;
> **set savepoint** $P_2$;
> **insert into** $r$ **values** $(x_3, v_3)$;
> **rollback to savepoint** $P_2$;
> **insert into** $r$ **values** $(x_4, v_4)$;
> **rollback to savepoint** $P_1$;
> **insert into** $r$ **values** $(x_5, v_5)$;
> **commit**.

The statement **rollback to savepoint** $P_2$ deletes from relation $r$ the inserted tuple $(x_3, v_3)$. The statement **rollback to savepoint** $P_1$ deletes from $r$ the inserted tuples $(x_4, v_4)$ and $(x_2, v_2)$. At the end of the transaction, an initially empty relation $r$ contains only the tuples $(x_1, v_1)$ and $(x_5, v_5)$.                                    □

We now add partial rollbacks to our transaction model. For that purpose, we define the following actions:

$B$
$\quad I[x_1, v_1]$
$\quad S[P_1]$
$\qquad I[x_2, v_2]$
$\qquad S[P_2]$
$\qquad\qquad I[x_3, v_3]$
$\qquad A[P_2]$
$\qquad\qquad\qquad I^{-1}[x_3, v_3]$
$\qquad C[P_2]$
$\qquad I[x_4, v_4]$
$\quad A[P_1]$
$\qquad\qquad I^{-1}[x_4, v_4]$
$\qquad\qquad I^{-1}[x_2, v_2]$
$\quad C[P_1]$
$\quad I[x_5, v_5]$
$C$

$\equiv$

$B$
$\quad I[x_1, v_1]$
$\quad S[P_1]$
$\qquad I[x_2, v_2]$
$\qquad I[x_4, v_4]$
$\quad A[P_1]$
$\qquad\qquad I^{-1}[x_4, v_4]$
$\qquad\qquad I^{-1}[x_2, v_2]$
$\quad C[P_1]$
$\quad I[x_5, v_5]$
$C$

$\equiv$

$B$
$\quad I[x_1, v_1]$
$\quad I[x_5, v_5]$
$C$

**Fig. 1.2** Nested partial rollbacks. Action $S[P]$ sets savepoint $P$, and partial rollback to $P$ is begun by action $A[P]$ and completed by action $C[P]$. Rolled-back segments of the transaction are shown *indented*. Because of the rollbacks, the overall effect on the logical database is the same as that of the transaction $BI[x_1, v_1]I[x_5, v_5]C$

7. $S[P]$: *set savepoint $P$.*
8. $A[P]$: *begin partial rollback to savepoint $P$.*
9. $C[P]$: *complete the partial rollback to savepoint $P$.*

Now in the forward-rolling phase $B\alpha$ of a transaction, string $\alpha$ may contain set-savepoint actions $S[P]$, completed rollbacks $S[P]\ldots A[P]\ldots C[P]$, and maybe one initialized but not yet completed partial rollback $S[P]\ldots A[P]\ldots$. Formally, the *forward-rolling phase* of a transaction can now be of any of the following three forms:

(a) A sequence $\alpha$ of actions $R$, $I$, $D$, $W$, and $S$
(b) An action sequence of form $\alpha S[P]\beta A[P]\beta^{-1}C[P]\gamma$, where $\alpha$, $\beta$, and $\gamma$ are of form (a) or (b)
(c) An action sequence of form $\alpha S[P]\beta\delta A[P]\delta^{-1}$, where $\alpha$, $\beta$, and $\delta$ are of form (a) or (b)

In case (b) the subsequence $S[P]\beta A[P]\beta^{-1}C[P]$ represents a completed partial rollback to savepoint $P$. In case (c) the subsequence $S[P]\beta\delta A[P]\delta^{-1}$ indicates that the transaction is rolling back to savepoint $P$.

The *undo sequence* for $\alpha$, denoted $\alpha^{-1}$ or *undo*$(\alpha)$, is now defined depending on its form: For a sequence $\alpha$ of form (a), the undo sequence $\alpha^{-1}$ is defined as before. For a sequence of form (b), the undo sequence is $\gamma^{-1}\alpha^{-1}$. For a sequence of form (c), the undo sequence is $\beta^{-1}C[P]\alpha^{-1}$.

Savepoints and partial rollbacks constitute an important database programming paradigm: transactions can be programmed freely to update the database immediately even if some subsequent event forces the update to be rolled back and another avenue to be followed so as to complete the transaction. In fact, with partial rollbacks, every transaction can be programmed to terminate with a **commit** request and never with a **rollback** request (i.e., total rollback). The effect of a total rollback can be achieved by setting a savepoint before the first update and then, at the end, by performing a partial rollback to that savepoint and committing the transaction.

## 1.10 Multiple Granularity

We may extend our key-range transaction model by adding *multiple granularity*, so that tuples can be grouped into relations. The relations of the database constitute a set that is totally ordered according to their identifiers. We denote this set by

$$\{(r_1, R_1), \ldots, (r_{n_i}, R_{n_i})\},$$

where $r_i$ is an identifier that uniquely identifies a relation in the database and $R_i$ is the relation schema ($= \underline{X}_i V_i$). Tuple $(x, v)$ in relation $r$ is then uniquely identified by the pair $(r, x)$.

The tuple-wise forward-rolling actions in the transaction model are now:

1. $R[r, x, \geq z, v]$: reading of the first tuple $(x, v)$ with $x \geq z$ from relation $r$.
2. $R[r, x, > z, v]$: reading of tuple $(x, v)$ next to $z$ from relation $r$.
3. $W[r, x, u, v]$: update of tuple $(x, u)$ in relation $r$.
4. $I[r, x, v]$: insertion of tuple $(x, v)$ into relation $r$.
5. $D[r, x, v]$: deletion of tuple $(x, v)$ from relation $r$.

New actions include:

(a) $R[r', \theta r, R']$: browsing the schema $R'$ of relation $r'$.
(b) $I[r, R]$: creation of a new relation $r(R)$ into the database, corresponding to the SQL statement **create table** $r(R)$.
(c) $D[r, R]$: deletion of an empty relation $r$ from the database, corresponding to the SQL statement **drop table** $r$ for an empty relation $r$.

Additional levels could be added to the granule hierarchy by grouping relations into databases (for different owners); new actions would then include ones corresponding to the SQL statements **create database** and **destroy database**.

## Problems

**1.1** The personnel database of an enterprise contains, among others, the relations created by the following SQL statements:

    **create table** *employee*(*empnr* **integer not null,**
        *name* **varchar(40) not null,** *address* **varchar(80) not null,**

*job* **varchar(20)**, *salary* **integer**, *deptnr* **integer not null**,
    **constraint** *pk* **primary key** (*empnr*),
    **constraint** *dfk* **foreign key** (*deptnr*) **references** *department*);
**create table** *department*(*deptnr* **integer not null**,
    *name* **varchar(20) not null**, *managernr* **integer**,
    **constraint** *pk* **primary key** (*deptnr*),
    **constraint** *efk* **foreign key** (*managernr*) **references** *employee*).

Consider the transaction on the database produced by the SQL program fragment:

**exec sql select max**(*empnr*) **into** *:e* **from** *employee*;
**exec sql select max**(*deptnr*) **into** *:d* **from** *department*;
**exec sql insert into** *department* **values** (*:d* + 1, 'Research', *:e*);
**exec sql insert into** *employee* **values** (*:e* + 1, 'Jones, Mary',
    'Sisselenkuja 2, Helsinki', 'research director', 3500, *:d* + 1);
**exec sql update** *department* **set** *managernr* = *:e* + 1
**where** *deptnr* = *:d* + 1;
**exec sql insert into** *employee* **values** (*:e* + 2, 'Smith, John',
    'Rouvienpolku 11, Helsinki', 'researcher', 2500, *:d* + 1);
**exec sql commit**.

(a) Give the string of tuple-wise actions (readings and insertions of single tuples) that constitutes the transaction. We assume that the tuples of the relations *employee* and *department* reside in the data pages in an arbitrary order and that there exist no index to the relations.

(b) Repeat (a) in the case in which there exists an ordered (B-tree or ISAM) index to the relation *employee* on attribute *empnr* and an ordered index to the relation *department* on attribute *deptnr*.

(c) Are the transactions created in (a) and (b) logically consistent? That is, do they preserve the integrity constraints of the database?

**1.2** The following SQL program fragments operate on relation $r(\underline{X}, V)$. Describe the transaction produced by the program fragments (1) in the read-write model of transactions and (2) in the key-range model of transactions.

(a) **update** $r$ **set** $V = V + 1$ **where** $X = x$;
    **update** $r$ **set** $V = V + 1$ **where** $X = y$; **commit**.

(b) **update** $r$ **set** $V = V + 1$ **where** $X = x$;
    **update** $r$ **set** $V = V + 1$ **where** $X = y$; **rollback**.

**1.3** Explain the meaning of the transaction

$$BI[r, x_1, v_1]S[P_1]I[r, x_2, v_2]S[P_2]I[r, x_3, v_3]A[P_2]$$
$$I^{-1}[r, x_3, v_3]C[P_2]I[r, x_4, v_4]A[P_1]$$
$$I^{-1}[r, x_4, v_4]I^{-1}[r, x_2, v_2]C[P_1]I[r, x_5, v_5]A$$
$$I^{-1}[r, x_5, v_5]I^{-1}[r, x_1, v_1]C.$$

Give SQL statements to generate this transaction.

**1.4** A banking database contains the relations

> *account*(*number*, *balance*),
> *holder*(*card number*, *account number*),
> *card*(*number*, *holder name*, *holder address*, *crypted password*),
> *transaction*(*site*, *date time*, *type*, *amount*, *account number*, *card number*),

where the relation *transaction* stores information about every completed or attempted withdrawal and deposit and about every balance lookup.

Give an embedded SQL program for a transaction for a withdrawal of $s$ euros using card $c$ with password $p$, where the withdrawal is allowed only if no overdraft will occur. The transaction also shows the balance that remains in the account after completing the withdrawal. We assume that the program includes the following statement:

**exec sql whenever sqlerror goto** $L$,

where $L$ is a program address to which the program control is transferred whenever an error status is returned by the execution of some SQL statement.

**1.5** Consider extending the read-write transaction model with an action $C[x]$, which *declares* the updates on $x$ as *committed*. This action can appear in the forward-rolling phase of the transaction, and it has the effect that even if the transaction eventually aborts and rolls back, the updates on $x$ done before $C[x]$ will not be undone. Accordingly, complete the transaction

$$BR[x, u]W[x, u, u']R[y, v]W[y, v, v']$$
$$C[x]R[z, w]W[z, w, w']W[y, v', v'']$$
$$C[y]W[x, u', u'']A$$

to a rolled back transaction. Consider situations in which this feature might be useful.

**1.6** Our key-range transaction model assumes that the tuples $(x, v)$ are only referenced via the unique primary keys $x$. Extend the model to include relations $r(\underline{X}, Y, V)$, where tuples $(x, y, v)$ can be referenced by either the primary key $x$ or by the (non-unique) secondary key $y$.

**1.7** We extend our key-range transaction model by a *cursor mechanism* that works as follows. When a transaction starts, it allocates a main-memory array (the *cursor*), private to the transaction, to store the tuples returned by all the read actions performed by the transaction, in the order of the actions. Each read action $R[x, v]$ appends the tuple $(x, v)$ to the next available entry in the cursor. In a more general setting, we would have several cursors, with special actions to open and close a cursor.

Now if the transaction performs a partial rollback to a savepoint, the contents of the cursor should be restored to the state that existed at the time the savepoint was set. Obviously, to that effect, we should define an undo action, $R^{-1}[x]$, for a read action $R[x]$. Elaborate this extension of our transaction model.

# Bibliographical Notes

The foundations of transaction-oriented database processing were laid by Eswaran et al. [1974, 1976] and Gray et al. [1976]. Transactions with partial rollbacks were already implemented in the System R relational database management system described by Astrahan et al. [1976], Gray et al. [1981], Blasgen et al. [1981, 1999], and Chamberlin et al. [1981]. The authors of System R state that the execution of a transaction should be atomic, durable, and consistent, where the last property is described as: "the transaction occurs as though it had executed on a system which sequentially executes only one transaction at a time," that is, in isolation [Gray et al.1981].

Gray [1980] presents a general formal model for transactions and their processing. Gray [1981] reviews the transaction concept and implementation approaches, analyzes the limitations of ordinary "flat transactions," and argues for the need of "nested transactions" as a means of modeling long-lived workflows. In their textbook on the implementation of transaction processing systems, Gray and Reuter [1993] also analyze thoroughly the transaction concept and review the history of the development of the transaction concept and transaction processing systems.

In many textbooks on database management and transaction processing, most notably in the classic works by Papadimitriou [1986] and Bernstein et al. [1987], the read-write transaction model (Sect. 1.7) was adopted as the basis for discussing transaction-oriented concepts. In this setting the database was considered to be a set of uninterpreted abstract data items $x$ that could be read (by action $R[x]$) and written (by action $W[x]$). Schek et al. [1993] consider the enhanced model in which the backward-rolling phase of an aborted transaction is represented explicitly as a string of undo actions; this feature is also included in the read-write model (also called the page model) used in the textbook by Weikum and Vossen [2002].

The key-range transaction model defined in this chapter and used previously by Sippu and Soisalon-Soininen [2001], Jaluta [2002], and Jaluta et al. [2003, 2005, 2006] was inspired by the model used by Mohan [1990a, 1996a] and Mohan and Levine [1992] to describe the ARIES/KVL and ARIES/IM algorithms. The transaction model considered by Gray et al. [1976] included the feature that a transaction can declare an update as committed before the transaction terminates, although this feature was only allowed for transactions run at the lowest isolation level (Problem 1.5).

# Chapter 2
# Operations on the Physical Database

The tuple collections of the logical database are stored in an underlying *physical database*, which consists of fixed-size database pages stored in non volatile random-access storage, usually on magnetic disk. For reading or updating tuples of the logical database, the pages that contain the tuples must be fetched from disk to the main-memory *buffer* of the database, from where the updated pages are later flushed back onto disk, replacing the old versions. The buffer-management component of the database management system takes care that frequently used database pages are kept available in the buffer as long as possible so as to reduce the need of expensive random reads and writes of the disk.

In this chapter we discuss issues of the physical database that are needed to understand the interplay between transaction management at the logical database level and the management of the underlying physical database structure under operations triggered by the logical database actions. These issues include the page-based organization of the database, the *fixing* of pages in the buffer for the time of accessing the page, integrity constraints on the physical database, the *latching* of pages for protecting the accesses against interference by other concurrent processes, and the need to implement *structure modifications* on the physical database as atomic operation sequences that commit independently of the triggering transactions.

## 2.1 Data Structures and Processes on the Server

Two *permanent (non volatile)* data collections are maintained on the transaction server:

1. *Data disks* containing the relations of the database, their indexes, and the system catalog or data dictionary.
2. *Log disks* containing the log records written for updates on the database.

© Springer International Publishing Switzerland 2014

S. Sippu, E. Soisalon-Soininen, *Transaction Processing*, Data-Centric Systems and Applications, DOI 10.1007/978-3-319-12292-2_2

When an instance of the database system is running, the following *volatile* data structures are maintained in the virtual memory shared by the server processes:

3. *Buffer pool* or *database buffer* for the buffering of database pages read from disk.
4. *Log buffer* for the buffering of the log file.
5. *Active-transaction table* for storing information about active transactions.
6. *Modified-page table* for storing information about buffered pages.
7. *Lock table* for storing the locks held by active transactions (when transactional concurrency control is based on locking).
8. Other volatile structures, such as a *query-plan cache* for storing (for reuse) compiled query execution plans.

The data disks and the database buffer are based on random access on database pages (via page identifiers), whereas the log disks and the log buffer are append-only sequential files. Each database update is performed on a page copied from the data disk into the database buffer, and the update is logged with a log record that is appended into the log buffer. The log record holds information that makes it possible to redo the update on an old version of the page (in the case when the update is lost due to a failure) and undo the update (at transaction rollback).

At transaction commit or when the log buffer is full, the contents of the log buffer are flushed onto the log disk, where the log records are appended next to the previously flushed log records. Thus every committed transaction has all its updates recorded permanently at least on the log disk.

The most important source of efficiency of transaction processing in a log-based system is that the random-access database residing on the data disks need not immediately reflect the updates of transactions: updated pages from the database buffer need not be flushed at transaction commit. For correctness of operation, an updated data page must not be flushed before flushing first the log records for the update.

In a database-system instance, several *database processes* operate on the shared database on disk and on the shared virtual-memory data structures:

1. *Server processes* and their threads (already mentioned above) that service requests from application processes and generate transactions
2. *Database-writer process* that flushes updated database pages from the buffer onto disk
3. *Log-writer process* that flushes log records from the log buffer onto the log disk when the log buffer becomes full or when a server process requests, at transaction commit, the log to be flushed onto disk
4. *Checkpoint process* that periodically takes a checkpoint in which certain volatile data is written onto disk
5. *Process-monitor process* that monitors other processes of the system and takes care of recovery of failed processes, including aborting and rolling back transactions that have terminated because of a failure of a server process
6. *Lock-manager process* that services requests to acquire or release locks on data items and takes care of detecting deadlocks

To reduce the number of messages between processes, in many systems the server processes lock data items by updating the shared lock table directly instead of sending requests to a dedicated lock-manager process. In that case the lock table must be protected by a mutual exclusion mechanism (semaphore) as any other virtual memory structures shared by many processes.

## 2.2 Database Pages and Files

The *physical database* is a collection of pages that contain records. A *page* is a fixed-size area on disk (e.g., 4 or 8 or 16 kilobytes), which occupies one disk block or, if the page is larger than a disk block, several consecutive disk blocks. A *disk block* is an area that consists of one or more consecutive sectors on one track of the surface of the disk. For example, if the size of a block is 8 kilobytes and the size of a sector is 512 bytes, the block consists of 16 consecutive sectors. A sector is the smallest addressable unit on the disk.

The operations of a physical database include *fetching* (reading) a page from disk to the database buffer in the main memory and *flushing* (writing) a page from the buffer to its location on disk. Both operations are performed in a single disk access, consisting of a seek of the correct track and the read or write of the sequence of consecutive disk blocks that constitute a page.

The logical database is implemented using the physical database by storing the tuples of the database in the pages. A page that contains tuples is called a *data page*. In addition to data pages, the physical database contains *index pages* for finding the data pages efficiently. The structure of the index pages is dependent on how the tuples are organized in the pages. For the purposes of database management, several other page types exist.

A tuple is normally stored as one record in a data page. If a tuple is larger than a page, it needs to be divided into several records which are chained together.

A page contains a header, the record area, and a record index. The *page header* contains at least the following information:

1. *Page id*: the unique identifier of the page.
2. *Page type*: "relation data page" (tuples from one relation), "cluster data page" (tuples from multiple relations), "index page" (part of an index to a relation), "free space" (an unallocated page), "data-dictionary page" (part of the data dictionary of the database), or "storage-map page" (page that describes the allocation status of other pages), etc.
3. *Internal identifier* for the relation, index, or other structure where the page belongs to.
4. *Record count*: the number of elements in the record index of this page.
5. *Free-space count*: information used to maintain areas of free space inside this page, such as the length of the longest contiguous free area and the total amount of free space on the page.

6. PAGE-LSN: The log sequence number (LSN) of the log record written for the latest
   update onto the page. When a buffered page is updated, the PAGE-LSN of the page
   is set to the LSN of the log record written for the update. In restart recovery,
   the PAGE-LSN in the disk version of a page is needed in determining how far
   the updates onto the page had reached at the time of the system failure (see
   Chap. 4). During normal processing, the PAGE-LSNs of the current versions of
   pages are utilized for accelerating undo actions (see Sect. 4.3) and traversals of
   index structures (see Chap. 7).
7. *Next page*: The page identifier of the "next page" in a linked structure of pages
   (depending on the graph structure).

The *record area* of a page contains the actual data for the page. In the case of a
data page, the contents are the tuples of the logical database that are placed on the
page. The record area is filled from beginning to end (starting after the header).

The *record index* of a page is an array $m$ at the end of the page. An element $m[i]$
of the array contains a byte offset pointing to the $i$th record in the record area. The
record index grows upward from the end: $m[1]$ is at the end of the page and $m[2]$
just before that. Figure 2.1 shows a tuple in page 924 at position $m[3]$.

The pages of a database are usually grouped into *files* or *segments*, each of which
consists of extents of one or more consecutive pages. Typically the grouping is based
on the types of the records. One relation in a relational database is often placed in
its own file. In some systems, the tuples of two or more relations that are related via
a foreign-key constraint can be stored in the same file—this organization is often
called a *cluster file*.

It is possible to map a file in a database management system directly to a file
in the file system provided by the operating system. Then, the database system can
use the services provided by the operating system, and the database management
software becomes smaller and usually simpler.

However, the file system does not always provide all the features that are
important for implementing database operations efficiently. A database management
system places stricter demands on the buffering of pages in the main memory than
a normal file system: the database management system has to ensure that, in certain

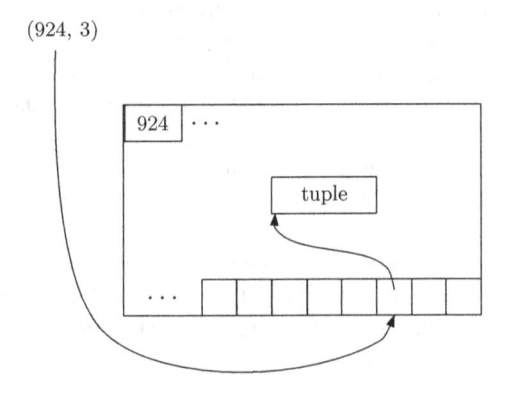

**Fig. 2.1** Referencing a tuple
in a data page

situations, a page is kept in main memory in the place that was reserved for it, or that a specific page can immediately be flushed from the buffer to disk.

It should also be possible for the database management system to make query evaluation more efficient by using read-ahead, that is, reading more than one page at a time from the disk to the buffer. This requires that clusters of multiple logically consecutive pages can be stored in physically consecutive areas on disk (i.e., on the same track). The file system provided by the operating system is not used in most database management systems. Instead, the areas of disk reserved for the database are handled directly by the database management system.

Every page in the database has a *page identifier* (or *page id*, for short) that uniquely identifies the page in the physical database. The page identifier contains the number of the file in the database and the number of the page inside the file. In a distributed database, a complete page identifier also includes an identifier for the site (node) in the system. The disk block that contains a page can be determined from the page identifier by using a file descriptor. The *file descriptor* points to the area reserved for the file on disk. For simplicity, we assume in this book that page identifiers are just page numbers.

The *record identifier* (or *record id*, *rid*) for the $i$th record in the record area of page $p$ is the pair $(p, i)$ where $p$ is the page identifier. The identifier $(p, i)$ of a tuple in record $i$ of a data page $p$ is also called a *tuple identifier* (or *tuple-id*, *tid*).

A lookup based on the tuple identifier is the fastest way to find a tuple:

1. Extract the page identifier $p$ from the tuple identifier $(p, i)$. Fix $p$ in the buffer; let $b$ be the address of the buffer frame containing $p$.
2. Fetch the tuple from the address $b + m[i]$.

The indirect addressing created by the record index $m$ enables the system to keep the same record identifier even when the record is moved inside the page.

In updating the record, it is possible that its length changes so that it no longer fits on the same page, and the record has to be moved to another page in the file. It is often advantageous to try to keep the record identifier even when the page changes: then we do not need to correct references to the record in, for example, indexes.

The solution is to use a *continuation record*: a small record is placed where the moved record was previously. The small record contains a *forwarding address* to the page that contains the updated record, that is, the identifier of the updated record (see Fig. 2.2). This adds one disk access to a tuple identifier lookup.

The above discussion on the internal structure of a page included methods for maintaining the free space inside the page. In the following we briefly discuss *free-space management* for the whole file.

Assume that a new tuple $t$ is inserted into a relation $r$, and it has to be decided into which page the tuple should go. In the case of a sparse index structure, such as the B-tree with the actual tuples in its leaf pages, the page $p$ that will receive the tuple $t$ is directly determined by the indexing attribute (primary key) of the structure, while in the case of an unordered sequential file (also called a *heap*) the tuple can go into any data page that has room for it. In the former case, if the B-tree leaf page $p$ is full and a page split needs to be done, we must find a completely empty page

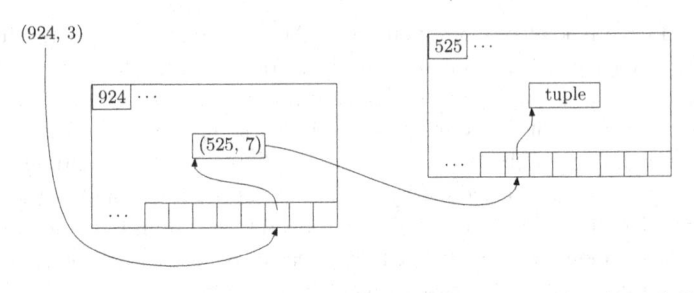

**Fig. 2.2** Representing a long tuple. The original identifier for the record is $(924, 3)$. There is a forwarding address to $(525, 7)$

$p'$, which is linked as a sibling of page $p$ and receives half of the tuples from $p$. In the latter case, we must find a page $p'$ in the heap with enough free space for the tuple $t$.

For allocating and deallocating pages of a B-tree, we can keep a *free-page chain*, where empty index and data pages are linked when they are detached from the B-tree. A newly allocated page is taken from the chain of free pages, if there is a page in the chain. Otherwise, the space allocated for the B-tree is extended by allocating a new contiguous area from disk and using the first page from this new area.

Another method for managing free space is to keep information on which pages are free and which are allocated in a *space map*. In this method, the first page (page 0) of a file is filled by a bit array with $e$ entries, containing the space map for the $e$ following pages, that is, pages $1, 2, \ldots, e$. The value 1 for the $i$th element (bit) in the space map denotes that page $i$ is allocated, while 0 means that page $i$ is free.

The space map can have about 32,000 entries on a 4-kilobyte page. If the file is larger than this, additional space maps are needed: page $e + 1$ is filled by a space map for the $e$ pages following that page, and so on. New pages can be allocated to the file in clusters of $e$ (or fewer) pages.

The free space of a heap-structured file can be maintained by a space map where the entries are approximated: The entries can be, for example, two-bit integers (i.e., values 0–3), so that the amount of free space is represented as:

0: the record area of the page is empty, or the page is unformatted.
1: 0–33% of the record area has been filled.
2: 33–66% of the record area has been filled.
3: over 66% of the record area has been filled.

A tuple $t$ to be added to a heap is converted to its internal representation (a record). The size of the record is counted, and the space maps of the file for $r$ are searched to find a page where $t$ fits.

## 2.3  Buffering and Fixing of Database Pages

When an instance of the database system is running, it is associated with a buffer in main memory. The *database buffer* or *buffer pool* is an array $B[1, 2, \ldots, N]$ of page-sized *buffer frames*. If there are pages of varying size in the database, there must be a buffer of its own for each different page size.

The number $N$ of buffer frames is orders of magnitude smaller than the number of pages in the database. Thus normally only a fraction of the set of pages in the database can reside in the buffer at a time. It is possible to adjust the size of the buffer, but it can seldom be more than a few thousand pages, whereas the database may contain millions or even billions of pages.

Before reading or writing the contents of a page $p$ in the database, the page must be fetched from disk into some free buffer frame $B[i]$. A server-process thread that accesses page $p$ *fixes*, or *pins*, $p$ into the buffer for the duration of the page access. If the page is not already in the buffer, the procedure call $fix(p)$ first fetches $p$ from disk into the buffer.

The buffer manager is not allowed to evict a fixed page from the buffer or to move it to another buffer frame. Thus, the buffer-frame address passed to the server-process thread that called *fix* stays valid for the duration of the fixing. After having finished processing the page, the process must *unfix*, or *unpin*, the page. A page can be evicted from the buffer only when it is no longer fixed for any process.

Pages are evicted from the buffer only when the buffer is full and a buffer frame needs to be assigned to a page $p$ to be fetched from disk (by a $fix(p)$ call). Then according to the *least-recently-used* (LRU) page-replacement policy, the page $q$ is evicted for which the longest time has passed since last being read or written; if that page is a modified page, it must first be flushed onto disk.

A buffered page is a *modified page* if its contents differ from the disk version of the page, that is, the page has been updated since it was last flushed onto disk. Otherwise, if the contents of the buffer and disk versions of the page are the same, the page is an *unmodified page*.

For each buffered page, the buffer manager maintains a *buffer control block* in the main memory, containing:

1. The page identifier of the page
2. The location of the page in the buffer (i.e., the address of the frame)
3. A *fixcount* that indicates how many process threads currently hold the page fixed
4. A *modified bit* that indicates whether or not the page has been updated since it was last fetched from disk
5. PAGE-LSN: the log sequence number (LSN) of the log record for the last update on the page (cf. Sect. 2.2)
6. A pointer to a latch (semaphore) that controls concurrent accesses to the page
7. Links needed to maintain the LRU chain

The addresses of the buffer control blocks are stored in a hash table indexed by the page identifiers. In this way it can be efficiently resolved whether or not a given page resides in the buffer and which buffer frame has been assigned to the page.

The implementation of an SQL query includes a series of pairs of calls to *fix* and *unfix* for the pages touched by the query. These fixings permit the query processor to operate on the pages touched by the query.

*Example 2.1* The execution of an SQL query

**select sum($V$) from $r$**

involves scanning the data pages of relation $r$ in their physical order. For every data page $p$, the following steps (among others) are performed:

1. *fix($p$)*.
2. Retrieve from the buffer frame of page $p$ the value of the $V$ attribute of every $r$-tuple and add it to the sum being counted.
3. *unfix($p$)*.

In fact, in order to determine the processing strategy for the query and to locate the data pages, some data-dictionary pages must first be fixed and inspected. Besides being fixed, each data page $p$ (as well as each data-dictionary page inspected) must also be *read-latched* for the duration of the inspection of the page contents, so that no other process thread can simultaneously update the page. Moreover, if the transaction is wanted to be run in full isolation from other transactions, the relation $r$ or all of its tuples must be *read-locked* for the transaction for commit duration. Latching is explained later in this chapter, and lock-based concurrency control is discussed in Chap. 6.                                                                    □

*Example 2.2* The execution of an SQL update statement

**update $r$ set $V = V + 1$ where $X = x$**

contains, among others, the following steps:

1. *fix($q$)* and *unfix($q$)* calls on pages $q$ in an index of relation $r$ based on attribute $X$, in order to locate the data page $p$ containing the $r$-tuple with $X = x$.
2. *fix($p$)*.
3. Locate the $r$-tuple with $X = x$ in page $p$ and increment the value of its attribute $V$ by 1.
4. Log the update for the transaction and stamp the LSN of the log record in the PAGE-LSN field of page $p$.
5. *unfix($p$)*.

Again, some data-dictionary pages must first be fixed (and read-latched) in order to find out that the index on $r$ exists. Besides fixing the page $p$ to be updated, the page must also be *write-latched* for the process thread for the duration of the update, so that no other process thread can simultaneously read or update the page. Moreover, in order to guarantee sufficient transactional isolation, the $r$-tuple to be updated must be *write-locked* for the transaction for commit duration.         □

---

**Algorithm 2.1** Function *fix*($p$)

---

**if** page $p$ is not in the buffer **then**
    **if** all buffer frames are occupied **then**
        select a buffered unfixed page $q$ to be evicted
        **if** page $q$ is modified **then**
            flush page $q$ to its disk address
        **end if**
        assign the buffer frame of page $q$ to page $p$
    **else**
        assign some unoccupied buffer frame to page $p$
    **end if**
    fetch page $p$ from its disk address to the frame
    create a buffer control block
    clear the modified-bit
    initialize fixcount as zero
**else**
    get the address of the buffer frame of $p$
**end if**
increment the fixcount
return the address of the frame

---

An implementation of the call *fix*($p$) is given as Algorithm 2.1. If all buffer frames are occupied and the pages fixed (i.e., all the fixcounts are non-zero), the process thread performing the call must wait until the fixcount of some buffered page comes down to zero. The call *unfix*($p$) decrements the fixcount of page $p$ by one.

## 2.4  Database States

The contents of the page on disk at the disk address of page $p$ is called the *disk version* of page $p$. The *buffer version* of a buffered page $p$ is the contents of $p$ in the buffer. The buffer version of a modified page differs from the disk version until the page is flushed from the buffer onto disk. In flushing onto disk, the disk version is overwritten by the buffer version. The *current version* of page $p$ is the buffer version if $p$ is buffered and the disk version otherwise.

The *disk version* of a physical database is the collection of disk versions of its pages. The *state* or *current version* of a physical database at a certain time is the contents of the current versions of its pages at that time.

The *disk version* of a logical database is the collection of tuples contained in the disk versions of the data pages of the underlying physical database. The *state* or *current version* of a logical database is the set of tuples in its relations, i.e., the tuples in the current versions of the data pages in the underlying physical database.

For reasons of efficiency, a modified page is not taken onto the disk immediately after the update operation has been performed nor even after the transaction that updated the page has committed. A page that is used often is kept in the buffer as long as possible, according to the LRU principle. Thus, the buffer version of a page may accumulate many updates, and the disk version can lag far behind the buffer version.

For example, some page may receive updates from a thousand transactions run in succession. It may then very well happen that the page is not once flushed onto disk in between, in which case the disk version lags a thousand updates behind the current version of the page.

## 2.5   Database Recovery and Checkpoints

In a system crash or shutdown, the contents of main memory disappear, including the buffers: the database buffer and the log buffer. Only the disk version of the database remains, with the log records that were taken to disk before the crash. The disk version usually does not reflect the state of the database at the time of the failure but may lag more or less behind.

Transactions have written log records for all the updates they have performed, and the *write-ahead-logging* (WAL) protocol ensures that log records for all the updates in the disk version of the database are found on the log disk (see Chap. 3). Moreover, the commit of a transaction includes writing a "commit" log record and flushing the log buffer onto the log disk, so that in the event of a system crash the log disk always contains log records up to and including the "commit" log record of the last transaction that committed before the failure.

In *restart recovery* (see Chap. 4), the database system is restarted, and the state of the database that existed at the time when the last log record found on the log disk was written is restored from the disk versions of the database pages and the log records: the pages that have updates recorded in the log are fetched from disk in the buffer, and the updates described by the log records are applied on the pages (Sects. 4.7 and 4.8). In addition, all forward-rolling transactions, that is, those for which no "abort" or "commit" log record is found on the log disk, are aborted and rolled back, and the rollback of all backward-rolling transactions is run into completion (Sect. 4.9).

To speed up restart recovery, modified pages are flushed from the buffer onto disk in *checkpoints* (Sect. 4.4) taken periodically (typically in every 5–10 min). Immediately after flushing onto disk, the disk version of the page has the same contents as the current version (i.e., the buffer version) of the page.

In a *complete checkpoint*, all modified pages in the buffer are flushed onto disk. After a complete checkpoint, the state of the disk version of the database is the same as the current state of the database, unless updating the database is permitted during the checkpoint (which it usually is, however).

Taking a complete checkpoint significantly slows down transaction processing. A more light-weight procedure for speeding up restart recovery is an *indirect*

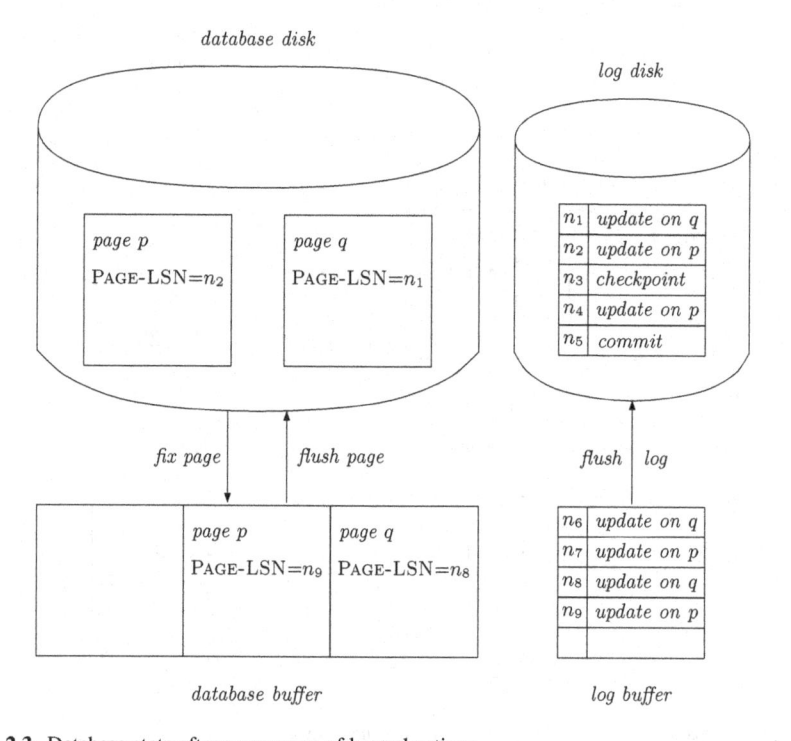

**Fig. 2.3** Database state after a sequence of logged actions

*checkpoint*, or a *fuzzy checkpoint*, in which only some of the modified pages are flushed. In a minimalistic fuzzy checkpoint, no database pages are flushed, but only some information about currently buffered pages are written to the log. This information includes page identifiers with LSNs indicating positions in the log after which there might be updates that are missing from the disk versions of pages; in restart recovery, this information is used to determine those pages whose disk versions might have some updates missing (see Chap. 4).

*Example 2.3* Figure 2.3 shows the states of two pages after a sequence of actions logged with LSNs $n_1$ to $n_9$. The updates up to LSN $n_2$ are present in the disk versions of pages $p$ and $q$ because the pages have been flushed from the buffer onto disk when taking the checkpoint logged with LSN $n_3$. Before the pages were flushed, following the write-ahead logging protocol, the log up to $n_2$ was flushed from the log buffer onto the log disk. Next time the log was flushed when a transaction was committed at LSN $n_5$. After that, active transactions updated pages $p$ and $q$, generating the log records with LSNs $n_6$ to $n_9$ appended to the log buffer.

The current versions of pages $p$ and $q$ are in the database buffer. The PAGE-LSN values stamped in the page headers indicate that the current version of page $p$ contains all the updates logged with LSNs $n_2$, $n_4$, $n_7$, and $n_9$, while the disk version of $p$ does not contain the three last updates, and that the current version of page $q$

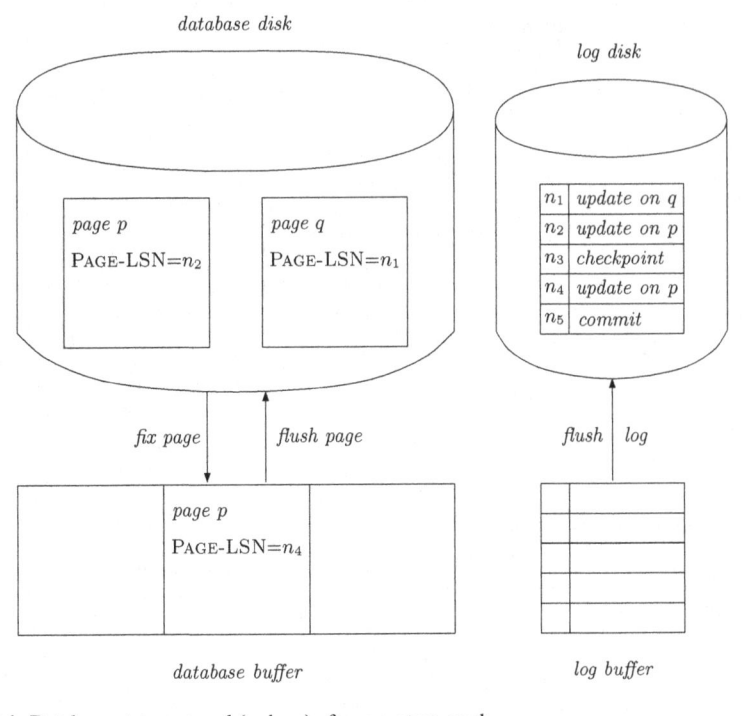

**Fig. 2.4** Database state restored (redone) after a system crash

contains all the updates logged with LSNs $n_1$, $n_6$, and $n_8$, while the disk version of $q$ does not contain the two last updates.

Assume now that the system crashes. The contents of main memory, including the database and log buffers, are then lost. In restart recovery, the database is restored to the state that existed at the time the most recent log record surviving on the log disk was written. This is done by applying logged updates onto the disk versions of pages. Thus, the disk version of page $p$ is fetched from disk and fixed in the buffer, the update logged with LSN $n_4$ is redone on the buffered page, the PAGE-LSN of the page is advanced to $n_4$, and the page is unfixed (Fig. 2.4). No page is flushed onto disk; so the disk version of the database remains unchanged. The recovery is then continued with aborting and rolling back the active transactions, that is, those that have no commit log record on the log disk.                                  □

## 2.6   Integrity of the Physical Database

The physical database (more properly, its state) is *integral* or *consistent* when the internal data structures in all its pages are consistent and the collection of pages makes up a valid graph structure of the type specified by the file structure. For example, a B-tree structure for a file defines precisely what kind of a graph structure the pages must form, what information there must be in each page of the structure, which balance conditions need to be fulfilled, and which algorithm is used to find records stored in the leaf pages of the B-tree.

The definition of integrity only pertains to the current version of the database. The disk version, on the contrary, does not need to be consistent. If each individual flushing of a page from the buffer onto disk and each individual fetching of a page from disk to the buffer are always performed correctly, the internal structure of every single page in the physical database is consistent when no logical database action is being performed. In this case we say that the pages are *page-action consistent*.

However, the collection of disk versions of the pages does not necessarily form a consistent structure. For example, when the structure of a B-tree changes due to the need to split a full page $p$ to accommodate an insert action, some of the pages involved in the structure modification (e.g., $p$ itself) may have been taken onto disk, while others (e.g., $p$'s parent and new sibling) may still reside only in the buffer. In the same way, the disk version of the logical database can continually be inconsistent.

When no logical database action by any active transaction is in progress, the current version of the physical database is consistent. In that case we say that (the current versions of) the physical and logical databases are *action-consistent*. When no transactions are active, so that every transaction is either committed or rolled back (the current versions of), the physical and logical databases are said to be *transaction consistent*.

Clearly, a transaction-consistent physical database is always action-consistent, and an action consistent physical database is always page-action consistent and consistent. If a complete checkpoint is taken when no transactions are active and no new transactions are allowed to the system in the interim, then even the disk versions of the physical and logical databases will be transaction consistent. A transaction-consistent logical database is consistent if all the transactions are logically consistent and have been run in full isolation.

## 2.7 Latching of Database Pages

The database must stay consistent under operations by several concurrent server-process threads. To keep the physical database consistent, a process thread that processes a database page must always *latch* the fixed page for the duration of the processing.

A *read latch* gives the process thread permission to read the page and prevents other processes and threads from concurrently writing the page. Multiple process threads can hold a read latch on the same page at the same time. A *write latch* gives permission to both read and write the page and prevents other process threads from reading or writing the page at the same time. The owner of a latch must *unlatch* the page (release the latch) after it has used the page.

A latch on a buffered page is implemented using a *semaphore* pointed to from the buffer control block of the page. Semaphores are the standard method for mutual exclusion of concurrent processes accessing shared main-memory structures. A semaphore is a main-memory structure that contains the mode of the semaphore (free, shared, exclusive), the number of process threads holding the semaphore (0 for free mode, 1 for exclusive mode, $\geq 1$ for shared mode), and a pointer to the list of process threads waiting for access to the semaphore. A read latch is implemented with a shared semaphore and a write latch with an exclusive semaphore.

Before each reference to page $p$ produced by an SQL query, the database management system automatically generates the pair of calls

$$fix\text{-}and\text{-}read\text{-}latch(p) \equiv fix(p)\&read\text{-}latch(p),$$

if the page is only read during this fixing, or

$$fix\text{-}and\text{-}write\text{-}latch(p) \equiv fix(p)\&write\text{-}latch(p),$$

if $p$ is to be updated during this fixing. The latch and fixing is freed by generating the pair of calls

$$unlatch\text{-}and\text{-}unfix(p) \equiv unlatch(p)\&unfix(p).$$

Actually, the pair of fix and latch calls and the pair of unlatch and unfix calls are most efficiently implemented as single calls, because we never need to hold fixed pages unlatched.

When a process or thread asks for a read latch onto a page that is currently write-latched by another process or thread or for a write latch onto a page that is currently read- or write-latched by another process or thread, the thread requesting for the latch is put to sleep, waiting for the other threads to release their latches. We assume that at any time, a process thread can be waiting for at most one semaphore and hence at most one latch.

A general policy is that no thread should hold a latch for a long time and that a thread may hold only a small constant amount of latches at any given time. This is because no wait-for graph is maintained for latches to detect possible deadlocks, and thus the latching protocol followed by the threads needs to prevent deadlocks. In most cases a thread only holds one latch at a time to protect its database action.

When a process wishes to read a tuple from a data page $p$, it calls *fix-and-read-latch*($p$), copies the tuple from the buffer frame of page $p$, and then calls *unlatch-and-unfix*($p$). When a process wishes to update a tuple on a data page $p$, it calls *fix-and-write-latch*($p$), updates the tuple in the buffer frame of page $p$, writes a log record for the update, stamps the LSN of this log record to the PAGE-LSN of $p$, and then calls *unlatch-and-unfix*($p$).

When traversing a chained database structure, a process thread needs to keep latches on two pages at a time in order to ensure that a correct path is traversed even if there are other process threads running simultaneously that may modify the pages.

*Example 2.4* Assume that pages $p_1, \ldots, p_n$ form an unidirectional chain so that the header of $p_i$ contains the page identifier of $p_{i+1}$, for $1 \leq i < n$, and the header of $p_n$

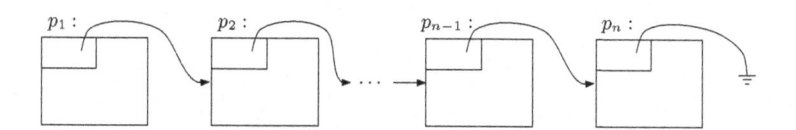

**Fig. 2.5** A unidirectional chain of pages

contains an empty link to terminate the chain (Fig. 2.5). A process thread holds the page id of page $p_1$, and it wants to read the pages in the order in which they are chained. To be sure that the link from page $p_i$ to page $p_{i+1}$ stays valid while the link is traversed, the process must use a latching protocol in which the latch on $p_i$ is released only after the latch on $p_{i+1}$ has been obtained.                                □

This latching protocol is called *latch-coupling* or *crabbing*. It is used, for instance, when searching for a record in a B-tree: on the search path from the root of the B-tree down to the leaf that contains the record, the parent page is kept latched until the latch on the child page has been acquired.

## 2.8  Modifications on the Physical Structure

When data is inserted into the database, new pages must be allocated to store that data, and those pages must be linked as part of the physical database structure. A split of a full B-tree page involves allocating a new empty page and linking it as a part of the B-tree structure. Deleting data from the database may similarly shrink the database structure, and emptied pages are detached from the structure. Such operations that modify the structure of the physical database but leave the logical database intact are called *structure modifications*.

The entire database structure must be maintained in a consistent state even when there are many structure modifications caused by concurrently running transactions. We have to require *atomicity* of the structure modifications.

During normal transaction processing, a safe method for guaranteeing the consistency of the physical database is to hold all the pages involved in a structure modification write-latched for the duration of the modification. For example, when a full child page $q$ of a B-tree page $p$ is split by allocating a new empty page $q'$, then all these three pages are held write-latched during the split, that is, until page $q'$ has been linked as a child of page $p$ and a half of the records in page $q$ has been moved to page $q'$ (and the modification has been logged).

*Example 2.5*  Assume that the logical database (or one of its relations) is implemented as a heap, that is, an unordered sequential file, for which new pages are allocated as new tuples are inserted into the database (Fig. 2.6). The pages in the heap are linked so that the next-page field in the header of a page contains the page id of the next page in the heap. In addition, we assume that the header of the first page $f$ in the heap also contains the identifier of the last page in the heap.

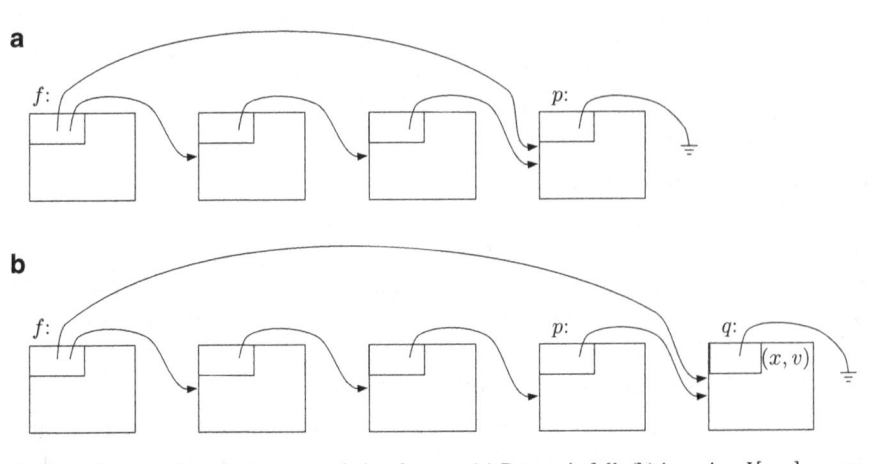

**Fig. 2.6** Organization of a heap as a chain of pages. (**a**) Page $p$ is full; (**b**) insertion $I[x, v]$ causes a new page $q$ to be allocated and linked into the end of the heap

Consider the case where an insert action $I[x, v]$ by transaction $T_1$ causes the allocation of a new page $q$ and linking it to the end of the heap, to be the follower of the last page $p$. Because the structure needs to be kept consistent during normal transaction processing, four pages must be kept write-latched for the duration of the allocation (and its logging): the page $s$ that contains the space map, the last page $p$, the page $q$ that is to be linked, and the first page $f$. Note that all these pages are updated in the allocation. Page $q$ that will receive the tuple $(x, v)$ must be formatted as an empty data page. Page $q$ is kept write-latched until the tuple $(x, v)$ has been inserted into page $q$ and the insertion has been logged.

Assume that another transaction now performs the action $I[y, w]$ (where $y \neq x$). This insertion also goes to page $q$ in the heap structure. When process thread that executes transaction $T_1$ releases its write latch acquired for the insertion $I[x, v]$, the process thread executing $T_2$ is free to write-latch page $q$ and to insert the tuple $(y, w)$ there.

Then $T_2$ commits while $T_1$ is still active and then wants to abort. In the backward-rolling phase, the undo action $I^{-1}[x, v]$ is executed so as to delete the tuple $(x, v)$ from page $q$. However, the allocation of page $q$ obviously must not be undone: $T_2$ has already made its own update on page $q$ and committed.                                    □

We conclude that a structure modification must be allowed to "commit" once it is successfully completed, regardless of the eventual outcome of the transaction that caused the structure modification.

Transaction processing thus involves operations at two levels. At the higher, or *logical level*, we have actions of transactions operating on the logical database, and at the lower, or *physical level*, we have operations on the pages of the physical database. The two levels are interconnected so that each action at the logical level, when performed on a given state of the physical database, is mapped to a sequence of page-wise operations at the physical level.

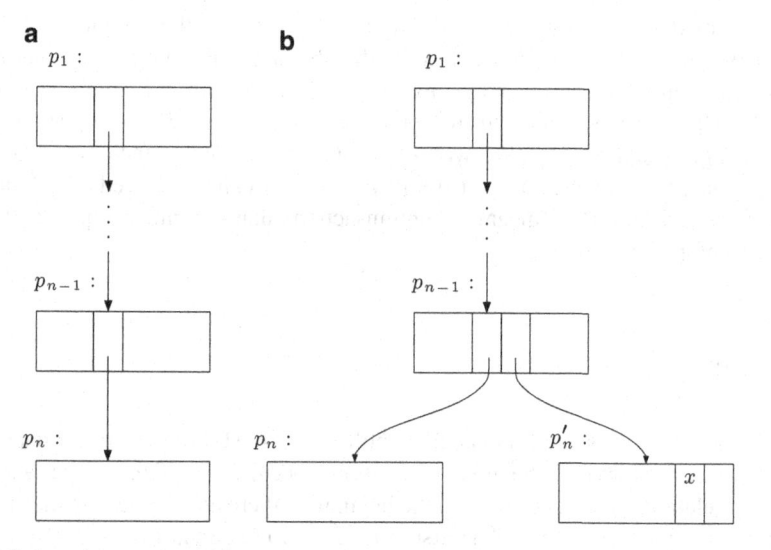

**Fig. 2.7** Inserting a tuple with key $x$ into a full leaf page $p_n$ of a B-tree; (**a**) before insertion, (**b**) after insertion

*Example 2.6*   Assuming that the physical database is a sparse B-tree index (Fig. 2.7), a read action $R[x]$ is mapped to a sequence

$$R[p_1]R[p_2]\ldots R[p_n]$$

of reads of pages $p_1, p_2, \ldots, p_n$ on the path from the root page $p_1$ down to the leaf page $p_n$ that contains the tuple with key $x$.

Similarly, an insert action $I[x]$ is mapped to a sequence

$$R[p_1]R[p_2]\ldots R[p_{n-1}]W[p_n]$$

of reads of non-leaf pages $p_1, p_2, \ldots, p_{n-1}$ and a read and write of the leaf page $p_n$, assuming that page $p_n$ has room for the tuple to be inserted.

If the page $p_n$ has no room for the tuple, the corresponding page-wise operation sequence might look like

$$R[p_1]R[p_2]\ldots R[p_{n-1}]W[p_n]W[p_n']W[p_{n-1}]W[p_n]W[p_n'],$$

where the first $W[p_n]$ is only a read and the subsequence $W[p_n']W[p_{n-1}]W[p_n]$ represents the split of page $p_n$, involving an allocation of a new page $p_n'$, moving the upper half of tuples in $p_n$ to $p_n'$ and linking $p_n'$ as a child of $p_{n-1}$; the final $W[p_n']$ represents the insertion of the tuple into page $p_n'$. The page split represented by the sequence $W[p_n']W[p_{n-1}]W[p_n]$ "commits" and thus persists as soon as it is completed, even if the logical-level transaction that performed the action $I[x]$ eventually aborts and rolls back.                                                                                    □

As is evident from the above examples, we can use the simple read-write transaction model with pages in place of the abstract uninterpreted data items to model transaction processing at the physical level: for page $p$, the action $R[p]$ fixes and read-latches page $p$ and reads its contents, and the action $W[p]$ fixes and write-latches page $p$ and reads and/or writes its contents. In the parlance of *multi-level transactions*, the "transactions" at the physical level can be viewed as a kind of *open nested transactions*, meaning subtransactions that commit independently of their parent transactions.

## Problems

**2.1** Assume that the size of the database buffer is 1,000 buffer frames for database pages of size 4 kilobytes. The tuples of relation $r$ occupy 10,000 data pages and the tuples of relation $s$ 500 data pages. In the beginning, there are no pages in the buffer, and no transactions are active. The first transaction to be run is $T_1$:

**exec sql update** $r$ **set** $A = A + 100$.
**exec sql commit**.

How many data pages of $r$ are fetched from disk into the buffer, and how many data pages are flushed from the buffer onto disk while $T_1$ is running? How does the disk version of the database differ from the current version of the database at the time $T_1$ commits?

Immediately after the commit of $T_1$, another transaction, $T_2$, is run:

**exec sql update** $s$ **set** $B = B + 200$;
**exec sql commit**.

How many data pages of $r$ and $s$ are fetched from disk, and how many data pages are flushed onto disk while $T_2$ is running? How does the disk version of the database differ from the current version of the database at the time $T_2$ commits? No checkpoints are taken.

Assuming a disk with average seek and rotation time 10 ms and sequential transfer rate of 10 megabytes per second, how long does it take to perform $T_1$ followed by $T_2$? You may assume that the pages of both relations are stored sequentially on the disk.

**2.2** Explain why a page-action-consistent database state is not necessarily action consistent and why an action-consistent state is not necessarily transaction consistent. Also explain why the disk version of a physical database is not necessarily action consistent even though no transactions are in progress.

**2.3** Recall from your operating systems course how LRU buffering is implemented.

**2.4** Assume that a server-process thread fails while keeping a page write-latched. Obviously, we have to regard the contents of the page in the buffer as *corrupted*, that is, non-page-action consistent. A corrupted page should not be flushed onto disk.

Instead, a page-action-consistent version of the page should somehow be recovered. How could this be possible? What should be done with the write latch? Observe that the write latch survives even if the thread fails.

**2.5** Assume that the physical structure of relation $r(\underline{A}, B, C)$ is a sparse B-tree index on $(B, A)$ and that the physical structure of relation $s(\underline{D}, E)$ is a sparse B-tree index on $D$, where the leaf pages (i.e., the data pages containing the tuples) of the B-tree are sideways linked in ascending key order: the *next-page* field in the page header of a leaf page contains the page id of the next leaf page in key order. Give the sequence of latchings and unlatchings of B-tree pages needed in executing the transaction generated from the program fragment:

**exec sql select** $B$, **count**($*$) **from** $r$, $s$ **where** $B = D$ **group by** $B$;
**exec sql commit**.

How many latches at most are held at a time?

**2.6** The *system catalog*, also called the *data dictionary*, of a relational database consists of several relations whose contents together contain the logical schema (i.e., the relational schema) and the physical schema (description of storage and index structures) of the database. A very simple system catalog might contain a relation $f$ with tuples such as $(r, X, b\text{-}tree, p)$ stating that for relation $r$ there is a B-tree index on attribute $X$ with root-page identifier $p$. Obviously, in executing an SQL statement such as

**update** $r$ **set** $V = V + 1$ **where** $X = x$

the tuple for $r$ must first be retrieved from the catalog relation $f$. This means that some pages of $f$ need to be latched. In what mode those pages are latched? How long are the pages kept latched? Actually, $f$ is not the only catalog relation that needs to be inspected in executing the operation. How would you represent the additional information needed here?

How are system catalog pages latched in performing the following SQL statements?

**create table** $r(X, V)$;
**alter table** $r$ **add primary key**($X$);
**create index** $I$ **on** $r(X)$.

## Bibliographical Notes

The basic organization of the physical database described in this chapter already appears in early database management systems such as System R [Astrahan et al. 1976] and INGRES [Stonebraker et al. 1976, Stonebraker, 1986]. In System R, the logical database is stored in logical segment spaces mapped to extents on disk storage, with each segment consisting of equal-sized pages mapped to disk blocks.

For addressing tuples in relations, System R and INGRES used tuple identifiers consisting of a page number within a segment and a byte offset from the bottom of a page denoting a slot that contains the byte location of the tuple in that page.

Stonebraker [1981] discusses various problems that arise in attempts to apply standard operating system services in implementing database management systems. Gray and Reuter [1993] give a thorough treatment of physical database organization; they show in detail how the physical structures mentioned in this chapter are organized and managed.

The idea of latching a page for the duration of a single action is essentially present in System R, where it is called physical locking [Astrahan et al. 1976]. The short-duration physical locks of System R allowed a page to contain uncommitted updates by several transactions at the same time, each protected by a commit-duration logical lock. Latches and their implementation as semaphores are discussed by Mohan [1990a, 1996a], Mohan et al. [1992a], Mohan and Levine [1992], and Gray and Reuter [1993], among others.

The way structure modifications are modeled in this book comes from the "nested top-level actions" of the ARIES recovery algorithm [Mohan et al. 1992a] and its adaptations to recoverable B-tree indexes [Mohan, 1990a, 1996a, Mohan and Levine, 1992]. Other solutions follow from schemes for multi-level transaction management [Weikum, 1991, Lomet, 1992, Weikum and Vossen, 2002]. Sippu and Soisalon-Soininen [2001] discuss a two-level model of search-tree transactions with tree-structure modifications as open nested subtransactions [Traiger, 1983, Gray and Reuter, 1993].

# Chapter 3
# Logging and Buffering

For the purposes of transaction rollback and restart recovery, a *transaction log* is maintained during normal transaction processing. The log is shared by all transactions, and it keeps, in chronological order, a record of each update on the database. The log record for an update action makes possible the redoing of the update on the previous version of an updated page when that update has been lost due to a failure. The log record for a forward-rolling update action also makes possible the undoing of the update in a backward-rolling transaction or in a transaction that must be aborted due to a failure. The log records are buffered in main memory before they are taken onto disk, but, unlike database pages, the log records are flushed onto the log disk whenever some transaction commits, so that each committed transaction is guaranteed to have every one of its updates recorded either on the disk version of the database or on the log disk.

In this chapter we show how the *physiological logging* protocol applied in the ARIES family of algorithms works in the case of the read-write and key-range transaction models. In fine-granular physiological logging, an update action on a tuple by a transaction is logged with a log record that carries, besides the arguments of the logical update action, also the identifier of the data page that received the update, so that the update can always be efficiently redone directly on the page. Undoing an update may also be possible directly on the page mentioned in the log record, provided that the page still is the correct target for the undo action; if not, the logical information in the log record is used to undo the update logically.

Besides logical update actions, also structure modifications need be logged. Unlike the solutions usually presented in connection with the ARIES recovery algorithm, we advocate the use of redo-only structure modifications, each of which is logged with a single redo-only log record that mentions all the few pages (a small constant number) involved in the modification. Such a log record is only used to redo a structure modification, if necessary, but the modification is never undone.

The log can serve its purposes only if a protocol called *write-ahead logging* (WAL) is followed in database and log buffering. The WAL protocol states that an

© Springer International Publishing Switzerland 2014
S. Sippu, E. Soisalon-Soininen, *Transaction Processing*, Data-Centric Systems
and Applications, DOI 10.1007/978-3-319-12292-2_3

updated database page can be flushed onto disk only if all the log records up to and including the log record for the last update on the page have already been flushed onto disk. Besides the WAL protocol and the flushing of the log as part of the commit protocol, no further restrictions are imposed on buffering, so that pages containing uncommitted updates can be "stolen" (i.e., flushed onto disk) and that no database page need be "forced" (i.e., flushed onto disk) at transaction commit.

## 3.1  Transaction Log

The *transaction log* keeps a record of all updates made by transactions to the database, that is, $W$ and $W^{-1}$ actions in the read-write model and $I$, $D$, $I^{-1}$, and $D^{-1}$ actions in the key-range model are recorded in the log. Also, when a transaction begins, commits, aborts, completes its rollback, sets a savepoint, begins rollback to a savepoint, or completes rollback to a savepoint (actions $B$, $C$, $A$, $S[P]$, $A[P]$, $C[P]$), this is logged. Other log records include those for structure modifications and checkpoints.

The log is an *entry-sequenced*, or *append-only*, file into which the log records of all transactions are appended in the chronological order of the updates. Log records can be deleted from the log only by truncating the log from the head, disposing of old log records that are no longer needed. The writes to the log file are buffered in a *log buffer*, whose contents are flushed when some transaction commits or the buffer becomes full or when the buffering policy (WAL) states that so must be done.

Because the log is shared by all transactions in the system, appending to the log must be protected by an exclusive latch, which is held for the time of appending the log record into the log buffer. Because the log buffer is flushed at every transaction commit (or completion of rollback), every transaction, be it forward-rolling, backward-rolling, committed or rolled back, is guaranteed to have on the log disk all its log records written before the last commit of some transaction.

As was mentioned earlier, each log record receives a unique *log sequence number*, the LSN, which is determined while holding the exclusive latch on the log tail. In a centralized database system, the log sequence numbers form a single increasing sequence: if action $o'$ of transaction $T'$ is performed after action $o$ of transaction $T$, then $\text{LSN}(o) < \text{LSN}(o')$.

In a shared-disks or page-server system, where sites (clients) independently generate log records for the actions they perform on the shared database, the log sequence numbers cannot be globally increasing. However, the LSNs of updates that are performed on the same page must always form a sequence that increases with time.

Log sequence numbers are often generated so that they also act as direct addresses to the log records. A structure that is often used is a pair containing the number of the log file and the byte offset. The byte offset represents the distance of the first byte of the log record, counting from the start of the file. The page where the log record is written can be found using the formula

page number = byte offset *div* page size.

Instead of the byte offset, the record identifier $(p, i)$ could be used, where $p$ is the logical page number $(1, 2, 3, \ldots)$ of page $p$ in the sequential log file and $i$ the number of the record on the page (an index to the internal record index in the page). Then $(p_1, i_1) < (p_2, i_2)$ if either $p_1 < p_2$ or $p_1 = p_2$ and $i_1 < i_2$.

The log record for a logical database action by transaction $T$ carries $T$'s unique *transaction identifier* (transaction-id) that is given by the system when the transaction starts (or when it performs its first database action). The general format of a log record for an action by transaction $T$ is:

$$n : \langle T, o[\bar{x}], n' \rangle,$$

where $T$ is the transaction identifier, $o[\bar{x}]$ represents the action being logged and its arguments, $n$ is the LSN of the log record, and $n'$ is an LSN value called the UNDO-NEXT-LSN that is needed if $T$ later performs a total or partial rollback.

For each active-transaction $T$, UNDO-NEXT-LSN($T$) is maintained as the LSN of the log record for the latest (not-yet-undone) forward-rolling update action by $T$. In the case that $T$ has not done any undo actions, UNDO-NEXT-LSN($T$) points to the log record for the latest update action by $T$ (Fig. 3.1a). In the case that $T$ has done undo actions, UNDO-NEXT-LSN($T$) skips over the log records for the forward-rolling update actions that have already been undone (Fig. 3.1b). Thus, UNDO-NEXT-LSN always points to the log record written for the action that is the next to be undone. In the case that $T$ has undone all its updates, UNDO-NEXT-LSN($T$) points back to the log record written for begin-transaction action ($B$) of $T$ (Fig. 3.1c).

A forward-rolling update action $o[\bar{x}]$ executed by transaction $T$ is logged in such a way that the action can be redone or undone if necessary. The undo action $o^{-1}[\bar{x}]$ is logged so that it can be redone: it must be possible to redo the action $o^{-1}[\bar{x}]$ using its log record, but undo is not necessary, since undo actions are never undone.

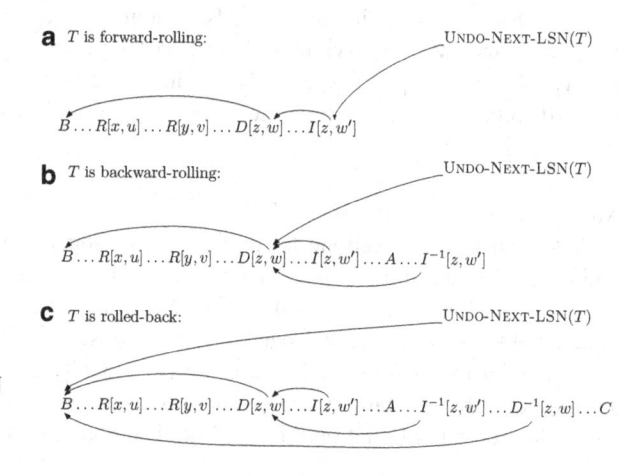

**a** $T$ is forward-rolling:     UNDO-NEXT-LSN($T$)

$B \ldots R[x, u] \ldots R[y, v] \ldots D[z, w] \ldots I[z, w']$

**b** $T$ is backward-rolling:     UNDO-NEXT-LSN($T$)

$B \ldots R[x, u] \ldots R[y, v] \ldots D[z, w] \ldots I[z, w'] \ldots A \ldots I^{-1}[z, w']$

**c** $T$ is rolled-back:     UNDO-NEXT-LSN($T$)

**Fig. 3.1** UNDO-NEXT-LSN chaining of log records written for transaction $T$

$B \ldots R[x, u] \ldots R[y, v] \ldots D[z, w] \ldots I[z, w'] \ldots A \ldots I^{-1}[z, w'] \ldots D^{-1}[z, w] \ldots C$

To perform a total rollback of $T$, the chain starting from the log record with LSN = UNDO-NEXT-LSN($T$) is traversed back to the begin-transaction log record of $T$; at each log record on the chain, the corresponding action is undone (either physically or logically). To perform a partial rollback to a savepoint, the UNDO-NEXT-LSN chain is traversed back to the log record written for the set-savepoint action.

In the redo pass of restart recovery (Sect. 4.8), the updates by transactions that were not written to disk before the failure are redone on disk versions of database pages. The redo pass is followed by the undo pass (Sect. 4.9), in which forward-rolling transactions are aborted and rolled back, and the rollback of backward-rolling transactions is completed by performing the undo actions of the not-yet-undone forward-rolling update actions.


## 3.2  Physiological Logging

Logging schemes can be classified into physical, logical, and physiological. In *physical logging*, or *value logging*, the log record for an update includes the page-id of the page being updated, the location of the bytes that were changed in the page, and the before and after images of the update. The *before image* gives the contents of the bytes before the update, and the *after image* gives the contents of the bytes after the update.

For example, when inserting tuple $(x, v)$ into page $p$ at record position $i$, the changed bytes at the position of the $i$th record in the record area of $p$ are written to the log, as well as the changed bytes in $p$'s record index (at entry $i$) and those in $p$'s header (in the field storing the total number of items stored in the record index). The changed byte sequences with their addresses (byte offsets from the start of the page) can all be written into one log record, if it fits on one page (as we will assume).

An update action logged with a physical log record can be redone by fixing and write-latching the page in question and writing the after image onto it. An operation can be undone by writing the before image.

In *logical logging*, the log records contain logical actions, that is, actions $W[x, u, v]$ and $W^{-1}[x, u, v]$ in the read-write model and, additionally, actions $I[x, v]$, $D[x, v]$, $I^{-1}[x, v]$, and $D^{-1}[x, v]$ in the key-range model. For example, a forward-rolling insertion of tuple $(x, v)$ by transaction $T$ would be logged with

$$n : \langle T, I, x, v, n' \rangle,$$

where $n'$ is the UNDO-NEXT-LSN of $T$.

A logical action is redone by performing the operation again (logically) and undone by performing the undo action (logically). Logical logging makes sense for a *main-memory database*, that is, a database that is kept in main memory all the time; its physical structure thus need not be composed of disk-page-size nodes. The contents of a main-memory database are periodically checkpointed into the log, and in the event of a failure, the database is initialized from the latest checkpoint whose log record is found on the log disk. The updates done by transactions after the checkpoint are redone logically on the database, after which the updates by active transactions are undone logically.

Modern disk-based database management systems usually employ *physiological logging*. Here, the log records are physical in terms of the page and logical in terms of the tuple. For example, the forward-rolling action $I[x, v]$ performed by transaction $T$ on leaf page $p$ of a sparse B-tree is logged with

$$n : \langle T, I, p, x, v, n' \rangle,$$

where $p$ is the page-id of page $p$.

In the recovery algorithm that we will present later in the next chapter (the ARIES algorithm), redo actions are always performed in exactly the order in which they are recorded in the log, which is also the original execution order of the actions on each page. Thus, the redo of a physiologically logged action can always be performed *physically*, that is, on the page whose page-id is recorded in the log record and even at the same record position where it was originally performed. Physical redo is efficient, because the target page can be found directly via the page-id: regardless of what the physical database structure is, it is not necessary to traverse the structure in search for the target of the action.

To redo an action $I[x, v]$ logged with $n : \langle T, I, p, x, v, n' \rangle$, page $p$ is fixed and write-latched, the tuple $(x, v)$ is inserted into page $p$, the PAGE-LSN of $p$ is advanced to $n$, and $p$ is unlatched and unfixed. To redo an action $D[x, v]$ logged with $n :$ $\langle T, D, p, x, v, n' \rangle$, page $p$ is fixed and write-latched, the tuple with key value $x$ in page $p$ (which is $(x, v)$) deleted, the PAGE-LSN is advanced, and $p$ is unlatched and unfixed.

However, the physical undo of a physiologically logged action is not always possible. This is because the server-process threads that perform the database actions of transactions release their write latches on the pages that they modify immediately after the modification is completed—that is, before the transaction commits—and they can undo their own actions independently of other transactions.

For example, consider a transaction $T$ that performs the action $D[x, v]$ on a leaf page $p$ of a B-tree. This action is logged with $n : \langle T, D, p, x, v, n' \rangle$. Next, the process thread that executes $T$ releases its write latch on $p$ and continues executing other actions. In the meantime, actions by other transactions fill page $p$ so that there is no room for more tuples, or alternatively, $p$ overflows and is split in two so that the newly allocated page $p'$ covers the key value $x$. Now $T$ is still active and wants to undo the action $D[x, v]$. The undo action $D^{-1}[x, v]$ cannot be performed physically on page $p$.

When an action cannot be undone physically, *logical undo* is used. Thus, the undo of the action $D[x, v]$ is first tried on the original page $p$: the page is fixed and write-latched, and if it still covers key value $x$ and there is room left for the tuple $(x, v)$, then the tuple is inserted into page $p$. Otherwise, the action is undone by executing the undo action $D^{-1}[x, v]$ logically: the search path for key value $x$ is traversed from the root page of the tree down to the leaf page $p'$ that covers $x$ and the tuple $(x, v)$ is inserted there.

## 3.3   Active-Transaction Table and Modified-Page Table

For the purposes of managing transactions during normal processing, the database management system records volatile information about currently active transactions. For each active-transaction $T$, the following information is kept in the *active-transaction table*, or *transaction table* for short:

1. The transaction identifier of $T$
2. The (type of the) transaction state of $T$: "forward-rolling" or "backward-rolling"
3. The UNDO-NEXT-LSN of $T$, that is, the LSN of the log record of the latest not-yet-undone forward-rolling update action by $T$
4. Other information relevant for an active transaction, such as a pointer to a chain of locks held by the transaction (see Sect. 6.1)

The entry for transaction $T$ is called the *transaction record* of $T$. The table is implemented using an efficient main-memory structure (hashing based on the transaction-id).

In practice, the transaction record for $T$ also contains PREV-LSN, the LSN of the latest logged action of $T$, but we omit it because it is not needed in the algorithms presented in this book. When $T$ contains no undo actions, the sequence of log records for update actions in the UNDO-NEXT-LSN chain of $T$ is equal to that in the PREV-LSN chain of $T$. In the general case the log records in the UNDO-NEXT-LSN chain form a subsequence of those in the PREV-LSN chain: rolled-back sequences are not included in the UNDO-NEXT-LSN chain, and some transaction-control actions are neither. For example, in Fig. 3.1a the UNDO-NEXT-LSN and PREV-LSN chains are equal, but in Fig. 3.1c the PREV-LSN chain links the log records for $C$, $D^{-1}[z, w]$, $I^{-1}[z, w']$, $A$, $I[z, w']$, $D[z, w]$, and $B$.

The active-transaction table is created and initialized to be empty when the system starts up. When a new transaction arrives in the system, a new transaction identifier $T$ is reserved for it, the *begin-transaction* log record $\langle T, B \rangle$ is written to the log, and the entry

$$(T, \text{forward-rolling}, \text{UNDO-NEXT-LSN} = n)$$

is inserted into the active-transaction table, where $n$ is the LSN of $T$'s begin-transaction log record, marking the end of the UNDO-NEXT-LSN chain. When a transaction commits or completes its rollback, its entry is deleted from the active-transaction table.

For the purposes of transaction rollback and restart recovery, it would only be necessary to record in the active-transaction table information about active transactions that have performed at least one update action, while it seems unnecessary to record their information about *read-only transactions* that only perform read actions. However, for the purposes of concurrency control, a transaction needs to have a unique transaction identifier when it acquires its first lock on a data item it wants to read or update.

In practice, the beginning of a new transaction is recorded in the log, and the transaction record is inserted into the active-transaction table when the transaction performs its first read or update action on the database. Accordingly, if a transaction never reads or writes a database, then neither its beginning or commit is recorded in the log or in the active-transaction table.

For the purposes of the redo pass of restart recovery, a main-memory structure called *modified-page table* (or *dirty-page table*), or *page table* for short, is maintained. This table complements the main-memory structures maintained by the buffer manager that contains information on which pages currently reside in the buffer, which processes have them fixed and latched, and which pages have been updated (recall the modified bit in the buffer control block of a buffered page). For each modified page $p$ in the buffer, the following information is stored in the modified-page table:

1. The page identifier of page $p$
2. The *recovery*-LSN value for page $p$, denoted REC-LSN($p$), which gives the earliest position of the log from which there may be updates on page $p$ that are not yet reflected in the disk version of the page

The table is implemented using an efficient main-memory structure, usually a hash table on the page-ids.

Whenever an unmodified page $p$ in the buffer is turned to a modified page due to the first update that makes the page differ from its disk version, the entry (page-id = $p$, REC-LSN = $n$) is inserted into the modified-page table, where $n$ is the LSN of the log record of that update. We denote $n$ by MODIFIED-LSN($p$). When $p$ is flushed onto disk, its entry is deleted from the modified-page table.

When no failures occur, the equality

$$\text{REC-LSN}(p) = \text{MODIFIED-LSN}(p)$$

is maintained. In general, in the presence of failures, we have

$$\text{REC-LSN}(p) \leq \text{MODIFIED-LSN}(p). \tag{3.1}$$

The REC-LSNs provide an easily maintained lower bound on the LSNs of log records that need to be applied in order to reconstruct the state of the database after a system crash or a process failure. When the current version of page $p$ has been lost or corrupted due to a failure, the redo pass of recovery (Sect. 4.8) restores the page from its disk version. This is done by scanning the log disk and redoing on the page every update on it logged with a LSN greater than or equal to REC-LSN($p$) if the update is not yet in the page, that is, the PAGE-LSN of the page being restored is less than the LSN of the log record being scanned.

## 3.4   Logging for the Read-Write and Key-Range Models

In the following we list the physiological log records for various actions in the read-write and key-range transaction models. In the log records below, $T$ denotes a transaction identifier, and $n$ is the LSN assigned to the log record.

The begin-transaction action $B$ of transaction $T$ is logged with the *begin-transaction* log record

$$n : \langle T, B \rangle. \tag{3.2}$$

The commitment or completion of rollback $C$ of transaction $T$ is logged with the *commit* log record

$$n : \langle T, C \rangle. \tag{3.3}$$

The abort $A$ of transaction $T$ is logged with the *abort* log record

$$n : \langle T, A \rangle. \tag{3.4}$$

The action $S[P]$ of setting savepoint $P$ in transaction $T$ is logged with the *set-savepoint* log record

$$n : \langle T, S[P], n' \rangle, \tag{3.5}$$

where $n'$ is the UNDO-NEXT-LSN of $T$ when the action is being performed. The UNDO-NEXT-LSN value is needed in order to correctly update UNDO-NEXT-LSN($T$) when a partial rollback to savepoint $P$ has been performed.

The action $A[P]$ of beginning partial rollback to savepoint $P$ in transaction $T$ is logged with the *begin-rollback-to-savepoint* log record

$$n : \langle T, A[P] \rangle. \tag{3.6}$$

The action $C[P]$ of completing partial rollback to savepoint $P$ in transaction $T$ is logged with the *complete-rollback-to-savepoint* log record

$$n : \langle T, C[P] \rangle. \tag{3.7}$$

The log record for a forward-rolling update action $(W, I, D)$ by transaction $T$ contains the name and arguments of the action, as well as the UNDO-NEXT-LSN value $n'$ set from UNDO-NEXT-LSN($T$) when the action is being performed; the value $n'$ is needed for updating UNDO-NEXT-LSN($T$) after possibly undoing the logged action in a partial or total rollback.

The write action $W[x, u, v]$ performed by transaction $T$ on data page $p$ is logged with the *write* log record

$$n : \langle T, W, p, x, u, v, n' \rangle. \tag{3.8}$$

The insert action $I[x, v]$ performed by transaction $T$ on data page $p$ is logged with the *insert* log record

$$n : \langle T, I, p, x, v, n' \rangle. \tag{3.9}$$

The delete action $D[x, v]$ performed by transaction $T$ on data page $p$ is logged with the *delete* log record

$$n : \langle T, D, p, x, v, n' \rangle. \tag{3.10}$$

The log records for the above three forward-rolling update actions are *redo-undo log records*: using them, the action can be both redone (physically) and undone (physically or, if this is not possible, logically).

The log record for a backward-rolling update action by transaction $T$ contains the name and (part of the) arguments of the action, as well as the UNDO-NEXT-LSN value $n'$.

The undo action $W^{-1}[x, u, v]$ performed by transaction $T$ on page $p$ is logged with the *undo-write* log record

$$n : \langle T, W^{-1}, p, x, u, n' \rangle, \tag{3.11}$$

when the log record for the respective forward-rolling write action $W[x, u, v]$ by $T$ is $n'' : \langle T, W, p', x, u, v, n' \rangle$ for some $n'' < n$ and $p'$. Note that it is not necessary to record the new value $v$ for redoing the action that reverts to the old value $u$. Also note that the page $p$ onto which the undo action is performed may be different from the page $p'$ onto which the forward-rolling action was performed, because the tuple $(x, v)$ may have migrated from page $p'$ to page $p$ in the interim due to structure modifications caused by $T$ or other transactions.

The undo action $I^{-1}[x, v]$ performed by transaction $T$ on page $p$ is logged with the *undo-insert* log record

$$n : \langle T, I^{-1}, p, x, n' \rangle, \tag{3.12}$$

when the log record for the respective forward-rolling insert action $I[x, v]$ by $T$ is $n'' : \langle T, I, p', x, v, n' \rangle$ for some $n'' < n$ and $p'$. Note that it is not necessary to record the entire inserted tuple in the log: this log record is only used to redo the undo of an insertion, which is a deletion of the tuple with key value $x$. If $p \neq p'$, the tuple $(x, v)$ has migrated from page $p'$ to page $p$ in the interim.

The undo action $D^{-1}[x, v]$ performed by transaction $T$ on page $p$ is logged with the *undo-delete* log record

$$n : \langle T, D^{-1}, p, x, v, n' \rangle, \tag{3.13}$$

when the log record for the respective forward-rolling delete action $D[x, v]$ by $T$ is $n'' : \langle T, D, p', x, v, n' \rangle$ for some $n'' < n$ and $p'$. In redoing the undo of a deletion of a tuple, the entire tuple is needed and hence must be included in the log record. If $p \neq p'$, structure modifications occurred in the interim have caused that the correct page to insert the tuple $(x, v)$ is no longer $p'$ but $p$. In a B-tree, when a full page is split into two pages, the key range covered by the page is split into two ranges, and when two sibling pages are merged into a single page, the key ranges covered by those pages are merged. Page $p$ may, for example, be the result of merging page $p'$ into its sibling page.

The log records for the above three undo actions are *redo-only log records*: using one, the undo action given by the log record can be redone (physically), but undo actions are never undone. The term *compensation log record* (CLR) is also used. Note that the UNDO-NEXT-LSN value $n'$ in the log record for the undo action is always the same as in the log record for the corresponding forward-rolling action.

*Example 3.1*  The committed transaction

$T_1: BR[x, u] R[y, v] D[z, w] I[z, w'] C$

generates the log records:

$n_1: \langle T_1, B \rangle$.
$n_2: \langle T_1, D, p, z, w, n_1 \rangle$.
$n_3: \langle T_1, I, p', z, w', n_2 \rangle$.
$n_4: \langle T_1, C \rangle$.

Here the tuple $(z, w)$ was deleted from page $p$, and the tuple $(z, w')$ was inserted into page $p'$. The log sequence numbers $n_1, n_2, n_3, n_4$ form an ascending sequence. When other concurrent transactions are present, log records generated by them are written in between $T_1$'s log records. The log records are chained backwards via the UNDO-NEXT-LSNs at update actions performed (Fig. 3.1a).                                  □

*Example 3.2*  The rolled-back transaction

$T_2: BR[x, u] R[y, v] D[z, w] I[z, w'] A I^{-1}[z, w'] D^{-1}[z, w] C$

generates the log records

$n_1: \langle T_2, B \rangle$.
$n_2: \langle T_2, D, p_1, z, w, n_1 \rangle$.
$n_3: \langle T_2, I, p_2, z, w', n_2 \rangle$.
$n_4: \langle T_2, A \rangle$.
$n_5: \langle T_2, I^{-1}, p_3, z, n_2 \rangle$.
$n_6: \langle T_2, D^{-1}, p_4, z, w, n_1 \rangle$.
$n_7: \langle T_2, C \rangle$.

The insertion of the tuple $(z, w')$ into page $p_2$ has been undone by deleting the tuple from its then current location in page $p_3$. (If $p_3 \neq p_2$, there must be between the log records with LSNs $n_3$ and $n_5$ a log record for a structure modification triggered by some other transaction that has moved the tuple from page $p_2$ to page $p_3$.) The deletion of $(z, w)$ from page $p_1$ has been undone by inserting $(z, w)$ into page $p_4$.

Observe that the UNDO-NEXT-LSN chain from the log record with LSN $n_5$ skips over the log records with LSNs $n_4$ and $n_3$ back to the log record with LSN $n_2$, which is the log record for the action to be undone next (Fig. 3.1b). Similarly, the UNDO-NEXT-LSN chain from the log record with LSN $n_6$ skips over the log records with LSNs $n_5, n_4, n_3$, and $n_2$ back to the log record with LSN $n_1$, which is the begin-transaction log record for $T_2$, thus indicating that the rollback is complete (Fig. 3.1c).

Also observe that the UNDO-NEXT-LSN value in the log record for an undo action is always the same as the UNDO-NEXT-LSN value in the log record for the action to be undone. □

Thus far we have assumed that the logical database consists of only one relation. If the logical database has more than one relation, an identifier for the relation $r$ must be included in the log records. The log records for the forward-rolling update actions are now:

$\langle T, W, p, r(x, u, v), n' \rangle$.
$\langle T, I, p, r(x, v), n' \rangle$.
$\langle T, D, p, r(x, v), n' \rangle$.

The log records for the undo actions are:

$\langle T, W^{-1}, p, r(x, u), n' \rangle$.
$\langle T, I^{-1}, p, r(x), n' \rangle$.
$\langle T, D^{-1}, p, r(x, v), n' \rangle$.

Naturally, in the most general case when the logical database is divided into several databases, the log records must also carry the identifier of the database on whose relation the update action is performed.

## 3.5  Logging Structure Modifications

As explained in Sect. 2.8, a structure modification on the physical database is an atomic unit of its own that "commits" independently of the outcome (commit or abort) of the transaction that caused the structure modification. To make possible the redoing of structure modifications that in the event of a process failure or a system crash are not reflected in the disk version of the database, all structure modifications must be logged.

For example, the allocation of a new page $q$ as part of a heap file structure could be logged as the following system-generated *structure-modification transaction*:

$\langle S, B \rangle$.
$\langle S, allocate\text{-}page\text{-}for\text{-}heap, s, f, p, q \rangle$.
$\langle S, C \rangle$.

Here $S$ is a new transaction identifier generated by the database management system. When the structure modification is recorded as committed, it can no longer be rolled back, that is, it behaves as a committed transaction.

However, for reasons of efficiency, we do not require full transaction semantics from a structure modification. The full semantics would require forcing the log onto disk, which we want to avoid.

If all of the actual physical database operations in a structure modification can be logged into one log record, it is not necessary to produce a transaction identifier or to log the begin or finish of the transaction. Thus, only the following record is logged for a structure modification that allocates a new page in a heap structure:

$$n : \langle allocate\text{-}page\text{-}for\text{-}heap, s, f, p, q \rangle. \tag{3.14}$$

The LSN $n$ of the log record is stamped into the PAGE-LSN fields of all the affected pages, $s, f, p, q$, while holding those pages write-latched.

*Example 3.3*  Assume that the actions of the transactions

$T_1 = BI[x, v]AI^{-1}[x, v]C$   and
$T_2 = BI[y, w]C$

are executed in the left-to-right order

$T_1 :$    $BI[x, v]$                      $AI^{-1}[x, v]C$
$T_2 :$                   $BI[y, w]C$

Also assume that the action $I[x, v]$ causes a new page to be allocated where the tuple $(x, v)$ is inserted. Then the following will be written to the log:

$n_1$: $\langle T_1, B \rangle$.
$n_2$: $\langle allocate\text{-}page\text{-}for\text{-}heap, s, f, p, q \rangle$.
$n_3$: $\langle T_1, I, q, x, v, n_1 \rangle$.
$n_4$: $\langle T_2, B \rangle$.
$n_5$: $\langle T_2, I, q, y, w, n_4 \rangle$.
$n_6$: $\langle T_2, C \rangle$.
$n_7$: $\langle T_1, A \rangle$.
$n_8$: $\langle T_1, I^{-1}, q, x, n_1 \rangle$.
$n_9$: $\langle T_1, C \rangle$.

The log record with LSN $n_2$ does not belong to any transaction. It is a redo-only log record that is only used to redo the effect of the allocation on one or more of the four pages $s, f, p, q$ affected should the effect of the allocation, due to a failure, be missing from some of those pages. The UNDO-NEXT-LSN value $n_1$ in $T_1$'s log record with LSN $n_3$ skips over that with LSN $n_2$ to $T_1$'s previous log record (with LSN $n_1$).

□

## 3.6 Updating a Data Page

We assume that log records are appended to the log by the call

$log(n, \langle \ldots \rangle)$.

The call write-latches the current tail of the log, determines the LSN for the log record, appends the log record to the log buffer, releases the latch, and assigns the LSN to the output parameter variable $n$.

In the following we consider briefly the implementation of forward-rolling actions $I[x, v]$ and $D[x, v]$, as regards data page updating, latching, and logging, omitting aspects related to transactional concurrency control (locking).

In executing $I[x, v]$, the data page $p$ that will receive the tuple $(x, v)$ is first located and write-latched. For example, if the physical database structure is a sparse B-tree, we have to locate the leaf page $p$ whose key range covers the key value $x$. We assume that locating the data page $p$ also includes any structure modifications (page splits) that must be done so as to arrange sufficient space in the page for the tuple to be inserted. When the server-process thread that services transaction $T$ has located, fixed, and write-latched data page $p$, the insertion is performed, after which page $p$ is unlatched:

1. Locate and write-latch the data page $p$ that will receive $(x, v)$.
2. *insert-into-page*$(T, p, x, v)$.
3. *unlatch-and-unfix*$(p)$.

Here the call *insert-into-page*$(T, p, x, v)$ (Algorithm 3.1) inserts $(x, v)$ into page $p$, logs the insertion with the redo-undo log record (3.9), stamps the LSN of the log record into the PAGE-LSN field of page $p$, and updates the UNDO-NEXT-LSN of transaction $T$ in the active-transaction table.

In executing $D[x, v]$, the data page $p$ that holds the tuple with key value $x$ is first located and write-latched. We assume that this also includes any structure modifications that must be done so as to arrange that a sufficient number of tuples will remain in page $p$ after the deletion of $(x, v)$ (in order to maintain the balance condition, if any). When the server-process thread that services transaction $T$ has located, fixed, and write-latched data page $p$, the deletion is performed, after which page $p$ is unlatched:

1. Locate and write-latch the data page $p$ that contains the tuple with key $x$.
2. *delete-from-page*$(T, p, x)$.
3. *unlatch-and-unfix*$(p)$.

Here the call *delete-from-page*$(T, p, x)$ (Algorithm 3.2) deletes the tuple with key $x$ from page $p$, logs the deletion with the redo-undo log record (3.10), stamps the LSN of the log record into the PAGE-LSN field of page $p$, and updates the UNDO-NEXT-LSN of transaction $T$.

---

**Algorithm 3.1** Procedure *insert-into-page*$(T, p, x, v)$

---

insert $(x, v)$ into data page $p$
$n' \leftarrow$ Undo-Next-LSN$(T)$
$log(n, \langle T, I, p, x, v, n' \rangle)$
Page-LSN$(p) \leftarrow n$
Undo-Next-LSN$(T) \leftarrow n$

---

---

**Algorithm 3.2** Procedure *delete-from-page*$(T, p, x)$

---

delete the tuple $(x, v)$ with key $x$ from data page $p$
$n' \leftarrow$ Undo-Next-LSN$(T)$
$log(n, \langle T, D, p, x, v, n' \rangle)$
Page-LSN$(p) \leftarrow n$
Undo-Next-LSN$(T) \leftarrow n$

---

## 3.7  Write-Ahead Logging

The entry-sequenced file that stores the log is kept in *stable storage*, that is, a permanent memory that is as reliable as possible. In practice, stable storage must be approximated with mirrored files stored on different disk drives. The log records are packed into buffer pages that are taken to the log file on disk when necessary.

To make the log file on stable storage reflect the state of the database as exactly as possible, modified pages of the log file should be flushed from the buffer onto disk immediately after writing a new log record. However, this would cause a disk access for every update, which is very inefficient. In practice, the log is forced to disk only when a transaction commits or completes its rollback or when the buffering policy requests it.

Whenever a transaction updates (by one of the actions $I$, $D$, $W$, $I^{-1}$, $D^{-1}$, or $W^{-1}$) a database page $p$ in the buffer, the LSN of the log record generated for this update is stamped into the Page-LSN field in the header of page $p$. Similarly is the LSN for a structure modification stamped into the Page-LSN fields of all those pages that are modified in the structure modification. For example, in Example 3.3 the LSN $n_2$ of the log record $\langle allocate\text{-}page\text{-}for\text{-}heap, s, f, p, q \rangle$ is stamped into the Page-LSN fields of all the four affected pages: $s$, $f$, $p$, and $q$.

Before page $p$ can be flushed from the buffer onto disk, the log manager must flush from the log buffer onto the log disk all log records whose LSNs are less than or equal to the Page-LSN of page $p$. This is called the WAL protocol.

In the event of a system failure, the contents of main memory are lost. Thus, the buffer versions of updated database pages as well as the log records in the log buffer are gone. What remains are the disk versions of the database pages and the log records that were flushed onto the log disk before the failure.

The protocol for page updating, which includes that the page must be kept write-latched during the generation of the log record and the stamping of its LSN into the Page-LSN of the page, guarantees that the log always "runs side by side" with the current version of the database, when no process failures or system crashes occur.

The WAL protocol in turn guarantees that the log disk always "runs ahead" of the disk version of the database, that is, the log disk always contains the log records for all the updates found in the disk versions of the pages (and often more). Thus, when recovering from a failure, the log contains the necessary information for redoing the updates that are recorded on the log disk but are missing from the disk version of the database and also for undoing the updates of active transactions that went onto disk before the failure. At the start of recovery, a transaction is considered to be active if it has no commit log record $(C)$ on the log disk.

## 3.8  The Commit Protocol

The *commit protocol* of a transaction (i.e., the execution of the $C$ action) includes generating the commit log record, flushing the log (called the principle of *force-log-at-commit*), releasing any resources (such as locks) reserved for and still held by the transaction, and deleting the transaction record from the active-transaction table. The commit protocol is given as Algorithm 3.3.

---

**Algorithm 3.3** Procedure *commit(T)*

---
$log(n, \langle T, C \rangle)$
*flush-the-log*()
Release all locks held by $T$
Delete the transaction record of $T$ from the active-transaction table
Return a notification of the commit of $T$ to the application process

---

The implementation of the complete-rollback action $C$ of an aborted transaction $T$ is the same as for the commit action except that in the last step, the application process that generates $T$ is notified that $T$ has completed its rollback instead of having committed.

The call *flush-the-log*() takes onto the log disk those log records from the log buffer that have not yet been written onto disk. In other words, the modified log file pages in the log buffer are flushed. If multiple pages are flushed, they are flushed in their logical order, so that the sequence of log records on the log disk is always a prefix of the sequence of log records appended into the log buffer, even if the system crashes during the flushing.

The above commit protocol is sufficient for a centralized database system, and also for local transactions in a distributed database system (where a transaction started at a given site is termed local if it only operates on data stored at that site). For a distributed transaction, that is, one that operates on data stored at several sites, the commit protocol is far more complicated: when a transaction operates on data stored at different sites, it gives rise to a set of local transactions (one for each of those sites), whose execution must be coordinated so that either all the local transactions commit or all abort and roll back. The management of distributed transactions is discussed in Chap. 13.

## 3.9   The Steal-and-No-Force Buffering Policy

As explained earlier, the buffer manager of the database management system services the page-fixing and page-latching calls (*fix-and-read-latch*($p$), *fix-and-write-latch*($p$), *unlatch-and-unfix*($p$)) of server-process threads that access the database and generally takes care that the database pages are easily available when needed. We also recall that the buffer has space for only a relatively small number of database pages at a time. The size of the buffer varies, but it is rarely larger than a few thousand pages.

Pages that are fixed in the buffer and the rules for writing log records prevent the buffer manager from removing pages freely from the buffer when it is full. A page that is currently fixed for a server-process thread that is executing a transaction cannot be evicted from the buffer. According to the WAL protocol, a modified database page can be taken to disk only after the log records for the updates have been taken to disk.

On the other hand, the commit protocol of a transaction (and the protocol for completing rollback) includes flushing the log to disk (force-log-at-commit). Usually, the database management system does not place further restrictions on buffer management. However, in some special-purpose database management systems, the buffer manager may apply a more limited protocol.

A data item (tuple or relation) of the logical database is *dirty*, if the transaction that updated it last is still active, and *clean*, if the transaction that updated it last has committed or completed its rollback. An updated database page is *dirty*, if it contains dirty data items, and *clean* otherwise.

When selecting an unfixed page for flushing from the buffer onto disk, the buffer manager can make a difference between dirty and clean pages or ignore the state of the page. In the latter case, the buffer manager applies the *steal policy*: it is allowed to "steal" the page from the buffer to disk, even though some transactions that updated the tuples on the page are still active. The other choice is the *no-steal policy*, according to which all dirty pages are kept in the buffer until the transactions that updated the pages have all committed or rolled back.

The no-steal policy may be claimed to make the undo pass of restart recovery faster: because dirty data is never taken onto disk, the recovery process might have less work in performing the undo actions of backward-rolling transactions that are aborted because of the failure. However, with the do-redo-undo recovery paradigm followed by the ARIES algorithm (see Sect. 4.6), the no-steal policy would yield little or no benefit, because a redo pass is run first in which all updates (be them by committed or rolled-back or active transactions) are redone (if found missing from the disk version), and only then an undo pass is run in which undo actions by aborted transactions are performed. Thus, even if the updates by aborted transactions do not appear in the disk version of the database, they may appear in the buffer version of the database reconstructed in the redo pass and hence must be undone in the undo pass that follows.

The no-steal policy might be of some use in the context of do-undo-redo recovery, in which the undo pass precedes the redo pass, so that the updates by aborted transactions lost due to the failure are not redone. But, as shown in Sect. 4.5, do-undo-redo recovery is problematic in fine-grained transaction processing in which a data page can simultaneously hold updates by several active transactions.

An obvious downside of the no-steal policy is that the buffer pool must be large enough to accommodate all the pages that have been updated by active transactions. Moreover, in fine-grained transaction processing a frequently updated page may continue to hold dirty data besides the committed data, so that, under the no-steal policy, the page cannot be flushed. This again implies that the disk version of the database lags more and more behind the current version of the database, so that more work must be done in the redo pass of recovery (see Sects. 4.6 and 4.8). This is why most database management systems apply the steal policy instead.

As for the question when should clean data be taken to disk, there are two different policies: force and no force. In the *force policy*, when a transaction requests to be committed or to complete its rollback, the buffer manager finds all pages that have been updated by the transaction but have not yet been flushed onto disk, and "forces," that is, flushes them onto disk (using WAL). Only after this, the transaction actually commits or completes its rollback, that is, the commit log record for the transaction is logged and the log is flushed onto the log disk. The *no-force policy* specifies that both dirty and clean modified data can stay in the buffer for as long as the buffer space is not needed for other purposes. Under fine-grained transaction processing, a page that holds data updated by a committing transaction may also hold data updated by other active transactions, in which case the force policy also takes dirty data onto disk, besides the clean data, or, more accurately, the data that is about to turn from dirty to clean.

The force policy obviously makes the redo pass of failure recovery faster. However, it slows down the commit action: updated pages are flushed onto disk as part of the commit protocol. An even more serious disadvantage is that a page that is updated repeatedly by different transactions is unnecessarily taken to disk each time one of the transactions commits. In the no-force policy, such a page (a "hotspot" page) would probably stay in the buffer between transactions and would not need to be taken to disk. This is why most database management systems use the no-force policy.

## 3.10  Reducing Disk Access

The buffer manager should try to minimize the number of disk accesses performed during transaction processing. When the buffer becomes full, the page to be evicted from the buffer should be selected intelligently. In managing virtual memory, the *least-recently-used* or LRU page-replacement strategy is often used: when all buffer frames are allocated, take to disk the page that has been unused for the longest time.

For virtual memory, LRU is a sufficient strategy, because the behavior of all kinds of programs is difficult to predict. For database pages, however, better strategies can be used: from the form of an SQL statement, the database management system can often predict which pages will be needed in executing the statement. The operating system needs to rely on the past to predict the future, while the database management system may know something about at least the near future.

*Example 3.4* Algorithm 3.4 computes the natural join $r \bowtie s$ of relations $r(A, B)$ and $s(B, C)$, when the join attribute $B$ is the key of neither relation, the relations are stored in separate files, and there are no indexes (locking operations have been left out of the algorithm).

---

**Algorithm 3.4** Computing the natural join $r(A, B) \bowtie s(B, C)$

---

```
for each data page p of r do
  fix-and-read-latch(p)
  for each data page q of s do
    fix-and-read-latch(q)
    for each r-tuple t in page p and each s-tuple u in page q do
      if t(B) = u(B) then
        output the joined tuple tu
      end if
    end for
    unlatch-and-unfix(q)
  end for
  unlatch-and-unfix(p)
end for
```

---

After page $p$ of $r$ has been used, it is no longer needed. Thus, a good buffering policy for $r$'s pages is "toss immediate." For each page $p$ of $r$, all pages $q$ of $s$ must be processed. After a page $q$ has been processed, it is needed only after all other pages of $s$ have been processed. A good buffering policy for the pages of $s$ is MRU (most recently used), that is, the opposite of LRU.

The page of $r$ that is being processed at a given time is fixed into the buffer during the processing of $s$, after which it can be designated to be an MRU page. Thus, MRU is an optimal strategy for the pages of both $r$ and $s$.                                    □

## Problems

**3.1** With physiological logging, what log records are produced by the transactions of Problem 1.2. Give the log records produced (a) for the read-write model, (b) the key-range model. You may assume that no structure modifications occur and that no other transactions are in progress.

**3.2** With physiological logging, what log records are produced by the transaction of Problem 1.3? You may assume that no structure modifications occur and that no other transactions are in progress.

**3.3** At the highest level of abstraction, logical logging would mean writing original SQL statements into the log, such as the single log record

$n : \langle T, \text{update } r \text{ set } V = V + 1 \text{ where } X > 50, m \rangle$

for the SQL statement

**update** $r$ **set** $V = V + 1$ **where** $X > 50$

done by transaction $T$. Are there any advantages in this approach? What are the disadvantages?

**3.4** With the no-steal buffering policy, no dirty (i.e., uncommitted) data never goes to disk, because all the pages updated by a transaction are kept in the buffer until the transaction has committed. Therefore, could we do without the before images in the log records? Note that the before images are needed only to roll back uncommitted transactions. Thus, a write operation $W[x, u, v]$ might be logged using a redo-only log record that contains the after image $v$ but not the before image $u$.

With the force buffering policy, a transaction is allowed to commit only after all pages updated by it have been forced to disk. Does this policy allow a similar simplification in what needs to be recorded in the log?

Is it possible for a database management system to apply both the no-steal and force policies, at the same time?

**3.5** Consider the extended read-write transaction model in which a transaction can use the action $C[x]$ to declare the updates on $x$ done so far as committed before the transaction terminates (see Problem 1.5). What log records must be generated so that the write actions declared as committed will not be undone if the transaction eventually aborts or performs a partial rollback?

**3.6** What changes are needed to the logging protocol for our key-range transaction model if the model is extended with the cursor mechanism outlined in Problem 1.7?

# Bibliographical Notes

The fine-granular physiological logging protocol with redo-undo log records for forward-rolling actions and redo-only compensation log records for undo actions in partial or total rollbacks and with PAGE-LSN stamping presented in this chapter is— apart from some minor simplifications—the one presented by Mohan et al. [1992a] for the ARIES recovery algorithm. The WAL and commit protocols are discussed by Gray [1978], Gray et al. [1981], Mohan et al. [1992a], and Gray and Reuter [1993], among others. The steal-and-no-force buffering policy has been advocated by Mohan et al. [1992a] and Gray and Reuter [1993], among others.

Logging was used in one form or another in early database management systems, with recovery algorithms based on paradigms different from that of ARIES. System R maintains "time-ordered lists of log entries, which record information about each change to recoverable data, and the entries for each transaction are chained together, and include the old and new values of all modified recoverable objects along with operation code and object identification" [Astrahan et al. 1976]. These log records are redo-undo log records for forward-rolling update actions, while undo actions are not logged. Also, modifications to dense indexes are not logged in System R, because the index entries are redundant in that they can be determined from the data tuples and information stored in the system catalog about the index. (See the discussion of recovery paradigms by Mohan et al. [1992a].)

In the B-tree index management algorithms (ARIES/KVL and ARIES/IM) designed to work with ARIES [Mohan, 1990a, 1996a, Mohan and Levine, 1992], structure modifications are logged differently from the way we propose in this book. The general approach for structure modifications in ARIES [Mohan et al. 1992a] is to log a structure modification involving several pages with as many log records as there are pages, each describing the modification on one page. Such a log record is a redo-undo log record as long as the multi-page structure modification is incomplete, but turns into a redo-only log record as soon as the modification is completed. This is achieved by ending the multi-page modification with a "dummy" CLR whose only purpose is to make the UNDO-NEXT-LSN chain to skip over all the one-page log records written for the multi-page modification. If the multi-page modification is left incomplete due to a process failure or a system crash, so that the dummy CLR is missing, the modification is undone. Since a structure modification is a physical operation, the undoing must also be physical; no logical undo is possible.

The discussion in Sect. 3.10 about the LRU and MRU buffering policies was inspired by Stonebraker [1981] and Silberschatz et al. [2006].

# Chapter 4
# Transaction Rollback and Restart Recovery

During normal processing, total or partial rollbacks occur when transactions themselves request such actions. In the event of a system crash or startup, *restart recovery* is performed before normal transaction processing can be resumed. This includes restoring the database state that existed at the time of the crash or shutdown from the disk version of the database and from the log records saved on the log disk, followed by the abort and rollback of all forward-rolling transactions and running the rollback of all backward-rolling transactions into completion.

In this chapter we show how the physiological redo-undo log records written during normal transaction processing are used to redo update actions whose effects are found missing from the disk version of the database and how the log records are used to perform the undo actions of backward-rolling transactions in our key-range transaction model. We also show how restart recovery can be speeded up by periodically written checkpoint log records that include snapshots of the active-transaction and modified-page tables.

The recovery algorithm to be presented is a slightly simplified version of the ARIES algorithm for a centralized database system. This algorithm consists of four phases, performed in the order shown: (1) the analysis pass for reconstructing the active-transaction and modified-page tables, (2) the redo pass for restoring the state of the database, (3) the undo pass for rolling back active transactions, and (4) the taking of a checkpoint. One of the most notable points in this organization of restart recovery is that the redo pass consists entirely of physically executed redo actions, thus restoring the state of the physical database that existed when the last log record saved on disk was written.

© Springer International Publishing Switzerland 2014
S. Sippu, E. Soisalon-Soininen, *Transaction Processing*, Data-Centric Systems and Applications, DOI 10.1007/978-3-319-12292-2_4

## 4.1   Transaction Aborts, Process Failures, and System Crashes

During transaction processing, a transaction can abort for one of the following reasons:

1. The application process itself requests the abort and rollback of the transaction by the SQL statement **rollback**.
2. The database management system aborts the transaction because of a deadlock or because the application process seems to have failed or the connection to it is cut off.
3. A *process failure* of the server-process thread that executes the database actions of the transaction.
4. A *system crash* of the database system due to a failure in the hardware or power supply or due to a software fault in the database management system or operating system.
5. The system is *shut down* by the database system administrator.

In case (1) the actions needed in aborting and rolling back the transaction are included in the execution of the **rollback** statement in the server-process thread that services the application process. In case (2) the backward-rolling phase of the transaction is performed on behalf of the transaction by the process-monitor process or by a new server-process thread allocated for the purpose. In case (3) the rollback is also performed by the process-monitor process or by a new server-process thread, but before that can happen, the possibly corrupted pages (if any) held write-latched by the failed process thread at the time of the failure must first be restored to an action-consistent state. This is done by fetching the disk versions of the pages and applying the log records to redo the missing updates on the pages.

In cases (4) and (5), all processes in the system fail, and the contents of main-memory disappear, so that all main memory structures needed for database management are lost; when the system is started up, *restart recovery* is performed so as to restore the state of the database that existed at the time of the crash or shutdown and to abort and roll back all forward-rolling transactions and to complete the rollback of all backward-rolling transactions. The restoring of the database state includes fetching from the disk all pages that resided as modified in the buffer at the time of the crash or shutdown and applying the log records to redo the missing updates. In case (5) there probably are no transactions active, so that no transaction rollbacks are needed.

In all the cases, transaction rollback can be performed by means of the log and the active-transaction table, whose contents survive in cases (1), (2), and (3) and are reconstructed from a logged snapshot and transaction-control log records in cases (4) and (5). The restoration of the database state in cases (3), (4), and (5) uses disk versions of modified pages, log records survived on disk, and a logged snapshot of the modified-page table (whose contents are also reconstructed, using log records for update actions).

## 4.2 Partial and Total Rollbacks

The execution of the SQL **rollback** statement in transaction $T$ is given as Algorithm 4.1 (for our key-range transaction model). First, the state of $T$ is changed to "backward-rolling" in the transaction record of $T$ in the active-transaction table, and the abort log record $\langle T, A \rangle$ is written to the log (Algorithm 4.2). Then the UNDO-NEXT-LSN chain for $T$ is traversed, starting from the UNDO-NEXT-LSN value stored in $T$'s transaction record in the active-transaction table, fetching from the log the records with the LSNs in the chain, and executing the undo actions of the update actions recorded with those log records (by calling one of *undo-insert* and *undo-delete*). Any set-savepoint log records encountered are just passed by. The rollback is finished when the begin-transaction log record $\langle T, B \rangle$ is encountered; then the transaction is recorded as having completed its rollback by executing the call *commit*($T$) (see Algorithm 3.3).

---

**Algorithm 4.1** Procedure *rollback*($T$)

---

  *abort*($T$)
  $r \leftarrow$ *get-log-record*(UNDO-NEXT-LSN($T$))
  **while** $r$ is not "$n : \langle T, B \rangle$" **do**
    **if** $r$ is "$n : \langle T, I, p, x, v, n' \rangle$" **then**
      *undo-insert*($n, T, p, x, n'$)
    **else if** $r$ is "$n : \langle T, D, p, x, v, n' \rangle$" **then**
      *undo-delete*($n, T, p, x, v, n'$)
    **else if** $r$ is "$n : \langle T, S[P], n' \rangle$" **then**
      UNDO-NEXT-LSN($T$) $\leftarrow n'$
    **end if**
    $r \leftarrow$ *get-log-record*(UNDO-NEXT-LSN($T$))
  **end while**
  *commit*($T$)

---

**Algorithm 4.2** Procedure *abort*($T$)

---

  *state*($T$) $\leftarrow$ "backward-rolling"
  *log*($n, \langle T, A \rangle$)

---

The call *get-log-record*($n$) returns the log record with LSN $n$. The call *undo-insert*($n, T, p, x, n'$) executes the undo action $I^{-1}[x, v]$, and the call *undo-delete*($n, T, p, x, v, n'$) executes the undo action $D^{-1}[x, v]$. The latter two calls both include the assignment UNDO-NEXT-LSN($T$) $\leftarrow n'$ in the active-transaction table; recall that $n'$ is the value of UNDO-NEXT-LSN($T$) at the point when the corresponding forward-rolling action was performed. Also note that in the case in which the log record $r$ encountered is one for a set-savepoint action ($\langle T, S[P], n' \rangle$), nothing is done besides the assignment of UNDO-NEXT-LSN($T$).

The execution of the SQL statement **rollback to savepoint** $P$ in transaction $T$ is similar (Algorithm 4.4), but it starts with logging a begin-rollback-to-savepoint log record (Algorithm 4.5), traverses the UNDO-NEXT-LSN chain back to the set-savepoint action (cf. Algorithm 4.3), and finishes the rollback with logging a complete-rollback-to-savepoint log record (Algorithm 4.6).

---

**Algorithm 4.3** Procedure *set-savepoint*$(T, P)$

---
$n' \leftarrow$ UNDO-NEXT-LSN$(T)$
$log(n, \langle T, S[P], n' \rangle)$
UNDO-NEXT-LSN$(T) \leftarrow n$

---

**Algorithm 4.4** Procedure *rollback-to-savepoint*$(T, P)$

---
*begin-rollback-to-savepoint*$(T, P)$
$r \leftarrow$ *get-log-record*(UNDO-NEXT-LSN$(T)$)
**while** $r$ is not "$n : \langle T, S[P], n' \rangle$" **do**
  **if** $r$ is "$n : \langle T, I, p, x, v, n' \rangle$" **then**
    *undo-insert*$(n, T, p, x, n')$
  **else if** $r$ is "$n : \langle T, D, p, x, v, n' \rangle$" **then**
    *undo-delete*$(n, T, p, x, v, n')$
  **else if** $r$ is "$n : \langle T, S[P'], n' \rangle$" for some $P' \neq P$ **then**
    UNDO-NEXT-LSN$(T) \leftarrow n'$
  **end if**
  $r \leftarrow$ *get-log-record*(UNDO-NEXT-LSN$(T)$)
**end while**
*complete-rollback-to-savepoint*$(T, P)$

---

**Algorithm 4.5** Procedure *begin-rollback-to-savepoint*$(T, P)$

---
$log(n, \langle T, A[P] \rangle)$

---

**Algorithm 4.6** Procedure *complete-rollback-to-savepoint*$(T, P)$

---
$log(n, \langle T, C[P] \rangle)$

---

## 4.3 Performing the Undo Actions

An undo action by a backward-rolling transaction or by a forward-rolling transaction that does a partial rollback is performed using as an argument the log record for the corresponding forward-rolling update action. The undo action $o^{-1}[\bar{x}]$ of an update action $o[\bar{x}]$ is first tried physically, that is, directly on the data page on which the action $o[\bar{x}]$ was performed. The page-id of that page is found in the log record for $o[\bar{x}]$. If such a *physical undo* is impossible, we resort to a *logical undo*: the execution of the undo action begins with a search of the database structure in order to locate the current target page of the action.

When an undo action $I^{-1}[x, v]$ must be performed logically, we assume that the data page $q$ that currently holds the tuple $(x, v)$ is located and fixed and write-latched by the call

$$find\text{-}page\text{-}for\text{-}undo\text{-}insert(q, x),$$

which is also assumed to take care of any structure modifications needed. These arrange that page $q$ will hold a sufficient number of tuples even after the deletion of $(x, v)$. Similarly, when an undo action $D^{-1}[x, v]$ is performed logically, the data page $q$ that covers key value $x$ is located and fixed and write-latched by the call

$$find\text{-}page\text{-}for\text{-}undo\text{-}delete(q, x, v),$$

which also takes care of any structure modifications needed to arrange sufficient space for the tuple $(x, v)$ to be inserted. Implementations for these calls are given in Chaps. 7 and 8 for the case in which the physical database is a sparse B-tree (see Sect. 7.6, in particular).

The call $undo\text{-}insert(n, T, p, x, n')$ (Algorithm 4.7) executes the undo action $I^{-1}[x, v]$ of an insert action $I[x, v]$ that was performed by transaction $T$ on data page $p$ and logged with LSN $n$. The new value of UNDO-NEXT-LSN($T$) becomes $n'$ after the undo action. A physical undo is always possible if the PAGE-LSN of page $p$ is still equal to the LSN found in the log record for $I[x, v]$, because in that case the contents of the page have not changed in between. On the other hand, even if then PAGE-LSN has advanced, $p$ may still be the correct target page; this is true when $p$ is still a data page and contains the tuple $(x, v)$ and a sufficient number of other tuples so that the page will not underflow when $(x, v)$ is deleted.

Otherwise, page $p$ is unlatched and unfixed, and the call $find\text{-}page\text{-}for\text{-}undo\text{-}insert(q, x)$ is used to locate the correct target page and to perform any structure modifications (page merges) needed to prevent the target data page from underflowing. Finally, when the correct target page has been found, the tuple $(x, v)$ is deleted, the deletion is logged with a redo-only log record (3.12), and the PAGE-LSN of the page and the UNDO-NEXT-LSN of the transaction are updated by the procedure $undo\text{-}insert\text{-}onto\text{-}page(T, p, x, n')$ (Algorithm 4.8).

---

**Algorithm 4.7** Procedure *undo-insert*(*n*, *T*, *p*, *x*, *n*′)

---

*fix-and-write-latch*(*p*)
**if** PAGE-LSN(*p*) $\neq$ *n* and, in addition, either *p* is not a data page or *p* does not contain the tuple
(*x*, *v*) or *p* would underflow due to the deletion of (*x*, *v*) **then**
    *unlatch-and-unfix*(*p*)
    *find-page-for-undo-insert*(*q*, *x*)
    *p* ← *q*
**end if**
*undo-insert-onto-page*(*T*, *p*, *x*, *n*′)
*unlatch-and-unfix*(*p*)

---

**Algorithm 4.8** Procedure *undo-insert-onto-page*(*T*, *p*, *x*, *n*′)

---

delete the tuple (*x*, *v*) from page *p*
$log(m, \langle T, I^{-1}, p, x, n' \rangle)$
PAGE-LSN(*p*) ← *m*
UNDO-NEXT-LSN(*T*) ← *n*′

---

Similarly to *undo-insert*, the call *undo-delete*(*n*, *T*, *p*, *x*, *v*, *n*′) (Algorithm 4.9)
executes the undo action $D^{-1}[x, v]$ of a delete action $D[x, v]$ performed by *T* on
*p* and logged with LSN *n*. Again a physical undo is always possible if the PAGE-
LSN of page *p* is still equal to the LSN found in the log record for $I[x, v]$. Even
if the PAGE-LSN has advanced, *p* is still the correct target page when *p* is a data
page and covers the key *x* and has room for the tuple (*x*, *v*). (In the case of a B-tree
page, we can be sure that the page covers *x* if *x* is between the least and greatest
keys currently in the page, although this test will miss some cases in which *p* still
covers *x*.)

Otherwise, page *p* is unlatched and unfixed, and the call *find-page-for-undo-
delete*(*q*, *x*, *v*) is used to locate the correct target page and to perform any structure
modifications (page splits) needed to prevent the target data page from overflowing.
The tuple (*x*, *v*) is inserted into target page, the insertion is logged with a redo-
only log record (3.13), and the PAGE-LSN of the page and the UNDO-NEXT-LSN of the
transaction are updated by the procedure *undo-delete-onto-page*(*T*, *p*, *x*, *n*′).

---

**Algorithm 4.9** Procedure *undo-delete*(*n*, *T*, *p*, *x*, *v*, *n*′)

---

*fix-and-write-latch*(*p*)
**if** PAGE-LSN(*p*) $\neq$ *n* and, in addition, either *p* is not a data page of *r* or *p* does not cover the
key value *x* or *p* does not have room for the tuple (*x*, *v*) **then**
    *unlatch-and-unfix*(*p*)
    *find-page-for-undo-delete*(*q*, *x*, *v*)
    *p* ← *q*
**end if**
*undo-delete-onto-page*(*T*, *p*, *x*, *v*, *n*′)
*unlatch-and-unfix*(*p*)

---

---

**Algorithm 4.10** Procedure *undo-delete-onto-page*($T, p, x, v, n'$)

insert the tuple $(x, v)$ into page $p$
$log(m, \langle T, D^{-1}, p, x, v, n' \rangle)$
PAGE-LSN($p$) $\leftarrow m$
UNDO-NEXT-LSN($T$) $\leftarrow n'$

---

We say that the *access-path-length*, or *path-length* for short, of a database action
(read or update) is the number of pages that need to be latched to locate the target
page of the action. The path-length of an action depends on the physical database
structure underlying the logical database. For example, if the relation $r(\underline{X}, V)$ is
organized as a sparse B-tree index (see Sect. 7.1), the path-length of an insert or
a delete action includes the number of pages latched during the B-tree traversal
needed to locate the target page of the action. Such traversals are included in the
implementations of read, insert, and delete actions (Sect. 6.7), as well as in the
implementations of undo-insert and undo-delete actions (the procedures *find-page-
for-undo-insert* and *find-page-for-undo-delete*).

As we shall see in Chaps. 6 and 7, a single root-to-leaf traversal along the B-
tree may not be sufficient, because two paths may have to be traversed to locate the
key next to a given key and because the acquisition of appropriate key-range locks
to protect the action may require even more traversals when the traversals are to
be done in a deadlock-free manner. Thus the ultimate access-path-length may be a
multiple of the height of the B-tree. On the other hand, in the best case the path-
length may be only one if the path previously traversed by a server-process thread
is remembered (see Sect. 7.2) and the traversal for the next action follows the same
path.

The path-length of an undo action may often be shorter than that of the
corresponding forward-rolling action, thanks to the possibility of doing physical
undo. When the page $p$ on which the forward-rolling insert or delete action
was performed can accommodate the undo-insert or undo-delete action, no other
database pages besides $p$ need be latched. Omitting any operations needed to locate
the log record of the forward-rolling action (by the procedure *get-log-record*), we
conclude that the path-length of the undo action is only one in the case of a physical
undo.

In the case that a physical undo is impossible, a logical undo must be performed,
which includes a traversal of the index structure. However, in such a traversal only,
the target page must be located, while no new locks are acquired, thus reducing the
number of retraversals needed (see Sect. 7.6). We conclude that the path-length of
a logical undo-insert action $I^{-1}[x, v]$ (resp. undo-delete action $D^{-1}[x, v]$) is never
greater than that of the forward-rolling delete action $D[x, v]$ (resp. insert action
$I[x, v]$), but probably less.

## 4.4  Checkpoints

When using the no-force buffering policy, there may be many pages in the buffer that are never taken to disk due to their high access rate. These pages can accumulate many updates by committed transactions. Thus, the disk versions of these pages may get well out of date: the PAGE-LSN of the disk version can be much less than the LSN of the latest update on the page (i.e., the PAGE-LSN of the buffer version).

In this situation, the REC-LSN values of the modified-page table point to early locations in the log. This also means that the point where the recovery process starts reading the log is far from the end: the recovery process starts from the REDO-LSN, which is at the minimum of the REC-LSNs or earlier. Thus, a lot of time is used for the redo actions.

The recovery process can be made faster by taking *checkpoints* every now and then (typically at 5–10 min intervals). A *complete checkpoint* can be taken as shown in Algorithm 4.11. The call

*flush-onto-disk*($p$)

takes page $p$ onto disk according to the WAL protocol (so that the log is first flushed) and marks page $p$ as unmodified by clearing the modified bit in the buffer control block of $p$.

---

**Algorithm 4.11** Procedure *take-complete-checkpoint*()

---

$log(n, \langle begin\text{-}complete\text{-}checkpoint\rangle)$
**for** every page $p$ with an entry in the modified-page table **do**
  *fix-and-read-latch*($p$)
  *flush-onto-disk*($p$)
  delete the entry for $p$ from the modified-page table
  *unlatch-and-unfix*($p$)
**end for**
$log(n', \langle end\text{-}complete\text{-}checkpoint\rangle)$
*flush-the-log*()
store BEGIN-CHECKPOINT-LSN $= n$ at a certain place on disk

---

Normal transaction processing can continue in parallel with taking a checkpoint. During taking the checkpoint new insertions into the modified-page table and the active-transaction table can occur; these are possibly not included in the logged tables.

The REDO-LSN value used as the starting point of the redo pass can be set to the LSN of the *begin-complete-checkpoint* log record of the last checkpoint that was completely done (i.e., whose *end-complete-checkpoint* record is found in the log).

Taking a complete checkpoint can be an expensive operation, and it can slow down normal transaction processing considerably. For instance, let the size of the buffer be 5,000 pages of 8 kilobytes each. Assume that 50% of the pages have been updated when the checkpoint is taken. Then we need to take 2,500 pages to disk. In the worst case, the buffer manager goes through the buffer sequentially and flushes every page marked as modified onto disk. Using an average seek time of 15 ms, this takes a total of 37.5 s. This operation slows down the normal operation of the system and is repeated every 5 or 10 min.

The situation can be made somewhat better by ordering the pages before taking them to disk to save in seeking the drive heads. We could also use a background process that takes modified pages to disk, so that there would be less pages to write when the actual checkpoint is taken.

A lighter-weight method than the complete checkpoint is to make an *indirect checkpoint* or *fuzzy checkpoint*, where only some of the modified pages are taken to disk. In a minimal fuzzy checkpoint, no pages are taken to disk, but copies of the modified-page table and the active-transaction table are written to the log. (The active-transaction table is logged so as to assist the undo pass of recovery.)

In a fuzzy checkpoint, the following is logged:

$\langle begin\text{-}checkpoint\rangle$.
$\langle active\text{-}transaction\text{-}table\{(T_1, s_1, n_1), \ldots, (T_k, s_k, n_k)\}\rangle$.
$\langle modified\text{-}page\text{-}table\{(p_1, r_1), \ldots, (p_m, r_m)\}\rangle$.
$\langle end\text{-}checkpoint\rangle$.

Here $s_i$ is the state of transaction $T_i$ (forward-rolling or backward-rolling), $n_i$ is the UNDO-NEXT-LSN of $T_i$, and $r_j$ is the REC-LSN of the modified page $p_j$.

In a fuzzy checkpoint, no pages are necessarily forced to disk, not even pages of the log file. When the *end-checkpoint* log record is eventually taken to disk, the LSN of the corresponding *begin-checkpoint* log record is written to a specific place in stable storage.

Transactions can perform updates and produce log records freely between those produced by the checkpoint. The modified-page table and active-transaction table must be protected using read latches when they are written to the log.

A page that is used often and for long duration may stay in the buffer beyond several checkpoints, which means that its REC-LSN freezes the REDO-LSN used as the starting point of the redo pass. Updated pages should be taken to disk using a background process. It is useful to mark modified pages that have not been taken to disk after the previous checkpoint to be taken to disk during the next checkpoint (before writing the modified-page table). In any case, it is important to make sure that the REDO-LSN value is advanced at every checkpoint. This can be done by flushing onto disk a couple of modified pages with the least REC-LSN values.

## 4.5  Problems with Do-Undo-Redo Recovery

The recovery algorithms used in some early database management systems such as System R were based on the *do-undo-redo* recovery paradigm, where forward-rolling actions of transactions were logged using redo-undo log records (the "do" phase of transaction processing) and where, in the event of a system crash, these log records were first used to undo the updates done by active transactions (the "undo" phase) and then to redo only the updates done by committed transactions that were missing from the disk version of the database (the "redo" phase). The motivation for this paradigm seems to be that since all the active transactions are eventually aborted and rolled back, it is a waste of time to redo the updates of active transactions but only those of the committed transactions. This recovery paradigm is presented in several textbooks on database management.

Unfortunately, the do-undo-redo paradigm has major drawbacks that make it difficult or impossible to apply in the transaction-processing environment considered in this book. First, and most importantly, the do-undo-redo paradigm is incompatible with physical redo and fine-grained transaction processing where a page can contain uncommitted updates by several transactions. It may be impossible to redo physically a committed update after rolling back an uncommitted update that was performed on the page earlier than the committed update (as shown in case (1) of the example below). As redoing can no longer be physical, we must resort to logical redo, implying that new log records need to be written for redo actions. Moreover, undo actions need to check whether or not they should be performed, and the PAGE-LSN values stamped on pages no longer correctly indicate the state of the pages.

*Example 4.1* Assume that transaction $T_1$ performs the action $D[x, v]$ to delete a tuple $(x, v)$ from a full data page $p$ and that then another transaction $T_2$ performs the action $I[y, w]$ to insert a tuple $(y, w)$ into page $p$, consuming the space created by the deletion of $(x, v)$. Then $T_2$ commits, but $T_1$ is still forward-rolling at the time the system crashes (Fig. 4.1). As $T_2$ committed before the crash, the log record of its insert action $I[y, w]$ and thus also the log record of the delete action $D[x, v]$ of $T_1$ were flushed onto the log disk before the crash. With the do-undo-redo recovery paradigm, $T_1$ is first aborted and rolled back, after which the updates of $T_2$ are redone.

We have three cases (as depicted in Fig. 4.1) to consider:

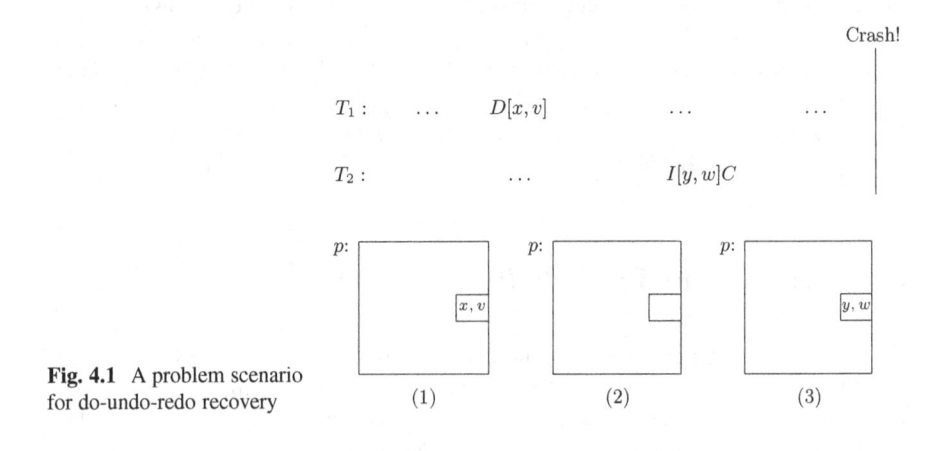

**Fig. 4.1** A problem scenario for do-undo-redo recovery

1. The modified page $p$ was not flushed onto disk before the crash, so that both updates are missing from the disk version of page $p$. In other words, page $p$ contains $(x, v)$, but not $(y, w)$.
2. Page $p$ was flushed onto disk between the two updates, so that the disk version of $p$ contains neither $(x, v)$ nor $(y, w)$.
3. Page $p$ was flushed onto disk after the two updates before the crash, so that the disk version of $p$ contains $(y, w)$, but not $(x, v)$.

In case (1) the effects of the deletion $D[x, v]$ by $T_1$ and the insertion $I[y, w]$ by $T_2$ are not present in the disk version of page $p$, so the deletion must not be undone, but the insertion, being an update by a committed transaction, must be redone. But redoing the insertion on the full page $p$ is not possible. Thus, we must resort to a *logical redo*: the physical database must be searched for another data page into which the tuple $(y, w)$ can be inserted. As regards the physical database, this new insertion is different from the original insertion and hence must be logged with a new physiological log record. Moreover, if the physical database is a B-tree, the page $p$ must be split so as to accommodate the insertion, and that split must also be logged.

In case (2) the effect of the deletion $D[x, v]$ by $T_1$ is present in the disk version of $p$, but the effect of the insertion $I[y, w]$ by $T_2$ is not. Thus the deletion must be undone and the insertion redone. The undo action $D^{-1}[x, v]$ is performed on page $p$. If the undo action is logged as usual, with a redo-only log record, and the LSN of that log record is stamped into the PAGE-LSN field of $p$, then the PAGE-LSN becomes greater than the LSN of the log record for $I[y, w]$ which should be redone. Thus, we can no longer rely on the PAGE-LSN value of a page in deciding which updates on that page should be redone.

In case (3) the effects of the deletion $D[x, v]$ and the insertion $I[y, w]$ are both present in the disk version of $p$, so the deletion must be undone, but the insertion must not be redone. As page $p$ is full, the undo action $D^{-1}[x, v]$ is performed logically. This is the only of the three cases that does not exhibit a problem. □

## 4.6 The ARIES Do-Redo-Undo Recovery Algorithm

The ARIES *recovery algorithm* (Algorithm for Recovery and Isolation Exploiting Semantics) first implemented in the IBM DB2 database management system is based on the *do-redo-undo* recovery paradigm, in which the entire (recent) history is first repeated, so that the missing updates by all transactions are first redone in their original chronological order and only then the updates by active transactions are undone with each undoing logged with a redo-only log record (Fig. 4.2). This approach ensures that the following properties hold in fine-grained transaction processing in which a data page can hold updates by several active transactions:

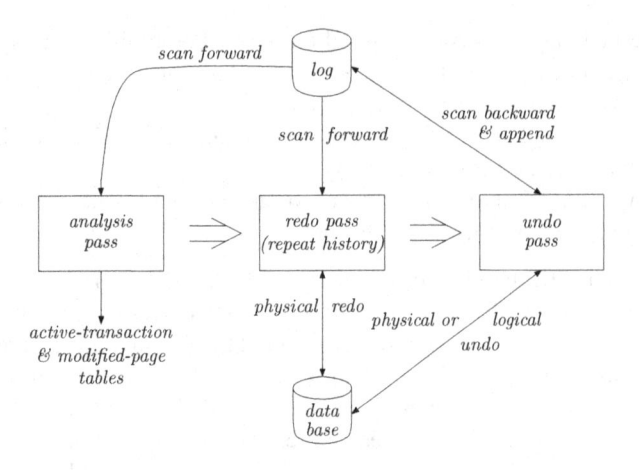

**Fig. 4.2** Analysis, redo, and undo passes of the ARIES recovery algorithm

(a) The PAGE-LSN value stamped on a version of a page $p$ always indicates that all updates with LSNs less than or equal to PAGE-LSN($p$) are present in page $p$, but updates with LSNs greater than PAGE-LSN($p$) are not, thus correctly indicating the state of the page.

(b) All redo actions can be performed physically on the same pages and at the same record positions where they were performed originally by checking the PAGE-LSN of the page. If PAGE-LSN is less than the LSN of the log record, then the logged update is missing from the page and must be redone.

(c) Any new log records need not, and must not, be written for the redo actions, because, by (b), the redo actions are exactly represented by the log records written for the original update actions.

(d) Undo actions are always legal so that there is no need to check whether or not they can be performed, provided that the transactions have been run sufficiently isolated (an issue to be discussed later).

*Example 4.2*  With the do-redo-undo recovery paradigm, restart recovery is performed as follows in the scenario of Example 4.1 (Fig. 4.1). First, the updates of both transactions are redone physically on the disk version of page $p$ fetched to the buffer:

1. When the effects of both updates are missing from the disk version of page $p$, the PAGE-LSN is less than the LSNs of both updates, and hence both updates are redone on $p$.

2. When the effect of the first update is in the disk version of $p$ but the second is missing, the PAGE-LSN is greater than or equal to the LSN of the first update, and hence only the second update is redone.

3. When both updates are in the disk version of $p$, the PAGE-LSN is greater than or equal to the LSN of the second update, and hence neither update is redone.

In all cases, when the updates have been redone, the database has been restored to the state that existed at the time when $T_2$ committed. Then $T_1$ is aborted and rolled back. The rollback includes performing (logically) the undo action $D^{-1}[x, v]$, which is legal in every case, and the legality need not be checked (see Algorithm 4.9).  □

Actually, the entire ARIES recovery algorithm consists of the following four steps, executed in the order shown, when the system restarts after a system crash:

1. The *analysis pass*: with a forward scan of the log starting from the most recently taken checkpoint, the active-transaction and modified-page tables are reconstructed.
2. The *redo pass*: with a forward scan of the log all updates missing from the disk version of the database are redone.
3. The *undo pass*: with a backward scan of the log, all forward-rolling transactions are aborted and rolled back, and the rollback of all backward-rolling transactions is run into completion.
4. Taking a checkpoint.

The ARIES algorithm requires that the buffer manager use the WAL protocol during both normal transaction processing and recovery. No other restrictions are imposed on the buffering policy: it is permitted to steal dirty pages from the buffer, and modified pages do not need to be taken to disk during transaction commitment. Thus, ARIES permits the buffering policies that are the most liberal: steal and no force.

## 4.7  Analysis Pass for Do-Redo-Undo Recovery

In the *analysis pass*, the log is scanned to find out which transactions were active and which pages possibly had updates in the buffer when the failure occurred. The input to the pass is the LSN of the *begin-checkpoint* record of the last (fuzzy) checkpoint that was completed (i.e., the value of BEGIN-CHECKPOINT-LSN). The output is the active-transaction table, the modified-page table, and the REDO-LSN value, which is the location where the redo pass must begin processing the log. Database pages are not accessed during the analysis pass.

In the analysis pass, the log is scanned from the *begin-checkpoint* record to the end of the log. In scanning the log, the call

*get-next-log-record*(n)

is used to fetch the log record with the LSN next to $n$. The active-transaction table and modified-page table are initialized from the log records written at the checkpoint.

While scanning the log, when an update to page $p$ is found and $p$ is not found in the modified-page table, the page-id of $p$ is stored in the table, and its REC-LSN is set to the LSN of the log record for the update. This is done because that update must have turned page $p$ from an unmodified page into a modified one. At the end of the log scan, the REDO-LSN is set to the minimum of the REC-LSN values in the modified-page table.

The active-transaction table is updated while scanning the log by inserting transaction records for transactions that begin after the start of the checkpoint (when a log record $\langle T, B \rangle$ is encountered), and deleting records for transactions that were committed or completed their rollback after the start of the checkpoint (when a log record $\langle T, C \rangle$ is encountered). The UNDO-NEXT-LSN values of active transactions are maintained accordingly.

The algorithm for the analysis pass is shown as Algorithm 4.12. The call *analyze-log-record*$(r)$ (Algorithm 4.13) updates the active-transaction table and the modified-page table as needed by the log record $r$. The processing of log records for structure modifications (in the call *analyze-smo*) is discussed in Sect. 4.11.

---

**Algorithm 4.12** Procedure *analysis-pass*()

---

$r \leftarrow$ *get-next-log-record* (BEGIN-CHECKPOINT-LSN)
**while** $r$ is an available log record with LSN $n$ **do**
    *analyze-log-record* $(r)$
    $r \leftarrow$ *get-next-log-record* $(n)$
**end while**
**return** REDO-LSN $= \min\{n \mid (p, \text{REC-LSN} = n)$ is in the modified-page table$\}$

---

**Algorithm 4.13** Procedure *analyze-log-record*()

---

**if** $r$ is of the form "$n$:$\langle$*active-transaction-table*, $\{\ldots\}\rangle$" **then**
    initialize the active-transaction table with the contents of $r$
**else if** $r$ is of the form "$n$:$\langle$*modified-page-table*, $\{\ldots\}\rangle$" **then**
    initialize the modified-page table with the contents of $r$
**else if** $r$ is of the form "$n$:$\langle T, B \rangle$" **then**
    insert the record $(T, \text{forward-rolling}, \text{UNDO-NEXT-LSN} = n)$ into the active-transaction table
**else if** $r$ is of the form "$n$:$\langle T, (I|D), p, \ldots \rangle$" **then**
    UNDO-NEXT-LSN$(T) \leftarrow n$
    **if** there is no entry for $p$ in the modified-page table **then**
       insert $(p, \text{REC-LSN} = n)$ into the modified-page table
    **end if**
**else if** $r$ is of the form "$n$:$\langle T, A \rangle$" **then**
    change the state of $T$ to "backward-rolling" in the active-transaction table
**else if** $r$ is of the form "$n$:$\langle T, (I|D)^{-1}, p, \ldots, n' \rangle$" **then**
    UNDO-NEXT-LSN$(T) \leftarrow n'$
    **if** there is no entry for $p$ in the modified-page table **then**
       insert $(p, \text{REC-LSN} = n)$ into the modified-page table
    **end if**
**else if** $r$ is of the form "$n$:$\langle T, C \rangle$" **then**
    delete the transaction record of $T$ from the active-transaction table
**else if** $r$ is of the form "$n$:$\langle T, S[P], n' \rangle$" **then**
    UNDO-NEXT-LSN$(T) \leftarrow n$
**else if** $r$ is a log record for a structure modification **then**
    *analyze-smo* $(r)$
**end if**

---

**Fig. 4.3** Log contents on the log disk at the time of a system crash

| |
|---|
| 101: $\langle T_1, B \rangle$ |
| 102: $\langle T_1, D, p_1, x_1, v_1, 101 \rangle$ |
| 103: $\langle begin\text{-}checkpoint \rangle$ |
| 104: $\langle active\text{-}transaction\text{-}table,$ |
| $\{(T_1, \text{forward-rolling}, \text{UNDO-NEXT-LSN} = 102)\} \rangle$ |
| 105: $\langle modified\text{-}page\text{-}table, \{(p_1, \text{REC-LSN} = 102)\} \rangle$ |
| 106: $\langle end\text{-}checkpoint \rangle$ |
| 107: $\langle T_1, I, p_1, x_1, v_1', 102 \rangle$ |
| 108: $\langle T_2, B \rangle$ |
| 109: $\langle T_1, C \rangle$ |
| 110: $\langle T_2, D, p_1, x_1, v_1', 108 \rangle$ |
| 111: $\langle T_3, B \rangle$ |
| 112: $\langle T_3, I, p_2, x_2, v_2, 111 \rangle$ |
| 113: $\langle T_2, I, p_1, x_3, v_3, 110 \rangle$ |
| 114: $\langle T_2, A \rangle$ |
| 115: $\langle T_2, I^{-1}, p_1, x_3, 110 \rangle$ |
| 116: $\langle T_4, B \rangle$ |
| 117: $\langle T_4, I, p_2, x_4, v_4, 116 \rangle$ |
| 118: $\langle T_4, C \rangle$ |

In the analysis pass, no pages of the database are read or written: only the log is read, and main-memory structures, namely, the active-transaction table and modified-page table, are reconstructed.

*Example 4.3* Assume that at the time of a system crash, the log contents surviving on the log disk are as shown in Fig. 4.3. The log contents are the result of running the sequence of actions

$$B_1 D_1[x_1, v_1] B_C \dots E_C I_1[x_1, v_1'] B_2 C_1 D_2[x_1, v_1'] B_3 I_3[x_2, v_2] I_2[x_3, v_3]$$
$$A_2 I_2^{-1}[x_3, v_3] B_4 I_4[x_4, v_4] C_4,$$

where the numerical subscript in an action indicates the transaction ($T_1$, $T_2$, $T_3$ or $T_4$) that performed the action and $B_C \dots E_C$ are the actions performed by the system in taking the fuzzy checkpoint. After performing the commit action $C_4$ of transaction $T_4$ (which includes flushing the log), some further actions by other transactions may have been performed before the crash, but as their log records have not been taken onto disk from the log buffer, no trace of them remains.

In the analysis pass, the log is scanned forward from the *begin-checkpoint* log record at LSN 103 to the end of the log. During the log scan, the following actions are performed at LSNs 103–118:

103: No action.

104: The active-transaction table is initialized from the logged copy as

$\{(T_1, \text{forward-rolling}, \text{UNDO-NEXT-LSN} = 102)\}$.

105: The modified-page table is initialized from the logged copy as

$\{(p_1, \text{REC-LSN} = 102)\}$.

106: No action.

107 (forward-rolling update by $T_1$ on page $p_1$): Set UNDO-NEXT-LSN$(T_1)$ ←
107 in the active-transaction table. The modified-page table need not be updated,
because page $p_1$ is already recorded there.

108 (begin of $T_2$): The record $(T_2$, forward-rolling, UNDO-NEXT-LSN $= 108)$ is
inserted into the active-transaction table.

109 (commit of $T_1$): The record for $T_1$ is deleted from the active-transaction table.
The contents of the table are now

$\{(T_2$, forward-rolling, UNDO-NEXT-LSN $= 108)\}$.

110 (update by $T_2$ on page $p_1$): Set UNDO-NEXT-LSN$(T_2)$ ← 110. The modified-
page table is not updated.

111 (begin of $T_3$): The record $(T_3$, forward-rolling, UNDO-NEXT-LSN $= 111)$ is
inserted into the active-transaction table.

112 (update by $T_3$ on page $p_2$): Set UNDO-NEXT-LSN$(T_3)$ ← 112 and insert the
record $(p_2$, REC-LSN $= 112)$ into the modified-page table.

113 (update by $T_2$ on page $p_1$): Set UNDO-NEXT-LSN$(T_2)$ ← 113. The modified-
page table is not updated.

114 (abort of $T_2$): Change the state of $T_2$ to "backward-rolling" in the active-
transaction table.

115 (undoing of an update by $T_2$ on $p_1$): Set UNDO-NEXT-LSN$(T_2)$ ← 110. The
modified-page table is not updated.

116 (begin of $T_4$): The record $(T_4$, forward-rolling, UNDO-NEXT-LSN $= 116)$ is
inserted into the active-transaction table.

117 (update by $T_4$ on page $p_2$): Set UNDO-NEXT-LSN$(T_4)$ ← 117. The modified-
page table is not updated.

118 (commit of $T_4$): The record for $T_4$ is deleted from the active-transaction table.

The log has now been scanned to the end. The contents of the reconstructed
active-transaction table are now

$\{(T_2$, backward-rolling, UNDO-NEXT-LSN $= 110)$,
$(T_3$, forward-rolling, UNDO-NEXT-LSN $= 112)\}$,

and the contents of the reconstructed modified-page table are:

$\{(p_1$, REC-LSN $= 102)$, $(p_2$, REC-LSN$=112)\}$.

The analysis pass ends with setting

REDO-LSN ← min$\{102, 112\} = 102$

as the starting point of the redo pass.                                        □

**Theorem 4.4** *The active-transaction table reconstructed in the analysis pass has
the contents that existed at the time when the last log record found on the log disk
was written. The modified-page table reconstructed in the analysis pass contains the
page-id of every page p that resided in the buffer as modified at the time when the
last log record found on the log disk was written, with a REC-LSN value satisfying
the following conditions:*

1. REC-LSN($p$) $\leq$ MODIFIED-LSN($p$), *where* MODIFIED-LSN($p$) *is the* LSN *of the log record of the most recent update found on the log disk that turned $p$ from an unmodified page into a modified page.*

2. *If $n$ is the* LSN *of an update on page $p$ with $n <$* REC-LSN($p$), *then $n \leq$* PAGE-LSN($p$), *where* PAGE-LSN($p$) *is the* PAGE-LSN *of the disk version of page $p$ at the time when the last log record found on the log disk was written.*

3. *The disk version of page $p$ at the time of the failure contains all updates on that page with* LSNs *less than* REC-LSN($p$).

*Proof* The active-transaction and modified-page tables are initialized from the contents of the tables logged when taking the checkpoint, as is apparent from Algorithm 4.13. The log surviving on the log disk is then scanned forward.

Upon encountering a log record "$n{:}\langle T, B \rangle$," the transaction record for a forward-rolling transaction $T$ with UNDO-NEXT-LSN $= n$ is inserted into the active-transaction table; upon encountering a log record "$n{:}\langle T, A \rangle$," the state of $T$ is changed to "backward-rolling," and upon encountering a log record "$n{:}\langle T, C \rangle$," the record for $T$ is deleted. Thus the active-transaction table being reconstructed indicates which transactions were forward-rolling or backward-rolling at any time after the checkpoint was taken and until the last log record surviving on the log disk was written.

Upon encountering a log record with LSN $n$ for an update action by $T$ or an undo action by $T$ on page $p$, the UNDO-NEXT-LSN value of $T$ in the active-transaction table is updated accordingly: the new value of UNDO-NEXT-LSN is the LSN $n$ of the redo-undo log record for a forward-rolling update action and the UNDO-NEXT-LSN $n'$ of the redo-only log record for an undo action. Also, the entry $(p,$ REC-LSN $= n)$ is inserted into the modified-page table if there is not yet any entry for $p$ in the table.

Obviously, if upon encountering a log record for an update on page $p$ there is no entry for page $p$ in the modified-page table being reconstructed, either page $p$ was not in the buffer at the time the checkpoint was taken or it resided there unmodified; in any case the update encountered must be one that turned $p$ from an unmodified page into a modified one, so that REC-LSN($p$) $=$ MODIFIED-LSN($p$) at that timepoint. For pages $p$ having an entry in the modified-page table logged at the checkpoint, the associated REC-LSN value is MODIFIED-LSN($p$) current at the time of taking the checkpoint if the entry was inserted into the table during normal processing, and, by induction, less than or equal to MODIFIED-LSN($p$) if the entry was inserted during an analysis pass.

In any case, an entry (page-id, REC-LSN) once inserted into the modified-page table being reconstructed in the analysis pass is not changed or deleted during the pass, because the analysis pass does not observe any flushings of pages from the buffer onto disk that may have occurred during normal processing before the failure. We conclude that condition (1) holds. By the definition of MODIFIED-LSN, this in turn implies condition (2). By the definition of PAGE-LSN, this in turn implies condition (3).                                                                              □

## 4.8   Redo Pass for Do-Redo-Undo Recovery

The *redo pass* "repeats history," that is, restores the state of the database as it was when the failure occurred. More specifically, the redo pass brings the physical database (and hence also the logical database) to the state that existed when the most recent log record surviving on the log disk was written.

The state of the database is reconstructed from the disk version of the physical database and the log found on the log disk. The log is scanned starting from the REDO-LSN to the end of the log, and the updates recorded in the log that are not found in the disk versions of the database pages are redone. Updates by active transactions are also redone, even though these transactions will eventually be rolled back in the undo pass of recovery.

While scanning the log, if a log record

$$n : \langle T, o, p, \bar{x}, n' \rangle$$

for a forward-rolling update action, or a log record

$$n : \langle T, o^{-1}, p, \bar{x}, n' \rangle$$

for a backward-rolling undo action, on page $p$ by a transaction $T$ is encountered, the LSN $n$ of the log record is compared with REC-LSN($p$). If $n <$ REC-LSN($p$), we know that the update logged with LSN $n$ is in the page $p$, so that the page need not be latched at all and the log record can be skipped (Fig. 4.4a).

On the contrary, if $n \geq$ REC-LSN($p$), the update may or may not be in the page. Therefore, the page must be fixed and write-latched and its PAGE-LSN be examined. If $n \leq$ PAGE-LSN($p$), the update is there, so the page is unlatched and unfixed, and the log record is skipped (Fig. 4.4b). Otherwise $n >$ PAGE-LSN($p$), indicating that the update is not in the page. In this case, we perform the logged action $o[\bar{x}]$ (or, respectively, $o^{-1}[\bar{x}]$) on page $p$, advance PAGE-LSN($p$) $\leftarrow n$, and unlatch and unfix page $p$ (Fig. 4.4c).

In the cases when $n \geq$ REC-LSN($p$) (Fig. 4.4b, c), page $p$, after being fixed and write-latched, is found unmodified, REC-LSN($p$) can be advanced by setting in the modified-page table:

$$\text{REC-LSN}(p) \leftarrow \max\{\text{REC-LSN}(p), \text{PAGE-LSN}(p) + 1\},$$

where the value "PAGE-LSN($p$) $+ 1$" represents the least LSN value greater than the value PAGE-LSN($p$). Because in this situation the contents of the disk and buffer versions of the page are the same and because the page thus has all updates with LSNs up to PAGE-LSN($p$), the above assignment is correct, preserving (3.1): REC-LSN($p$) $\leq$ MODIFIED-LSN($p$).

This updating of the REC-LSNs is essential for accelerating the redo pass, because PAGE-LSN($p$) may be much greater than the LSN of the log record being scanned. Once the disk version of page $p$ has been fetched from disk to the buffer and its PAGE-LSN examined and found to be greater than the REC-LSN($p$) currently recorded in the modified-page table, REC-LSN($p$) is advanced to PAGE-LSN($p$)+1, so that page

**a**  log record being scanned:  $n : \langle \ldots, p, \ldots \rangle$

$\qquad$ REC-LSN($p$) $\longrightarrow n_1 : \langle \ldots, p, \ldots \rangle$

$\qquad$ PAGE-LSN($p$) $\longrightarrow n_2 : \langle \ldots, p, \ldots \rangle$

$\qquad n <$ REC-LSN($p$): page $p$ need not be fixed and latched.

**b**  $\qquad$ REC-LSN($p$) $\longrightarrow n_1 : \langle \ldots, p, \ldots \rangle$

log record being scanned:  $n : \langle \ldots, p, \ldots \rangle$

$\qquad$ PAGE-LSN($p$) $\longrightarrow n_2 : \langle \ldots, p, \ldots \rangle$

$\qquad n \geq$ REC-LSN($p$): page $p$ has to be fixed and latched, but the update must not be redone, because $n \leq$ PAGE-LSN($p$).

**c**  $\qquad$ REC-LSN($p$) $\longrightarrow n_1 : \langle \ldots, p, \ldots \rangle$

$\qquad$ PAGE-LSN($p$) $\longrightarrow n_2 : \langle \ldots, p, \ldots \rangle$

log record being scanned:  $n : \langle \ldots, p, \ldots \rangle$

$\qquad n \geq$ REC-LSN($p$): page $p$ has to be fixed and latched, and the update has to be redone, because $n >$ PAGE-LSN($p$).

**Fig. 4.4** Different scenarios in the redo pass

$p$ does not need to be fixed and write-latched again when processing any further log records for updates on $p$ with LSNs less than PAGE-LSN($p$) $+ 1$. Moreover, if another recovery needs to be started soon after this one, its redo pass will be faster, since the checkpoint taken at the end of the recovery will have a copy of the modified-page table with more up-to-date REC-LSNs.

The normal steal-and-no-force policy is followed during the redo pass. As the redo pass starts with an empty database buffer, no pages need to be flushed onto disk if all the pages that need redoing fit in the buffer at the same time. In this case the disk version of the database remains as it was at the time of the crash, and the redone updates appear only in the buffer version of the database resulting from the redo pass. However, if a page is flushed onto disk during the redo pass, its entry is deleted from the modified-page table (as is done during normal processing), and if an update on that page is encountered in the redo pass when scanning a log record with LSN $n$, then an entry for the page with REC-LSN $= n$ is inserted into the modified-page table.

Nothing is written to the log during the redo pass. This is because actions are always redone physically on the same pages (and even in the same record positions) where they were originally done, and thus the records already in the log describe the redo actions exactly. As all the surviving log records are on the log disk and no new log records are generated, the log buffer remains empty during the redo pass. Thus, the write-ahead-logging protocol need not be followed during the redo pass.

The algorithm for the redo pass is shown as Algorithm 4.14. The call *redo(r)* (see Algorithm 4.15) redoes the update logged with log record *r*. The handling of log records for structure modifications (the call *redo-smo(r)*) is described in Sect. 4.11.

---

**Algorithm 4.14** Procedure *redo-pass*()

---

$r \leftarrow$ *get-log-record* (REDO-LSN)
**while** $r$ is a log record **do**
  **if** $r$ is of the form "$n : \langle T, o, p, \bar{x}, n' \rangle$" or "$n : \langle T, o^{-1}, p, \bar{x}, n' \rangle$" **then**
    **if** there is no entry for page $p$ in the modified-page table **then**
      insert $(p, \text{REC-LSN} = n)$ into the modified-page table
    **end if**
    **if** REC-LSN$(p) \leq n$ **then**
      *fix-and-write-latch*$(p)$
      **if** page $p$ is unmodified **then**
        REC-LSN$(p) \leftarrow \max\{\text{REC-LSN}(p), \text{PAGE-LSN}(p) + 1\}$
      **end if**
      **if** PAGE-LSN$(p) < n$ **then**
        *redo(r)*
        PAGE-LSN$(p) \leftarrow n$
      **end if**
      *unlatch-and-unfix*$(p)$
    **end if**
  **else if** $r$ is a log record for a structure modification **then**
    *redo-smo(r)*
  **end if**
  $r \leftarrow$ *get-next-log-record* $(n)$
**end while**

---

 

---

**Algorithm 4.15** Procedure *redo(r)*

---

**if** $r$ is of the form "$n : \langle T, I, p, x, v, n' \rangle$" **then**
  insert the tuple $(x, v)$ into page $p$
**else if** $r$ is of the form "$n : \langle T, D, p, x, v, n' \rangle$" **then**
  delete the tuple $(x, v)$ from page $p$
**else if** $r$ is of the form "$n : \langle T, I^{-1}, p, x, n' \rangle$" **then**
  delete the tuple with key $x$ from page $p$
**else if** $r$ is of the form "$n : \langle T, D^{-1}, p, x, v, n' \rangle$" **then**
  insert the tuple $(x, v)$ into page $p$
**end if**

---

*Example 4.5* Continuing with our previous example, in the redo pass of recovery, the log is scanned from the record at REDO-LSN $= 102$ up to the end of the log.

Assume that in the disk version of page $p_1$, PAGE-LSN$(p_1) = 102$, so that page $p_1$ has updates up to and including LSN 102. Thus, page $p_1$ has been flushed onto disk between the actions logged with LSNs 105 and 107. Note that $p_1$ was still a modified page when the modified-page table was logged at LSN 105. We also assume that in the disk version of page $p_2$, PAGE-LSN($p_2$) < 101, meaning that the page was not flushed onto disk after the actions logged with LSNs 112 and 117. Also note that during taking the checkpoint, page $p_2$, if buffered, was unmodified, since it does not appear in the logged modified-page table.

During the log scan, the following actions are performed at LSNs 102–118:

102: Because REC-LSN($p_1$) = 102 ≤ 102, the page $p_1$ must be fixed and write-latched. The fixing brings page $p_1$ from disk to the buffer. The PAGE-LSN of the page can now be examined. As the page is unmodified, we set

$$\text{REC-LSN}(p_1) \leftarrow \max\{\text{REC-LSN}(p_1), \text{PAGE-LSN}(p_1) + 1\}$$
$$= \max\{102, 102 + 1\} = 103$$

in the modified-page table. Because PAGE-LSN($p_1$) = 102 ≥ 102, the LSN of the log record, the result of the logged action $D[x_1, v_1]$ is already on the page. Page $p_1$ is unlatched and unfixed.

103–106: No action.

107: Because REC-LSN($p_1$) = 103 ≤ 107, page $p_1$ is fixed and write-latched. As the page is still unmodified, we again set REC-LSN($p_1$) ← 102 + 1 = 103. Because PAGE-LSN($p_1$) = 102 < 107, the result of the action $I[x_1, v_1']$ is missing from the page. The tuple $(x_1, v_1')$ is inserted into the page, and the PAGE-LSN is advanced: PAGE-LSN($p_1$) ← 107. Page $p_1$ is unlatched and unfixed.

108–109: No action.

110: Because REC-LSN($p_1$) = 103 ≤ 110, page $p_1$ is fixed and write-latched. As the page is no longer unmodified, its REC-LSN is not updated. Because PAGE-LSN($p_1$) = 107 < 110, the result of the action $D[x_1, v_1']$ is missing from page $p_1$. The tuple $(x_1, v_1')$ is deleted from page $p_1$ and PAGE-LSN($p_1$) ← 110. Page $p_1$ is unlatched and unfixed.

111: No action.

112: Because REC-LSN($p_2$) = 112 ≤ 112, page $p_2$ is fixed and write-latched. The fixing brings page $p_2$ from disk to the buffer. As the page is unmodified, we set

$$\text{REC-LSN}(p_2) \leftarrow \max\{\text{REC-LSN}(p_2), \text{PAGE-LSN}(p_2) + 1\} = 112.$$

Recall that PAGE-LSN($p_2$) < 101. Because PAGE-LSN($p_2$) < 112, the result of the action $I[x_2, v_2]$ is missing from the page. The tuple $(x_2, v_2)$ is inserted into the page and PAGE-LSN($p_2$) ← 112. Page $p_2$ is unlatched and unfixed.

113: Because REC-LSN($p_1$) = 103 ≤ 113, page $p_1$ is fixed and write-latched. Because PAGE-LSN($p_1$) = 110 < 113, the result of the action $I[x_3, v_3]$ is missing from the page. The tuple $(x_3, v_3)$ is inserted into the page and PAGE-LSN($p_1$) ← 113. Page $p_1$ is unfixed and unlatched.

114: No action.

115: Because REC-LSN($p_1$) = 103 ≤ 115, page $p_1$ is fixed and write-latched. Because PAGE-LSN($p_1$) = 113 < 115, the result of the action $I^{-1}[x_3, v_3]$ is missing from the page. The tuple $(x_3, v_3)$ is deleted from the page and PAGE-LSN($p_1$) ← 115. Page $p_1$ is unlatched and unfixed.

| log |
|---|
| 101: $\langle T_1, B \rangle$ |
| 102: $\langle T_1, D, p_1, x_1, v_1, 101 \rangle$ |
| 103: $\langle begin\text{-}checkpoint \rangle$ |
| 104: $\langle active\text{-}transaction\text{-}table, \{$ |
| $(T_1, \text{forward-rolling},$ |
| $\text{UNDO-NEXT-LSN} = 102)\})$ |
| 105: $\langle modified\text{-}page\text{-}table, \{$ |
| $(p_1, \text{REC-LSN} = 102)\})$ |
| 106: $\langle end\text{-}checkpoint \rangle$ |
| 107: $\langle T_1, I, p_1, x_1, v_1', 102 \rangle$ |
| 108: $\langle T_2, B \rangle$ |
| 109: $\langle T_1, C \rangle$ |
| 110: $\langle T_2, D, p_1, x_1, v_1', 108 \rangle$ |
| 111: $\langle T_3, B \rangle$ |
| 112: $\langle T_3, I, p_2, x_2, v_2, 111 \rangle$ |
| 113: $\langle T_2, I, p_1, x_3, v_3, 110 \rangle$ |
| 114: $\langle T_2, A \rangle$ |
| 115: $\langle T_2, I^{-1}, p_1, x_3, 110 \rangle$ |
| 116: $\langle T_4, B \rangle$ |
| 117: $\langle T_4, I, p_2, x_4, v_4, 116 \rangle$ |
| 118: $\langle T_4, C \rangle$ |

| active-transaction table | | |
|---|---|---|
| tr-id | state | UNDO-NEXT-LSN |
| $T_2$ | backward-rolling | 110 |
| $T_3$ | forward-rolling | 112 |

| modified-page table | |
|---|---|
| page-id | REC-LSN |
| $p_1$ | 102 |
| $p_2$ | 112 |

| database buffer | | |
|---|---|---|
| page $p_1$ | page $p_2$ | |
| PAGE-LSN=115 | PAGE-LSN=117 | $\cdots$ |
| | $(x_2, v_2)$ | |
| | $(x_4, v_4)$ | |

**Fig. 4.5** Database state after the redo pass of ARIES recovery

116: No action.

117: Because $\text{REC-LSN}(p_2) = 112 \leq 117$, page $p_2$ is fixed and write-latched. As the page is no longer unmodified, its REC-LSN is not updated. Because $\text{PAGE-LSN}(p_2) = 112 < 117$, the result of the action $I[x_4, v_4]$ is missing from the page. The tuple $(x_4, v_4)$ is inserted into the page and $\text{PAGE-LSN}(p_2) \leftarrow 117$. Page $p_2$ is unfixed and unlatched.

118: No action.

The redo pass is now complete. In page $p_2$ there are the tuples $(x_2, v_2)$ and $(x_4, v_4)$, while tuples with keys $x_1$ and $x_3$ are missing from the database, as was the case at the time of the crash (Fig. 4.5).                                   □

**Theorem 4.6** *The analysis pass followed by the redo pass brings the physical database (and hence also the logical database) to the state that existed when the most recent log record found on the log disk was written. The resulting database state is action consistent, provided that each structure modification has been logged with a single redo-only log record.*

*Proof* By Theorem 4.4, the REDO-LSN value produced by the analysis pass is a correct LSN value for starting the forward scan of the log in the redo pass. The claim then follows from the fact that the updates logged with log records with LSNs greater than or equal to REDO-LSN up to the end of the log are redone on the disk versions of pages fetched to the buffer in the original chronological order of the updates when those updates are missing, and from the fact that the PAGE-LSN of each page records the LSN of the log record of the last update on the page.

We also note that the updating of the REC-LSNs during the redo pass is correct because REC-LSN($p$) is only updated when processing a log record for an update on a currently unmodified page $p$. Thus we can safely set REC-LSN($p$) ← PAGE-LSN($p$) + 1, not violating condition (3.1).

As each log record for an update action by a transaction describes an entire action and such was also assumed to be the case with structure modifications, the resulting database must be action consistent, because the redo pass then performed complete operations and because the modification on a single page described by the redo-only log record for a multi-page structure modification can be selectively redone on that page (as will be shown in Sect. 4.11).                                   □

The number of disk accesses performed during the redo pass is usually much less than the number of disk accesses performed during normal transaction processing from the timepoint of logging the log record with LSN =REDO-LSN up to the event of the system crash. First, during the redo pass, no page is fetched from disk just for reading, that is, no read-latching is performed. Second, a page is write-latched only if there is a log record with LSN ≥ REDO-LSN that describes an update on it. Moreover, thanks to the maintenance of the REC-LSN values, it may not be necessary to write-latch and inspect every such page at every logged update on it.

This can be contrasted with normal transaction processing where an update action may involve latching several pages, because of the need to locate the target page of the update via an access path such a B-tree index. As discussed above in Sect. 4.3 (and presented in detail in Chaps. 6 and 7), a forward-rolling insert or delete action may involve several root-to-leaf traversals of a B-tree index, thus implying an access-path-length that is a multiple of the height of the B-tree. A logical undo action also involves traversing of the index. Instead, the path-length is only one for the redoing of a forward-rolling insert or delete action, or an undo-insert or undo-delete action, given the log record of the action.

## 4.9  Undo Pass for Do-Redo-Undo Recovery

In the *undo pass*, forward-rolling transactions are rolled back and the rollback of backward-rolling transactions is completed. First, every forward-rolling transaction $T$ is aborted by performing the call *abort*($T$) (Algorithm 4.2), which sets the state of $T$ to "backward-rolling" in the transaction record of $T$ in the active-transaction table and appends the log record $\langle T, A \rangle$ to the log buffer. Next the log is scanned in reverse chronological order, starting from the end and going backwards, via the UNDO-NEXT-LSN chains, towards the beginning, examining the log records for forward-rolling updates. The next log record to be examined is always the one whose LSN is the same as the currently greatest UNDO-NEXT-LSN value in the active-transaction table.

Whenever an update log record

$$\langle T, o, p, \bar{x}, n \rangle$$

is encountered in the log scan, the page $p$ is fixed and write-latched, and the undo action $o^{-1}[\bar{x}]$ is executed on the page $p'$ that currently covers the key of the target tuple of the action (given in $\bar{x}$).

The undo action is executed in the same manner as the undo action of a single backward-rolling transaction or of a forward-rolling transaction doing a partial rollback, during normal transaction processing. As explained in Sect. 4.3, the undo action is first attempted physically, but if this fails, it is done logically. If the page $p$ mentioned in the log record as the original target page of the forward-rolling action $o[\bar{x}]$ still covers the key of the tuple and the undo action $o^{-1}[\bar{x}]$ can be executed on page $p$, then $p$ is the target page $p'$ of the undo action, and the undo action is physical; otherwise $p$ is unlatched and unfixed, and the undo action is executed logically (see Algorithms 4.7 and 4.9).

As explained in Sect. 3.4, the undo action is logged with a redo-only log record with

$$m'\colon \langle T, o^{-1}, p', \bar{x}', n \rangle,$$

where $\bar{x}'$ is the part of the arguments $\bar{x}$ needed for redoing the undo action (recall the log records (3.11)–(3.13)). The LSN $m'$ is stamped into the PAGE-LSN field of page $p'$, and the page is unlatched and unfixed. The UNDO-NEXT-LSN of $T$ is set back to $n$ in the active-transaction table. The next log record to be examined is the one whose LSN is now the greatest UNDO-NEXT-LSN of the transaction records in the active-transaction table.

Whenever a log record $\langle T, B \rangle$ is encountered, we conclude that transaction $T$ has been completely rolled back; hence, $\langle T, C \rangle$ is written to the log, the log is flushed onto disk, and the transaction record for $T$ is deleted from the active-transaction table. The undo pass ends when there are no longer any transactions left in the active-transaction table.

The normal steal-and-no-force and write-ahead-logging policies are applied during the undo pass. The algorithm for the undo pass is given as Algorithm 4.16.

---

**Algorithm 4.16** Procedure *undo-pass*()

---

**for** each forward-rolling transaction $T$ in the active-transaction table **do**
   *abort*($T$)
**end for**
**while** the active-transaction table is nonempty **do**
   $n \leftarrow \max\{\text{UNDO-NEXT-LSN}(T) \mid T$ is in the active-transaction table$\}$
   $r \leftarrow$ *get-log-record*($n$)
   **if** $r$ is "$n : \langle T, B \rangle$" **then**
      *log*($m, \langle T, C \rangle$)
      *flush-the-log*()
      delete the record of $T$ from the active-transaction table
   **else if** $r$ is "$n : \langle T, I, p, x, v, n' \rangle$" **then**
      *undo-insert*($n, T, p, x, n'$)
   **else if** $r$ is "$n : \langle T, D, p, x, v, n' \rangle$" **then**
      *undo-delete*($n, T, p, x, v, n'$)
   **else if** $r$ is "$n : \langle T, S[P], n' \rangle$" **then**
      UNDO-NEXT-LSN($T$) $\leftarrow n'$
   **end if**
**end while**

---

*Example 4.7* In the case of our running example, the undo pass is performed as follows. First, the only forward-rolling transaction, $T_3$, is turned into backward-rolling by appending the abort log record

   119: $\langle T_3, A \rangle$

to the log buffer and by setting the state of $T_3$ as "backward-rolling" in the active-transaction table. The contents of the active-transaction table are now:

   $\{(T_2, \text{backward-rolling}, \text{UNDO-NEXT-LSN} = 110),$
   $(T_3, \text{backward-rolling}, \text{UNDO-NEXT-LSN} = 112)\}.$

Then the log is scanned backwards starting from the log record with LSN 112 $= \max\{110, 112\}$ (the maximum of the UNDO-NEXT-LSNs in the active-transaction table).

The log record with LSN 112 was written for an action $I[x_2, v_2]$ performed by $T_3$ on page $p_2$. Thus, page $p_2$ is fixed and write-latched. We observe that the tuple with key $x_2$ still resides in page $p_2$ and that the page does not underflow if the tuple is deleted. The undo action $I^{-1}[x_2, v_2]$ is thus performed physically by deleting the tuple $(x_2, v_2)$ from page $p_2$. The redo-only log record

   120: $\langle T_3, I^{-1}, p_2, x_2, 111 \rangle$

is appended to the log buffer, PAGE-LSN($p_2$) $\leftarrow$ 120 is set, page $p_2$ is unlatched and unfixed, and UNDO-NEXT-LSN($T_3$) $\leftarrow$ 111 is set in the active-transaction table.

Now the maximum of the UNDO-NEXT-LSNs in the active-transaction table is 111. Thus, the next log record to be processed is the one with LSN 111. This is the begin-transaction log record for $T_3$. So the rollback of $T_3$ has been completed, the commit log record

121: $\langle T_3, C \rangle$

is appended to the log buffer, the log is flushed, and the transaction record of $T_3$ is deleted from the active-transaction table. The contents of the active-transaction table are now

$\{(T_2, \text{backward-rolling}, \text{UNDO-NEXT-LSN} = 110)\}.$

The next log record to be processed is the one with LSN 110, written for the action $D[x_1, v_1']$ performed by $T_2$ on page $p_1$. Page $p_1$ is fixed and write-latched. We observe that page $p_1$ still covers key $x_1$ and that there is room for the tuple $(x_1, v_1')$ in the page. The undo action $D^{-1}[x_1, v_1']$ is performed by inserting the tuple $(x_1, v_1')$ into page $p_1$, the redo-only log record

122: $\langle T_2, D^{-1}, p_1, x_1, v_1', 108 \rangle$

is appended to the log buffer, PAGE-LSN($p_1$) $\leftarrow$ 122 is set, page $p_1$ is unlatched and unfixed, and UNDO-NEXT-LSN($T_2$) $\leftarrow$ 108 is set in the active-transaction table.

The contents of the active-transaction table are now

$\{(T_2, \text{backward-rolling}, \text{UNDO-NEXT-LSN} = 108)\}.$

The next log record to be processed is thus the one with LSN 108, the begin-transaction log record for $T_2$. As the rollback of $T_2$ is thus complete, the log record

123: $\langle T_2, C \rangle$

is appended, the log is flushed, and the transaction record of $T_2$ is deleted from the active-transaction table.

This completes the undo pass, because the active-transaction table is now empty. The logical database has been brought to a state that contains only the updates by the committed transactions $T_1$ and $T_4$. □

**Theorem 4.8** *The analysis pass followed by the redo pass followed by the undo pass brings the database to a transaction-consistent state that contains all the updates by transactions that committed before the crash and none of the updates by other transactions, provided that the sequence of undo actions to be performed in the undo pass can be run on the logical database resulting from the redo pass.*

*Proof* By Theorem 4.6, the redo pass brings the database into an action-consistent state that is the result of applying the sequence of update actions whose log records are found on the disk at the time of the crash. As every transaction that committed before the crash has all its log records on the log disk before the crash, the database state resulting from the redo pass thus contains all the updates by committed transactions. The undo pass rolls back all the transactions that were active, that is, either forward-rolling or backward-rolling, at the time of the crash. Because of the assumption that each undo action involved in the rollbacks can be performed on the database, each individual active transaction will be rolled back correctly, removing its effect from the database. At the end of the undo pass, the database thus contains only the updates by a set of committed transactions and is therefore transaction-consistent.                                                                             □

The provision mentioned in Theorem 4.8 is needed, because if the transactions have not been run sufficiently isolated, the undo pass may not be possible to execute.

*Example 4.9*  Assume that action $D[x, u]$ by transaction $T_1$ is followed by action $I[x, v]$ by transaction $T_2$, after which $T_2$ commits and a crash occurs when $T_1$ is still forward-rolling:

$$B_1 D_1[x, u] B_2 I_2[x, v] C_2 \text{ crash!}$$

After the redo pass, the current database contains the tuple $(x, v)$. Now $T_1$, which should roll back, cannot perform its undo action $D^{-1}[x, u]$ on a database that contains $(x, v)$, because of the key constraint.

The action $I[x, v]$ by $T_2$ is termed a "dirty write," the most serious of the isolation anomalies to be discussed in Chap. 5: the action overwrites an uncommitted (or "dirty") update. To make recovery possible in all circumstances, dirty writes must be prevented.                                                                              □

As the very last phase of restart recovery, a checkpoint is taken. After this, the system is ready to process new transactions.

*Example 4.10*  In the case of our running example, the following is written to the log at the end of restart recovery:

124: ⟨*begin-checkpoint*⟩
125: ⟨*active-transaction-table*, {}⟩
126: ⟨*modified-page-table*, {$(p_1, \text{REC-LSN} = 103), (p_2, \text{REC-LSN} = 112)$}⟩
127: ⟨*end-checkpoint*⟩

As a conclusion, we note that the database state after recovery is as shown in Fig. 4.6. Observe that the active-transaction table is always empty in this case, unless we allow new transactions to enter the system while the undo pass of recovery is still ongoing—an issue to be discussed in Sect. 9.7.                                         □

| log |
|-----|
| 101: $\langle T_1, B \rangle$ |
| 102: $\langle T_1, D, p_1, x_1, v_1, 101 \rangle$ |
| 103: $\langle begin\text{-}checkpoint \rangle$ |
| 104: $\langle active\text{-}transaction\text{-}table, \{$ |
| $\quad (T_1, \text{forward-rolling},$ |
| $\quad \text{UNDO-NEXT-LSN} = 102) \} \rangle$ |
| 105: $\langle modified\text{-}page\text{-}table, \{$ |
| $\quad (p_1, \text{REC-LSN} = 102) \} \rangle$ |
| 106: $\langle end\text{-}checkpoint \rangle$ |
| 107: $\langle T_1, I, p_1, x_1, v_1', 102 \rangle$ |
| 108: $\langle T_2, B \rangle$ |
| 109: $\langle T_1, C \rangle$ |
| 110: $\langle T_2, D, p_1, x_1, v_1', 108 \rangle$ |
| 111: $\langle T_3, B \rangle$ |
| 112: $\langle T_3, I, p_2, x_2, v_2, 111 \rangle$ |
| 113: $\langle T_2, I, p_1, x_3, v_3, 110 \rangle$ |
| 114: $\langle T_2, A \rangle$ |
| 115: $\langle T_2, I^{-1}, p_1, x_3, 110 \rangle$ |
| 116: $\langle T_4, B \rangle$ |
| 117: $\langle T_4, I, p_2, x_4, v_4, 116 \rangle$ |
| 118: $\langle T_4, C \rangle$ |
| 119: $\langle T_3, A \rangle$ |
| 120: $\langle T_3, I^{-1}, p_2, x_2, 111 \rangle$ |
| 121: $\langle T_3, C \rangle$ |
| 122: $\langle T_2, D^{-1}, p_1, x_1, v_1', 108 \rangle$ |
| 123: $\langle T_2, C \rangle$ |
| 124: $\langle begin\text{-}checkpoint \rangle$ |
| 125: $\langle active\text{-}transaction\text{-}table, \{ \} \rangle$ |
| 126: $\langle modified\text{-}page\text{-}table, \{$ |
| $\quad (p_1, \text{REC-LSN} = 103),$ |
| $\quad (p_2, \text{REC-LSN} = 112) \} \rangle$ |
| 127: $\langle end\text{-}checkpoint \rangle$ |

| active-transaction table | | |
|-----|-----|-----|
| tr-id | state | UNDO-NEXT-LSN |
| | | |

| modified-page table | |
|-----|-----|
| page-id | REC-LSN |
| $p_1$ | 102 |
| $p_2$ | 112 |

| database buffer | | |
|-----|-----|-----|
| page $p_1$ | page $p_2$ | |
| PAGE-LSN=122 | PAGE-LSN=120 | $\cdots$ |
| $(x_1, v_1')$ | $(x_4, v_4)$ | |

**Fig. 4.6**  Database state at the end of ARIES recovery

## 4.10  Recovery from a Process Failure

Assume that the server-process thread that executes a transaction $T$ fails, terminating abnormally for some reason. If some pages were write-latched by that thread at the time of the failure, we have to regard the contents of the buffer versions of those pages as corrupted.

Fortunately, the redo pass of the ARIES algorithm can be applied *selectively*, that is, as restricted to the corrupted pages only. First, the LSN of the latest log record is saved. Then, a recovery process is started, and the write latches held by the failed thread are transferred to that process the corrupted pages are purged from the buffer without flushing them onto disk and then fixed again, thus bringing the disk versions of the pages into the buffer.

The REDO-LSN is set to the minimum of the REC-LSNs of the corrupted pages, and the log (including the log records on disk and in the log buffer) is scanned from the REDO-LSN to the saved LSN, redoing on the write-latched pages all the missing updates with LSNs greater than or equal to REDO-LSN and less than or equal to the saved LSN. In this way, we restore all the updates on the corrupted pages logged before the failure caused the corruption of the pages in the buffer.

After the selective redo pass, transaction $T$ is aborted and rolled back in the usual way by the recovery process, as if $T$ itself had performed the SQL **rollback** statement.

## 4.11 Recovering Structure Modifications

Consider a structure modification that has been logged with a single redo-only log record of the form

$$n : \langle S, p_1, \ldots, p_k, V_1, \ldots, V_l \rangle,$$

where $S$ is the name of the structure modification, $p_1, \ldots, p_k$ are the page-ids of the pages updated in the modification, and $V_1, \ldots, V_l$ are other arguments needed to represent the effects of the modification on the affected pages, such as sets of records moved from a page to another. We call such a modification a *redo-only structure modification*.

Examples of log records for redo-only structure modifications include

$$n: \langle allocate\text{-}page\text{-}for\text{-}heap, s, f, p, q \rangle$$

representing the allocation of a new page $q$ and linking it into a chain of pages, and

$$n: \langle page\text{-}split, p, q, x, q', s, V, h \rangle$$

representing a split of a full B-tree page $q$ of height $h$, where $p$, $q$, $q'$ and $s$ are page-ids, the set $V$ contains the records moved from page $q$ to its new sibling page $q'$, $x$ is the minimum key of the those records, $(x, q')$ is the child-router record inserted into the parent page $p$, and page $s$ is the space-map page used to allocate the page $q'$.

In the analysis pass (Algorithms 4.12 and 4.13), the log records for structure modifications are processed by the call *analyze-smo*($r$) (Algorithm 4.17).

---

**Algorithm 4.17** Procedure *analyze-smo*($r$)

---

**if** $r$ is of the form "$n : \langle S, p_1, \ldots, p_k, V_1, \ldots, V_l \rangle$" **then**
    **for** $i = 1, \ldots, k$ **do**
        **if** there is no entry for $p_i$ in the modified-page table **then**
            insert $(p_i, \text{REC-LSN} = n)$ into the modified-page table
        **end if**
    **end for**
**end if**

---

In the redo pass (Algorithm 4.14), the logged structure modifications are redone, one page at a time, as the update actions by transactions, when it turns out that the result of the modification is not present in the page. The log records for structure modifications are processed by the call *redo-smo*($r$) (Algorithm 4.18).

We assume that the effect of a structure modification on each single page involved is encoded in the log record in such a way that the modification on that page can be redone using only that page possibly with some other, non-page arguments in the log record. In other words, we assume that the log record allows for *selective redoing*, so that the effects of the modification on any single page involved can be accomplished without inspecting the contents of the other pages mentioned in the log record.

For example, using the log record $n: \langle page\text{-}split, p, q, x, q', s, V, h \rangle$, (1) the effect of the split on page $q$ can be redone by deleting from $q$ all records with keys greater than or equal to $x$, (2) the effect on $q'$ by formatting $q'$ as an empty B-tree page and inserting into it the records from the set $V$, (3) the effect on the space-map page $s$ by setting the bit corresponding to $q'$, and (4) the effect on the parent page $p$ by inserting into it the child-router record $(x, q')$.

---

**Algorithm 4.18** Procedure *redo-smo*($r$)

---

**if** $r$ is of the form "$n : \langle S, p_1, \ldots, p_k, V_1, \ldots, V_l \rangle$" **then**
    **for** all $i = 1, \ldots, k$ **do**
        **if** there is no entry for page $p_i$ in the modified-page table **then**
            insert $(p_i, \text{REC-LSN} = n)$ into the modified-page table
        **end if**
        **if** $\text{REC-LSN}(p_i) \leq n$ **then**
            *fix-and-write-latch*($p_i$)
            **if** page $p_i$ is unmodified **then**
                $\text{REC-LSN}(p_i) \leftarrow \max\{\text{REC-LSN}(p_i), \text{PAGE-LSN}(p_i) + 1\}$
            **end if**
            **if** $\text{PAGE-LSN}(p_i) < n$ **then**
                using the arguments $V_1, \ldots, V_l$ from the log record, apply the effect of structure modification $S$ onto page $p_i$
                $\text{PAGE-LSN}(p_i) \leftarrow n$
            **end if**
            *unlatch-and-unfix*($p_i$)
        **end if**
    **end for**
**end if**

---

When every structure modification can be logged in its entirety with a single log record, the WAL protocol guarantees that after the redo pass of recovery, the physical database is consistent (and the logical database action-consistent) and there is never any need to roll back any unfinished structure modifications. Recall that we require that all the pages involved in the modification be kept simultaneously write-latched during the modification. Then it is sufficient that the log record for the modification is redo only.

This suggests that we define all structure modifications in such a way that this property is maintained. As the principle of latching requires that any process thread may keep only a small number of pages latched simultaneously, this means that in the case of a balanced tree-like index such as a B-tree, a structure modification must be defined to involve only a small fixed number of pages (typically from two adjacent levels of the tree) and still having the property of retaining the physical consistency of the index.

However, there exist implementations of structure modifications that involve several levels of a B-tree index and are logged using several (redo-undo) log records (each involving two adjacent levels of the tree), where the pages involved in the modification of one level may be unlatched and stolen from the buffer before the modifications at the other levels are completed. Such a multi-level redo-undo modification may need to be undone, when the entire modification has not been completed before the crash.

There are problems with redo-undo structure modifications. A structure modification is a physical operation, meaning that undoing such an operation must also be physical. Therefore undoings of structure modifications cannot be mixed with logical undoings of transactions' actions. One solution is to implement the undo pass of restart recovery in two backward scans of the log: in the first scan, the unfinished structure modifications are undone, thus bringing the physical database into a consistent state, and then in the second scan the active transactions are rolled back.

In the ARIES/KVL and ARIES/IM locking and index management methods designed to work with ARIES, a structure modification involving multiple levels of a B-tree, such as a sequence of page splits or merges, can be left unfinished in the event of a system crash and must be undone in the undo pass of restart recovery. However, the undo pass is still implemented by a single backward scan of the log. During normal transaction processing, a special *tree latch* is used to ensure that only a single structure modification can be running at a time. This in turn guarantees in the undo pass the correctness of the search path for any tuple update that must be undone logically.

The handling of structure modifications on B-trees is treated in more detail in Chap. 8.

## Problems

**4.1** At the time of a system failure, the log surviving on the log disk contains the following records:

101: $\langle$begin-checkpoint$\rangle$
102: $\langle$active-transaction-table, {}$\rangle$
103: $\langle$modified-page-table, {}$\rangle$
104: $\langle$end-checkpoint$\rangle$
105: $\langle T_1, B \rangle$
106: $\langle T_1, I, p, x_1, v_1, 105 \rangle$
107: $\langle T_2, B \rangle$
108: $\langle T_2, I, p, x_2, v_2, 107 \rangle$
109: $\langle$allocate-page-for-heap, $s, f, p, q\rangle$
110: $\langle T_1, I, q, x_3, v_3, 106 \rangle$
111: $\langle T_2, S[P], 108 \rangle$
112: $\langle T_2, I, q, x_4, v_4, 111 \rangle$
113: $\langle T_1, I, q, x_5, v_5, 110 \rangle$
114: $\langle T_2, I, q, x_6, v_6, 112 \rangle$
115: $\langle T_2, A[P] \rangle$
116: $\langle T_2, I^{-1}, q, x_6, 112 \rangle$
117: $\langle T_3, B \rangle$
118: $\langle T_3, I, q, x_7, v_7, 117 \rangle$
119: $\langle T_1, C \rangle$

None of the pages updated ($s$, $f$, $p$, $q$) were flushed onto disk after the updates logged with LSNs 106–118. The structure modification logged with LSN 109 uses space-map page $s$ to allocate a new page $q$ and assigns the page-id $p$ into the last-page link field in the first page, $f$, of the heap and in the next-page link field in the last page, $p$, of the heap. The allocation is triggered by an attempt by transaction $T_1$ to insert a tuple, $(x_3, v_3)$, into page $p$, which turns out to be full; the insertion goes to the allocated page $q$.

The ARIES recovery algorithm is used to recover the system from the failure.

(a) What actions are involved in the analysis pass of ARIES? At what LSN does the log scan begin? What are the contents of the active-transaction table and the modified-page table reconstructed in the analysis pass? What is the value set to the REDO-LSN?

(b) What actions are involved in the redo pass of ARIES? At what LSN does the log scan begin? When are pages latched and unlatched and what is done on those pages? What are the PAGE-LSN values of the current versions of the pages $s$, $f$, $p$, and $q$ at the end of the redo pass? No pages are flushed onto disk during the redo pass.

(c) What actions are involved in the undo pass of ARIES? We assume that all undo actions can be performed physically. What log records are generated? What are the PAGE-LSN values of the current versions of the pages $s$, $f$, $p$, and $q$ at the end of the redo pass?

(d) Give the log records written in the fuzzy (indirect) checkpoint taken at the end of the recovery.

**4.2** Repeat items (b) and (d) in the previous problem in the case that pages $f$, $p$, and $q$ were flushed onto disk during normal processing between the writing of the log records with LSNs 112 and 113 and that page $q$ is flushed onto disk during the redo pass between the scanning of the log records with LSNs 114 and 115.

**4.3** As stated in Theorem 4.4 about the analysis pass of ARIES recovery, the active-transaction table reconstructed in the analysis pass exactly represents the states of active transactions at the time when the last log record survived on the log disk was written, while the reconstructed modified-page table may not represent exactly the set of modified pages with their MODIFIED-LSNs at the same timepoint. Show that the reconstructed modified-page table may contain pages whose REC-LSNs are less than their MODIFIED-LSNs, and even pages that are unmodified, at the time of the system crash. Do such pages cause any harm?

**4.4** In the redo pass of the ARIES algorithm, an update is redone on page $p$ only if the PAGE-LSN of $p$ is less than the LSN of the log record for the update. In the undo pass, the PAGE-LSN of page $p$ is not inspected in order to verify that the update to be undone really is on $p$, that is, the PAGE-LSN of $p$ is greater than or equal to the LSN of the log record for the update. Why is that verification unnecessary?

**4.5** What actions should be taken for a new failure that occurs during restart recovery when the ARIES algorithm is in the middle of its (a) analysis pass, (b) redo pass, (c) undo pass?

**4.6** During normal transaction processing, a great deal of log records are created. At which location (LSN) is it possible to truncate the head of the log file and dispose of the oldest log records?

**4.7** In the undo pass of recovery, the log is flushed every time the rollback of some transaction is completed (Algorithm 4.16). The log is also flushed during normal transaction when a transaction completes its rollback (Sect. 3.8, Algorithm 3.3). The flushing of the log is not absolutely necessary in these cases. Explain why. However, it is advantageous to do so. Why?

**4.8** Case (2) of Example 4.1 suggests that with the do-undo-redo recovery paradigm, we should not log undo actions, in order to avoid advancing the PAGE-LSN of a page above the LSNs of log records for actions to be redone later on that page. Can you make this approach viable? What problems do you observe? With the do-redo-undo recovery paradigm of ARIES, it is absolutely necessary to log the undo actions. Which of the functions of ARIES would not work if the undo actions were not logged?

**4.9** Consider the extended read-write transaction model in which a transaction can use the action $C[x]$ to declare the updates on $x$ done so far as committed before the transaction terminates (see Problems 1.5 and 3.5). An update declared as committed becomes immediately visible to other transactions and will not be undone even if the transaction eventually aborts and rolls back its uncommitted updates. What changes would be needed in the algorithms of this chapter and Chap. 3 to implement this feature? Is it possible with ARIES?


# Bibliographical Notes

The do-undo-redo recovery paradigm was implemented in the System R database management system of Astrahan et al. [1976], where it was used in connection with *shadow paging* [Lorie, 1977, Gray et al. 1981]. With shadow paging, the first time a page is modified after a checkpoint, a new disk page is associated with it, and later, when the page (the current version) is flushed onto disk, it goes to the new disk address. The old version, called the shadow page, is not discarded until the next checkpoint is taken. Restart recovery is performed from the shadow pages and the log, using the do-undo-redo paradigm. In checkpointing, all actions of transactions are quiesced, and all the modified pages in the buffer are flushed, resulting in an action-consistent disk version of the database, which is then made the new shadow version. In System R, the undo actions are not logged nor is there any logging of modifications on index and space-management information. Also, the shadow-page technique makes it unnecessary to have the PAGE-LSN concept to track the state of pages [Mohan et al. 1992a, Mohan, 1999].

The pitfalls associated with the do-undo-redo recovery paradigm of System R are discussed in detail by Mohan et al. [1992a]. They show that with the WAL buffering policy and fine-grained (i.e., tuple-level) concurrency control, the do-undo-redo recovery paradigm is incorrect. Mohan et al. [1992a] also present a detailed review of different recovery paradigms and a survey of their implementations in various commercial database management systems and research prototypes.

Weikum and Vossen [2002] consider a recovery paradigm called "redo-winners," in which the redo pass is done first, but only the updates by "winner" (i.e., committed) transactions are redone, after which the "loser" (i.e., uncommitted) transactions are rolled back. This paradigm, although avoiding some of the pitfalls related to the do-undo-redo recovery paradigm, still cannot perform physical redo in all cases.

The do-redo-undo or "repeating history" recovery paradigm, together with associated features such as the PAGE-LSN for tracking the state of pages, the WAL buffering policy, redo-only logging of undo actions, and fine-grained concurrency control, are the fundamental basis of the ARIES family of algorithms [Mohan et al. 1992a]. The version of the ARIES recovery algorithm presented in this chapter is basically the one presented by Mohan et al. [1992a] but slightly simplified and modified for the purposes of our key-range transaction model. Also, issues related to distributed transactions have been deferred to Chap. 13. Our approach to handling structure modifications is different from that presented for ARIES by Mohan [1990a, 1996a], Mohan et al. [1992a], and Mohan and Levine [1992]: we advocate redo-only structure modifications instead of redo-undo ones.

The original ARIES algorithm was developed in the 1980s and publicly documented in a research report in 1989 [Mohan et al. 1992a, Mohan, 1999]. Later, enhanced versions and extensions of the basic ARIES algorithm were developed; these include ARIES-RRH (restricted repeating of history) [Mohan and Pirahesh, 1991], ARIES/NT (nested transactions) [Rothermel and Mohan, 1989], ARIES/KVL (key-value locking) [Mohan, 1990a], ARIES/IM (index management) [Mohan and Levine, 1992], ARIES/LHS (linear hashing with separators) [Mohan, 1993a], and ARIES/CSA (client-server architecture) [Mohan and Narang, 1994]. The ARIES algorithm and most of the extensions have been implemented in several industrial-strength database management systems, including IBM DB2 [Mohan, 1999].

# Chapter 5
# Transactional Isolation

The latching protocol applied when accessing pages in the physical database maintains the integrity of the physical database during transaction processing, so that, for example, a B-tree index structure is kept structurally consistent and balanced. When the physical database is consistent, the logical database consisting of the tuples in the data pages is action consistent, meaning that the logical database is the result of a sequence of completely executed logical database actions.

Action consistency however does not imply the integrity of the logical database. First, an action-consistent logical database may not be transaction-consistent, that is, the result of completely executed (i.e., committed or rolled-back) transactions: some of the data items may be in an uncommitted state. Second, a transaction-consistent database may still not be consistent if the transactions have not been executed in sufficient isolation.

From Sect. 1.6, we recall that *isolation*, the third ACID property of transactions, implies that a transaction has the impression that it is running alone although in reality many transactions are running concurrently. The actions by different transactions must be executed in such a mutual order that the actions by different transactions do not interfere with each other, such as an action by a transaction accessing a data item updated by another active transaction, which may cause the integrity of the database to be lost. When transactions are run in complete isolation, the execution is necessarily equivalent to some serial execution order of the transactions in which only one transaction is active at a time.

In this chapter we study transactional isolation by means of *isolation anomalies* that represent situations in which isolation is violated in a transaction history, that is, in a mutual execution order of the actions of concurrently running transactions. A transaction may be set to run at a designated *isolation level*, meaning that some of the less serious isolation anomalies are allowed for the transaction, but the more serious ones are disallowed. At the highest isolation level, called *serializability*, all anomalies are disallowed. When we want to be sure that the integrity of the logical

© Springer International Publishing Switzerland 2014

S. Sippu, E. Soisalon-Soininen, *Transaction Processing*, Data-Centric Systems and Applications, DOI 10.1007/978-3-319-12292-2_5

database is maintained under all circumstances, we run all updating transactions at the serializable isolation level.

## 5.1 Transaction Histories

In fine-granular transaction processing, a transaction $T_2$ can usually read or write a data item $y$ in data page $p$ even though another, still active transaction $T_1$ has read or written another data item $x$ in the same page $p$. This is made possible by the latching protocol described earlier: according to this protocol, a page is kept latched only for the duration of the read or write action. Also, transaction $T_2$ should be allowed to read a data item $x$ that has been read by another, still active transaction $T_1$, provided that no active transaction writes $x$.

The latching protocol thus allows many different ways in which the action sequences of concurrently running transactions can be interleaved. Some of these interleavings are correct and preserve the integrity of the database, but some may result in inconsistencies in the logical database or prevent recovery from failures.

First we note that in a specific interleaving, an individual transaction may see a database state that is different from that seen by the transaction executed by the same database program in some other interleaving. For example, a sequence of database-program steps that scans the entire database generates a transaction that contains as many read actions as there are tuples in the database state seen at the moments the individual actions are executed. That sequence of actions (i.e., the transaction generated) depends on the interleaving of the actions with updating transactions that may be running concurrently.

*Example 5.1* Consider a set of three transactions of the following forms:

$T_1 = BR[x_1, >0, u_1] R[x_2, >x_1, u_2] \ldots R[\infty, >x_n, 0]C$.
$T_2 = BI[4, 4]C$.
$T_3 = BD[2, 2]C$.

Assume that these transactions are run on the database

$\{(0,0), (1, 1), (2, 2), (3, 3)\}$.

Obviously, the exact sequence of actions for the transaction of the form $T_1$ depends, besides on the contents of the database, also on how the actions are interleaved with the actions of transactions $T_2$ and $T_3$. The committed state of $T_1$ can take four different forms, depending on how the actions of $T_1$ are interleaved with the actions of $T_2$ and $T_3$:

1. $(T_1, BR[1, 1] R[2, 2] R[3, 3] R[\infty, 0]C)$.
2. $(T_1, BR[1, 1] R[2, 2] R[3, 3] R[4, 4] R[\infty, 0]C)$.
3. $(T_1, BR[1, 1] R[3, 3] R[\infty, 0]C)$.
4. $(T_1, BR[1, 1] R[3, 3] R[4, 4] R[\infty, 0]C)$.

(Recall the formal definition of a transaction as a pair of a transaction identifier and a transaction state.)

Case (1) occurs in interleavings in which the action $I[4,4]$ of $T_2$ is executed after the action $R[\infty,0]$ of $T_1$ and the action $D[2,2]$ of $T_3$ is executed after the action $R[x_2,>1,u_2]$ of $T_1$, such as in the following interleaving:

$T_1:$ $BR[1,1]R[2,2]R[3,3]R[\infty,0]C$
$T_2:$ $\qquad\qquad\qquad\qquad\qquad\qquad BI[4,4]C$
$T_3:$ $\qquad\qquad\qquad\qquad\qquad\qquad\qquad\qquad\qquad BD[2,2]C$

Case (2) occurs when $I[4,4]$ is executed before $R[x_4,> 3,u_4]$, but $D[2,2]$ is executed after $R[x_2,>1,u_2]$, such as in the following:

$T_1:$ $BR[1,1]R[2,2]R[3,3]\qquad\qquad\qquad R[4,4]R[\infty,0]C$
$T_2:$ $\qquad\qquad\qquad\qquad BI[4,4]C$
$T_3:$ $\qquad\qquad\qquad\qquad\qquad\qquad\qquad\qquad\qquad BD[2,2]C$

Case (3) occurs when $I[4,4]$ is executed after $R[\infty,0]$, but $D[2,2]$ is executed before $R[x_2,>1,u_2]$. Case (4) occurs when $I[4,4]$ is executed before $R[x_4,>3,u_4]$ and $D[2,2]$ is executed before $R[x_2,>2,u_2]$. $\qquad\qquad\qquad\square$

Let $\{T_1,\ldots,T_n\}$ be a set of transactions. This set can contain forward-rolling, committed, backward-rolling, and rolled-back transactions. A *transaction schedule* or *transaction history* $H$ of the set of transactions $\{T_1,\ldots,T_n\}$ is a *shuffle* of the action strings $T_1,\ldots,T_n$. In other words, $H$ is an action sequence that satisfies the following conditions:

1. $H$ is an action sequence composed exactly of the actions present in the transactions $T_1,\ldots,T_n$.
2. The actions of each transaction $T_i$ are present in $H$ in the same order as in $T_i$, that is, $H$ is of the form $\alpha_1\beta_1\alpha_2\beta_2\ldots\beta_m\alpha_{m+1}$, where $\beta_1\beta_2\ldots\beta_m = T_i$.

A *serial history* $H$ is a permutation of entire transactions of which at most one, the last, is active. In a serial history all the actions of each transaction $T_i$ are consecutive, so that $H$ is of the form $T_{i_1}\ldots T_{i_n}$, where $\{i_1,\ldots,i_n\} = \{1,\ldots,n\}$, and each transaction $T_{i_j}$, $j = 1,\ldots,n-1$, is committed or rolled back. A history that is not serial is called a *nonserial history*.

*Example 5.2* For the transactions of Example 5.1, the possible serial histories are in the four cases:

1. $T_1T_2T_3$ or $T_1T_3T_2$.
2. $T_2T_1T_3$.
3. $T_3T_1T_2$.
4. $T_2T_3T_1$ or $T_3T_2T_1$.

All the six histories result in the same database:

$\{(0,0),(1,1),(3,3),(4,4)\}.$

Moreover, the two histories in case (1), as well as those in case (4), can be regarded as "equivalent" because in each case $T_1$ is the same transaction reading the same set of tuples.                                                                                      □

In a serial history, all the transactions are executed one by one from the beginning (action $B$) to the end (action $C$). Such a history must be viewed as correct, as long as we do not impose any constraint on the timepoint when a transaction is expected to be run or on the specific state of the logical database on which a transaction is meant to be run. Recall from the discussion about the ACID property "C" in Sect. 1.6 that the sequence of application-program steps that generates a transaction is assumed to be correctly programmed for any serial execution. Indeed, when all committed transactions are programmed to retain the consistency of an initially consistent database, a serial history is guaranteed to retain the consistency of an initially consistent database.

Two histories $H$ and $H'$ of the same transaction set $\{T_1, \ldots, T_n\}$ on database $D$ are *equivalent* if they produce on $D$ the same database state. Observe that the requirement that $H$ and $H'$ are histories of the same transaction set includes that each transaction has exactly the same sequence of actions and with same input and output both in $H$ and in $H'$. Thus, in Example 5.2, histories from two different cases are not equivalent because transaction $T_1$ varies from one case to another.

There usually exist many nonserial histories that are equivalent to some serial history, thus suggesting that transactions can be run concurrently in a nonserial way. In classical concurrency control theory, a history that is equivalent to a serial history is called *serializable* (a property that is weaker than the one with the same name to be defined in Sect. 5.5).

*Example 5.3* One of the many possible nonserial histories of the transactions in Example 5.1 is the history given for case (2). Using subscripting to distinguish between actions belonging to different transactions, that history can be represented alternatively as

$$B_1 R_1[1, 1] R_1[2, 2] R_1[3, 3] B_2 I_2[4, 4] C_2 R_1[4, 4] R_1[\infty, 0] C_1 B_3 D_3[2, 2] C_3,$$

where an action $o[\bar{x}]$ by transaction $T_i$ is denoted by $o_i[\bar{x}]$. Clearly, the history is equivalent to the serial history for case (2) in Example 5.2:

$T_1$ :                       $BR[1, 1] R[2, 2] R[3, 3] R[4, 4] R[\infty, 0] C$
$T_2$ :     $BI[4, 4] C$
$T_3$ :                                                                   $BD[2, 2] C$

Thus, the history is serializable in the classical sense. As will be seen soon, it is also free of any isolation anomalies; hence it is acceptable.                          □

In terms of the classical read-write transaction model, the *isolation* of transactions means that a transaction $T$ is not free to read or write data items written by another transaction $T'$ or to overwrite data items read by $T'$ while $T'$ is still active. In our key-range transaction model, $T$ is also not free to read a key range from which $T'$ has deleted a data item or to insert data items into a key range previously read by $T'$.

In every serial history, transactions are trivially isolated from each other, because the transaction that reads or writes first is committed or rolled back before the other transaction performs its action. However, equivalence to a serial history does not imply isolation.

*Example 5.4* The history

$$T_1 : \quad BR[1,1]R[2,2] \qquad\qquad R[3,3] \qquad\qquad R[4,4]R[\infty,0]C$$
$$T_2 : \qquad\qquad\qquad\qquad\qquad\qquad\qquad BI[4,4]C$$
$$T_3 : \qquad\qquad\qquad\qquad BD[2,2]C$$

is equivalent to the serial history, $T_2 T_1 T_3$, of case (2) in Example 5.2, but the transactions are not isolated: the action $R[2,>1,2]$ of $T_1$ is an "unrepeatable read": the action cannot be repeated in $T_1$ after the action $D[2,2]$ of $T_3$. In this specific case, the breach of isolation does not cause any harm. But if $T_1$ performed some update based on the output of the action $R[2,2]$, and if $T_3$, instead of committing immediately, aborted and rolled back or inserted a tuple $(2,u)$ with $u \neq 2$ and then committed, the integrity of the database might be lost. $\square$

## 5.2 Uncommitted Updates

A data item $x$ is *dirty* or *uncommitted* at a certain moment if it was last updated (inserted, written, or deleted) by a transaction $T$ that is still active, and it is possible that $T$ will still change (i.e., delete, write, or insert) $x$. Such a future change is possible if $T$ is forward-rolling (and can thus perform any number of new forward-rolling database actions) or $T$ is backward-rolling but has not yet restored the original value to $x$ that existed when $T$ first overwrote it.

In the following, we define precisely the concept of an uncommitted update in relation to our transaction models. For that purpose, let $H$ be a transaction history and $T$ one of its active transactions, so that $H$ does not contain the $C$ action for $T$, and let $x$ be a key that is updated in $H$, so that it is the key of some tuple inserted, written, or deleted in some action in $H$. The key $x$ has a *dirty update* or an *uncommitted update* by $T$ in $H$, if one of the following conditions is true:

1. The last update on $x$ in $H$ is by a forward-rolling update action $o[x]$ (i.e., one of $I[x]$, $W[x]$ or $D[x]$) by $T$.
2. The last update on $x$ in $H$ is by a backward-rolling action $o^{-1}[x]$ (i.e., one of $I^{-1}[x]$, $W^{-1}[x]$ or $D^{-1}[x]$) by $T$ such that somewhere before the corresponding forward-rolling action $o[x]$ ($I[x]$, $W[x]$ or $D[x]$) by $T$ there is still some forward-rolling update action $o'[x]$ by $T$ that has not been undone in $H$.

In case (2), the update on $x$ by $o^{-1}[x]$ does not yet restore on the tuple with key $x$ the value that existed at the time when $T$ performed its first update on it. As soon as the first update on $x$ by $T$ has been undone, $x$ is considered to be committed, because the tuple has been restored to its most recent committed state.

*Example 5.5* Consider the following history for a rolled-back transaction $T_1$:

$$H = BI[y, w]D[x, u]I[x, v]AI^{-1}[x, v]D^{-1}[x, u]I^{-1}[y, w]C.$$

Key $y$ has an uncommitted update by $T_1$ in the following prefix of $H$:

$$BI[y, w].$$

In the following prefixes of $H$, keys $y$ and $x$ have uncommitted updates by $T_1$:

$$BI[y, w]D[x, u].$$
$$BI[y, w]D[x, u]I[x, v].$$
$$BI[y, w]D[x, u]I[x, v]A.$$
$$BI[y, w]D[x, u]I[x, v]AI^{-1}[x, v].$$

Key $y$ has an uncommitted update by $T_1$ in the prefix

$$BI[y, w]D[x, u]I[x, v]AI^{-1}[x, v]D^{-1}[x, u].$$

However, $x$ is already committed here, because the last update on $x$ is by the undo action $D^{-1}[x, u]$, and the corresponding forward-rolling action $D[x, u]$ is the first update on $x$ by $T_1$. Both $x$ and $y$ are committed in the prefix

$$H = BI[y, w]D[x, u]I[x, v]AI^{-1}[x, v]D^{-1}[x, u]I^{-1}[y, w],$$

although the transaction is still active, namely, backward-rolling, because its rollback has not yet been completed with the commit action $C$. Of course, the transaction is de facto rolled back, because $C$ is the only action it can (and must) do next.                                                                                                   □

By the above definition, a partial rollback of a forward-rolling transaction can change an uncommitted key $x$ to committed, after which the transaction can again update $x$, after which a new partial rollback can change $x$ to committed.

*Example 5.6* In the following, as the history advances, the key $x$ changes from uncommitted to committed, then to uncommitted, then to committed, then to uncommitted, and finally to committed.

$$BS[P]W[x].$$
$$BS[P]W[x]A[P]W^{-1}[x].$$
$$BS[P]W[x]A[P]W^{-1}[x]C[P]S[Q]W[x].$$
$$BS[P]W[x]A[P]W^{-1}[x]C[P]S[Q]W[x]A[Q]W^{-1}[x].$$
$$BS[P]W[x]A[P]W^{-1}[x]C[P]S[Q]W[x]A[Q]W^{-1}[x]C[Q]W[x].$$
$$BS[P]W[x]A[P]W^{-1}[x]C[P]S[Q]W[x]A[Q]W^{-1}[x]C[Q]W[x]C.$$

□

## 5.3  Isolation Anomalies

In a concurrent history, the isolation of transactions can be violated in several ways. The isolation violations can be classified into three *isolation anomalies*: dirty writes, dirty reads, and unrepeatable reads. Without assuming any specific transaction model, these isolation anomalies can be characterized as follows:

1. A transaction $T_2$ *does a dirty write* if $T_2$ updates data items satisfying predicate $P$ when some data item satisfying $P$ has an uncommitted update by another transaction $T_1$.
2. A transaction $T_2$ *does a dirty read* if $T_2$ reads data items satisfying predicate $P$ when some data item satisfying $P$ has an uncommitted update by another transaction $T_1$.
3. A transaction $T_1$ *does an unrepeatable read* if $T_1$ reads data items satisfying predicate $P$, after which another transaction $T_2$ updates some data item satisfying $P$ while $T_1$ is still forward-rolling. The read action by $T_1$ is termed unrepeatable because repeating it after the commit of $T_2$ will most probably produce a different result.

*Example 5.7* The update action $(W_2[x], I_2[x], D_2[x])$ by transaction $T_2$ is a dirty write in the histories

$$B_1 \ldots W_1[x] \ldots B_2 \ldots W_2[x] \ldots C_1 \ldots ,$$
$$B_1 \ldots W_1[x] \ldots B_2 \ldots D_2[x] \ldots C_1 \ldots ,$$
$$B_1 \ldots I_1[x] \ldots B_2 \ldots W_2[x] \ldots C_1 \ldots ,$$
$$B_1 \ldots I_1[x] \ldots B_2 \ldots D_2[x] \ldots C_1 \ldots ,$$
$$B_1 \ldots D_1[x] \ldots B_2 \ldots I_2[x] \ldots C_1 \ldots ,$$

assuming that the update action of transaction $T_1$ is not undone before the update action of $T_2$. The predicate $P$ in the definition is here "$X = x$," where $X$ is the primary-key attribute.

On the other hand, the history

$$B_1 \ldots S_1[P]W_1[x] \ldots B_2 \ldots A_1[P] \ldots W_1^{-1}[x]C_1[P] \ldots W_2[x] \ldots C_1 \ldots$$

does not exhibit a dirty write, because $W_1[x]$ is undone before $W_2[x]$. □

*Example 5.8* The action $R_2[x]$ by transaction $T_2$ is a dirty read in the histories

$$B_1 \ldots W_1[x] \ldots B_2 \ldots R_2[x] \ldots C_1 \ldots ,$$
$$B_1 \ldots I_1[x] \ldots B_2 \ldots R_2[x] \ldots C_1 \ldots ,$$

assuming that the update action of transaction $T_1$ is not undone before the read action of $T_2$. The predicate $P$ in the definition is "$X = x$." The histories represent the simple type of a dirty read in which the key read is the key that has the uncommitted update. □

*Example 5.9* The action $R_2[x, >z]$ by transaction $T_2$ is a dirty read in the history

$$B_1 \ldots D_1[y] \ldots B_2 \ldots R_2[x, >z] \ldots C_1 \ldots ,$$

if $x > y > z$, assuming that the delete action of transaction $T_1$ is not undone before the read action of $T_2$. The predicate $P$ in the definition is "$x \geq X > z$."

The above history exhibits a dirty read of type *phantom phenomenon*. The tuple with key $y$ that disappears from the key range $(z, x]$ due to the delete action by $T_1$ is called a *phantom tuple*. □

*Example 5.10* The action $R_1[x]$ by transaction $T_1$ is an unrepeatable read in the histories

$$B_1 \ldots R_1[x] \ldots B_2 \ldots W_2[x] \ldots C_1 \ldots ,$$
$$B_1 \ldots R_1[x] \ldots B_2 \ldots D_2[x] \ldots C_1 \ldots ,$$

assuming that $T_1$ is still forward-rolling when transaction $T_2$ performs its update action. The predicate $P$ in the definition is "$X = x$."

The read action $R_1[x]$ is called unrepeatable, because repeating it in $T_1$ after the commit of $T_2$ may be impossible: after $W_2[x]$ the action $R[x]$ is likely to produce a tuple with a value different from that obtained in the first reading, and after $D_2[x]$, the action $R[x, \geq x]$ is impossible, and the action $R[x', \geq z]$ will produce a tuple with key $x'$ different from $x$. It is a natural requirement that if $T_1$ does not itself update the tuple with key $x$ in the interim, then the read action $R[x]$ should be repeatable within $T_1$.

The above histories represent the simple type of an unrepeatable read in which the key updated is the key read. $\quad\square$

*Example 5.11* The action $R_1[x, >z]$ by transaction $T_1$ is an unrepeatable read in the history

$$B_1 \ldots R_1[x, >z] \ldots B_2 \ldots I_2[y] \ldots C_1 \ldots ,$$

if $x > y > z$, assuming that $T_1$ is still forward-rolling when transaction $T_2$ performs its insert action. The predicate $P$ in the definition is "$x \geq X > z$."

The above history exhibits an unrepeatable read of type *phantom phenomenon*. The tuple with key $y$ that appears into the key range $(z, x]$ due to the insert action by $T_2$ is called a *phantom tuple*. $\quad\square$

We say that a transaction $T$ in history $H = \alpha\beta$ is *forward-rolling* (resp. *aborted, backward-rolling, committed, rolled-back, active,* or *terminated*) *at* $\alpha$, if the prefix $T'$ of $T$ contained in $\alpha$ is forward-rolling (resp. aborted, backward-rolling, committed, rolled-back, active, or terminated) in the history $\alpha$.

For our transaction models (the read-write and key-range model), the isolation anomalies are defined as follows:

1. An update action $o_2[x]$ on key $x$ by transaction $T_2$ is a *dirty write* in history $\alpha o_2[x]\beta$, if $x$ has an uncommitted update by another transaction $T_1$ in $\alpha$.
2. A read action $R_2[x, \theta z]$ by transaction $T_2$ is a *dirty read* in history $\alpha R_2[x, \theta z]\beta$, if some key $y$ with $x \geq y \,\theta\, z$ has an uncommitted update by another transaction $T_1$ in $\alpha$. The dirty read is a *simple dirty read* if here $x = y$ and a *phantom dirty read* (or a dirty read of type *phantom phenomenon*) if $x > y$.
3. A read action $R_1[x, \theta z]$ by transaction $T_1$ is an *unrepeatable read* in history $\alpha R_1[x, \theta z]\beta o_2[y]\gamma$, if $T_1$ is forward-rolling at $\alpha R_1[x, \theta z]\beta$ and $o_2[y]$ is an update action by another transaction $T_2$ on key $y$ with $x \geq y \,\theta\, z$. The unrepeatable read is a *simple unrepeatable read* if here $x = y$ and a *phantom unrepeatable read* (or an unrepeatable read of type *phantom phenomenon*) if $x > y$.

In the definition of unrepeatable reads, it should be noted that a read action $R_1[x, \theta z]$ is not regarded as unrepeatable in the history

$$\alpha R_1[x, \theta z] \beta o_2[y] \gamma$$

if $T_1$ is backward-rolling at $\alpha R_1[x, \theta z] \beta$. This is because $T_1$ will eventually roll back all its updates, including those possibly based on the unrepeatably read key.

The update actions exhibiting one of the three isolation anomalies defined above can be forward-rolling insert, delete, or write actions or undo actions. However, if such an action is an undo action, the anomaly may also be exhibited by a forward-rolling action.

*Example 5.12* In the history

$$B_1 \ldots W_1^{-1}[x] \ldots B_2 \ldots W_2^{-1}[x] \ldots C_2 \ldots C_1$$

the action $W_2^{-1}[x]$ is a dirty write, if $W_1^{-1}[x]$ is the last update on $x$ in $B_1 \ldots W_1^{-1}[x] \ldots B_2 \ldots$ and does not restore $x$ to its original, committed state. But then the history must be of the form

$$B_1 \ldots o_1[x] \ldots W_1[x] \ldots W_1^{-1}[x] \ldots B_2 \ldots W_2[x] \ldots W_2^{-1}[x] \ldots C_2 \ldots C_1,$$

where $W_1[x]$ and $W_2[x]$ are the forward-rolling actions that are undone by the actions $W_1^{-1}[x]$ and $W_2^{-1}[x]$, respectively, and $o_1[x]$ is a forward-rolling update action on $x$. The pair of forward-rolling update actions $o_1[x]$ and $W_2[x]$ now exhibit a dirty write. □

The following lemma states that the appearance of dirty writes in a history can always be characterized by a pair of conflicting forward-rolling update actions.

**Lemma 5.13** *Let $H$ be a history of transactions in the read-write or key-range transaction model. If $H$ contains a dirty write, then $H$ is of the form*

$$H = \alpha o_1[x] \beta o_2[x] \gamma,$$

*where $o_1[x]$ is a forward-rolling insert action $I_1[x]$, delete action $D_1[x]$, or write action $W_1[x]$ by some transaction $T_1$ active at $\alpha o_1[x] \beta$; $o_2[x]$ is a forward-rolling insert action $I_2[x]$, delete action $D_2[x]$ or write action $W_2[x]$ by some transaction $T_2$ other than $T_1$; and the action sequence $\beta$ does not contain the undo action for $o_1[x]$.*

*Proof* By the definitions of a dirty write and an uncommitted update, we can write $H$ as

$$H = \alpha o_1[x] \beta o_2[x] \gamma,$$

where the action $o_1[x]$ by transaction $T_1$ is the last update on key $x$ in $\alpha o_1[x] \beta$, $T_1$ is active at $\alpha o_1[x] \beta$, the action $o_2[x]$ is an update on $x$ by some transaction $T_2$ other than $T_1$, and, if $o_1[x]$ is an undo action, then it does not restore $x$ to its committed state, so that there is in $\alpha$ a forward-rolling update action $o_1'[x]$ on $x$ by $T_1$ that has no corresponding undo action in $\alpha o_1[x] \beta$.

Thus, if $o_1[x]$ is an undo action, we can write $H$ as

$$H = \alpha' o_1'[x] \alpha_2 o_1[x] \beta o_2[x] \gamma = \alpha' o_1'[x] \beta' o_2[x] \gamma,$$

where $o_1'[x]$ is a forward-rolling update action for which there is no corresponding undo action in $\beta'$.

Now if $o_2[x]$ is an undo action, we can further write $H$ either as

$$H = \alpha' o_1'[x]\beta'' o_2'[x]\beta_2 o_2[x]\gamma = \alpha' o_1'[x]\beta'' o_2'[x]\gamma'$$

or as

$$H = \alpha'' o_2'[x]\beta'' o_1'[x]\beta' o_2[x]\gamma = \alpha'' o_2'[x]\beta'' o_1'[x]\gamma',$$

depending on whether the forward-rolling action $o_2'[x]$ corresponding to the undo action $o_2[x]$ is in $\beta'$ or in $\alpha'$. In the former case the undo action for $o_1'[x]$, if any, is not in $\beta''$, since it is not in $\beta' = \beta'' o_2'[x]\beta_2$. In the latter case $T_2$ is active at $\alpha'' o_2'[x]\beta''$ and the undo action $o_2[x]$ for $o_2'[x]$ is not in $\beta''$, since it is in $\gamma'$. Thus both of these cases exhibit the required form of the history $H$.                      □

The appearance of dirty reads in a history can always be characterized by a conflicting pair of a forward-rolling update action and a read action:

**Lemma 5.14** *Let $H$ be a history of transactions in the read-write or key-range transaction model. If $H$ contains a dirty read, then $H$ is of the form*

$$H = \alpha o_1[y]\beta R_2[x, \theta z]\gamma,$$

*where $x \geq y \, \theta \, z$, $o_1[y]$ is a forward-rolling insert action $I_1[y]$, delete action $D_1[y]$, or write action $W_1[y]$ by some transaction $T_1$ active at $\alpha o_1[y]\beta$, $R_2[x, \theta z]$ is a read action by some transaction $T_2$ other than $T_1$, and the action sequence $\beta$ does not contain the undo action for $o_1[y]$.*

*Proof* By the definitions of a dirty read and an uncommitted update, we can write $H$ as

$$H = \alpha o_1[y]\beta R_2[x, \theta z]\gamma,$$

where $x \geq y \, \theta \, z$, the action $o_1[y]$ by transaction $T_1$ is the last update on key $y$ in $\alpha o_1[y]\beta$, $T_1$ is active at $\alpha o_1[y]\beta$, the action $R_2[x, \theta z]$ is a read action by some transaction $T_2$ other than $T_1$, and, if $o_1[y]$ is an undo action, then it does not restore $y$ to its committed state, so that there is in $\alpha$ a forward-rolling update action $o_1'[y]$ on $y$ by $T_1$ that has no corresponding undo action in $\alpha o_1[y]\beta$.

Thus, if $o_1[y]$ is an undo action, we can write $H$ as

$$H = \alpha' o_1'[y]\alpha_2 o_1[y]\beta R_2[x, \theta z]\gamma = \alpha' o_1'[y]\beta' R_2[x, \theta z]\gamma,$$

where $o_1'[y]$ is a forward-rolling update action for which there is no corresponding undo action in $\beta'$, as required.                                                     □

A history that contains an unrepeatable read but no dirty reads can be characterized by a conflicting pair of a read action and a forward-rolling update action:

**Lemma 5.15** *Let $H$ be a history of transactions in the read-write or key-range transaction model. If $H$ contains an unrepeatable read, then $H$ either contains a dirty read or is of the form*

$$H = \alpha R_1[x, \theta z]\beta o_2[y]\gamma,$$

*where* $x \geq y \ \theta \ z$, $R_1[x, \theta z]$ *is a read action by a transaction* $T_1$ *forward-rolling at* $\alpha R_1[x, \theta z]\beta$ *and* $o_2[y]$ *is a forward-rolling insert action* $I_2[y]$, *delete action* $D_2[y]$, *or write action* $W_2[y]$ *by some transaction* $T_2$ *other than* $T_1$.

*Proof* By the definition of an unrepeatable read, $H$ is of the form

$$H = \alpha R_1[x, \theta z]\beta o_2[y]\gamma,$$

where $x \geq y \ \theta \ z$, $R_1[x, \theta z]$ is a read action by a transaction $T_1$ forward-rolling at $\alpha R_1[x, \theta z]\beta$, and $o_2[y]$ is an update action by some transaction $T_2$ other than $T_1$.

If $o_2[y]$ is an undo action, we can write $H$ either as

$$H = \alpha R_1[x, \theta z]\beta' o_2'[y]\beta_2 o_2[y]\gamma = \alpha R_1[x, \theta z]\beta' o_2'[y]\gamma'$$

or as

$$H = \alpha' o_2'[y]\beta' R_1[x, \theta z]\beta o_2[y]\gamma = \alpha' o_2'[y]\beta' R_1[x, \theta z]\gamma',$$

depending on whether the forward-rolling update action $o_2'[y]$ corresponding to $o_2[y]$ is in $\beta$ or in $\alpha$. The former of these is of the desired form, while the latter exhibits a dirty read.                                                                          □

## 5.4  Isolation Anomalies and Database Integrity

We will show that each of the three isolation anomalies can destroy the integrity of the database even if the database application programs generate only logically consistent transactions when run serially on an initially consistent logical database. In each case we give a concurrent history of just two transactions, both of which are committed (or rolled back). In each case, only one isolation anomaly is present in the history, and when the history is run on a consistent database, the result is an inconsistent database.

The following example demonstrates that a dirty write can easily break the integrity of a database.

*Example 5.16* The integrity constraint for the database states that the database must have tuples with keys $x$ and $y$, and the values must always be the same. Let transaction $T_1$ be generated from the program

**exec sql update** $r$ **set** $V = 1$ **where** $X = x$;
**exec sql update** $r$ **set** $V = 1$ **where** $Y = y$;
**exec sql commit**,

and let transaction $T_2$ be generated from the program

**exec sql update** $r$ **set** $V = 2$ **where** $X = x$;
**exec sql update** $r$ **set** $V = 2$ **where** $Y = y$;
**exec sql commit**.

When the programs are run serially, one by one, on an initially consistent database $r(\underline{X}, V)$, the transactions generated are of the forms:

$$T_1 = BW[x, u, 1]W[y, u, 1]C$$
$$T_2 = BW[x, v, 2]W[y, v, 2]C,$$

where $v = 1$ if $T_1$ is run first and $u = 2$ if $T_2$ is run first. Both transactions are logically consistent with respect to the integrity constraint: both transactions retain the consistency of the logical database. Hence both serial histories $T_1T_2$ and $T_2T_1$ keep the database consistent.

However, it is possible that the following nonserial history is generated:

$T_1$:    $BW[x, 0, 1]$                                          $W[y, 2, 1]C$
$T_2$:                    $BW[x, 1, 2]W[y, 0, 2]C$

This history can be run on all databases that include tuples $(x, 0)$ and $(y, 0)$, and will produce a database where the tuples with keys $x$ and $y$ are $(x, 2)$ and $(y, 1)$. Thus, this history breaks the integrity of the database. The action $W[x, 1, 2]$ by $T_2$ is a dirty write, and this is the only isolation anomaly in this history.                    □

The next two examples show how a dirty read can break the consistency of a database.

*Example 5.17*  An integrity constraint states that the database must have tuples with keys $x$ and $y$, and the values in both tuples must be positive. When run serially on a consistent database $r(\underline{X}, V)$, the programs

**exec sql update** $r$ **set** $V = 0$ **where** $X = x$;
**exec sql update** $r$ **set** $V = 1$ **where** $X = x$;
**exec sql commit**

and

**exec sql select** $V$ **into** : w **from** $r$ where $X = x$;
**exec sql update** $r$ **set** $V = $ : w **where** $Y = y$;
**exec sql commit**

generate two logically consistent transactions:

$$T_1 = BW[x, u, 0]W[x, 0, 1]C;$$
$$T_2 = BR[x, w]W[y, v, w]C.$$

On the other hand, the nonserial history

$T_1$:    $BW[x, 1, 0]$                                          $W[x, 0, 1]C$
$T_2$:                    $BR[x, 0]W[y, 1, 0]C$

is also possible; this history can be run on every database that includes tuples $(x, 1)$ and $(y, 1)$ and produces a database where the tuples with keys $x$ and $y$ are $(x, 1)$ and $(y, 0)$. Thus, the history breaks the integrity of the database. The only isolation anomaly in the history is the dirty read $R[x, 0]$ by $T_2$. Transaction $T_2$ reads a value that is not yet a final, committed value and uses it to update the database.                    □

*Example 5.18* The integrity constraint is the same as in the previous example (positive values for the tuples with keys $x$ and $y$). Transaction $T_2$ is as previously:

$$T_2 = BR[x, w]W[y, v, w]C.$$

The other transaction,

$$T_3 = BW[x, u, 0]AW^{-1}[x, u, 0]C,$$

is generated from the program

**exec sql update** $r$ **set** $V = 0$ **where** $X = x$;
**exec sql rollback**.

$T_3$ is a rolled-back transaction and thus trivially logically consistent with respect to any integrity constraint whatsoever. However, the following history of transactions $T_2$ and $T_3$ breaks the integrity of the database:

$T_2$: $\qquad\qquad\qquad BR[x, 0]W[y, 1, 0]C$
$T_3$: $\quad BW[x, 1, 0] \qquad\qquad\qquad\qquad\qquad AW^{-1}[x, 1, 0]C$

The only isolation anomaly in the history is the dirty read $R[x, 0]$ by $T_2$. Here, too, $T_2$ reads a value that is not final and uses it to update the database. $\qquad\square$

The next two examples show how an unrepeatable read can break the integrity of the database.

*Example 5.19* The integrity constraint of the database states that $u + v \geq 0$ must always hold for the tuples $(x, u)$ and $(y, v)$ with keys $x$ and $y$ ("the sum of the balances of accounts $x$ and $y$ must never be negative"). The transactions

$$T_1 = BR[y, v]W[x, u, -v]C \quad \text{and}$$
$$T_2 = BR[x, u]W[y, v, -u]C$$

both bring the sum of the values down to zero, thus preserving the integrity constraint. Transaction $T_1$ (resp. $T_2$) can be interpreted as "a withdrawal of the sum of balances $u + v$ from account $x$ (resp. $y$)." The history

$T_1$: $\quad BR[y, v] \qquad\qquad\qquad\qquad W[x, u, -v]C$
$T_2$: $\qquad\qquad\qquad BR[x, u]W[y, v, -u]C$

can be run on every database that contains the tuples $(x, u)$ and $(y, v)$ and produces a database with the tuples $(x, -v)$ and $(x, -u)$, thus breaking the integrity of the database when $u + v > 0$ initially. The only isolation anomaly in the history is the unrepeatable read $R[y, v]$ by $T_1$.

To keep the integrity of the database, transaction $T_1$ should have read the current value of $y$ before updating $x$. Repeating the read action $R[y, v]$ in $T_1$ after $T_2$ has committed would have resulted in the action $R[y, -u]$, so that $W[x, u, -v]$ would have been realized as $W[x, u, u]$. $\qquad\square$

*Example 5.20* The integrity constraint of the database states that the value of the tuple with key 0 must be the sum of the values of the tuples with positive keys. The integrity constraint holds, for example, for the database

$D = \{(0, 6), (2, 2), (4, 4)\}.$

A transaction of the form

$$T_1 = BR[x_1, >0, u_1] R[x_2, >x_1, u_2] \ldots R[\infty, >x_n, 0] W[0, v, \sum_{i=1}^{n} u_i] C$$

is generated from the program

> **exec sql select sum($V$) into** : s **from** $r$ **where** $X > 0$;
> **exec sql update** $r$ **set** $V = $ : s **where** $X = 0$;
> **exec sql commit**,

and a transaction

$$T_2 = BI[1, 1] I[3, 3] R[0, v] W[0, v, v + 4] C$$

from the program

> **exec sql insert into** $r$ **values** $(1, 1)$;
> **exec sql insert into** $r$ **values** $(3, 3)$;
> **exec sql select** $V$ **into** : v **from** $r$ where $X = 0$;
> **exec sql update** $r$ **set** $V = $ : v $+ 4$ where $X = 0$;
> **exec sql commit**.

Both transactions are logically consistent with respect to the integrity constraint. The transactions exemplify a case in which one transaction ($T_1$) scans a relation in key order while simultaneously another transaction ($T_2$) inserts tuples in the range being scanned.

On the above database $D$, the history

$T_1$: $BR[2, 2]$ $\hspace{4cm}$ $R[3, 3] R[4, 4] R[\infty, 0] W[0, 10, 9] C$
$T_2$: $\hspace{2.5cm}$ $BI[1, 1] I[3, 3] R[0, 6] W[0, 6, 10] C$

can be run on $D$ and produces the database

$$D' = \{(0, 9), (1, 1), (2, 2), (3, 3), (4, 4)\},$$

which does not satisfy the integrity constraint. The only isolation anomaly in this history is the unrepeatable read $R[2, > 0, 2]$ by $T_1$, which is of type phantom phenomenon.

The sum calculated by $T_1$ includes the value of the second tuple inserted by $T_2$, but not that of the first. Accordingly, the history is neither equivalent to the serial history $T_1 T_2 = $

$T_1$: $\hspace{0.5cm}$ $BR[2, 2] R[4, 4] R[\infty, 0] W[0, 6, 6] C$
$T_2$: $\hspace{5cm}$ $BI[1, 1] I[3, 3] R[0, 6] W[0, 6, 10] C,$

nor to the serial history $T_2 T_1 = $

$T_1$: $\hspace{5cm}$ $BR[1, 1] \ldots R[\infty, 0] W[0, 10, 10] C$
$T_2$: $\hspace{0.3cm}$ $BI[1, 1] I[3, 3] R[0, 6] W[0, 6, 10] C.$

Both of these serial histories produce on $D$ the database

$D'' = \{(0, 10), (1, 1), (2, 2), (3, 3), (4, 4)\},$

which satisfies the integrity constraint.                                                □

The definition of unrepeatable reads presented in the previous section also encompasses read actions that are included in action sequences rolled back in partial rollbacks completed before executing the offending update action by the other transaction. There is good reason to restrict the definition of unrepeatable reads so as to exclude such read actions.

*Example 5.21* According to the definition, the action $R[x]$ is an unrepeatable read in the history

$T_1$:    $B \ldots S[P] \ldots R[x] \ldots A[P] \ldots C[P] \ldots$                 $\ldots W[y]C$
$T_2$:                                                   $BW[x]C$

However, it is unlikely that the unrepeatable read would cause any violation of the integrity of the database. This is because it would exhibit bad transaction-programming practice if an update done after $C[P]$ in $T_1$, such as the update $W[y]$, would somehow depend on information read by $R[x]$.                 □

The above example suggests that we should impose the following additional requirement on a logically consistent transaction: no update action following a completed partial rollback should use information read by a read action included in the action sequence rolled back in that partial rollback.

## 5.5  SQL Isolation Levels

The SQL language has a statement **set transaction**, which can be used in an application program to set the *isolation level*, or *degree of isolation*, of the next transaction to be started.

The isolation anomalies that form the basis of the definition of isolation levels are not exactly the same as the ones we defined above: the dirty and unrepeatable reads in SQL parlance do not include phantom reads.

There are four possible isolation levels in SQL (ordered by complexity, the first ones being the least restrictive and also the simplest to implement): read uncommitted, read committed, repeatable read, and serializable. The meaning of these isolation levels is stated below according to the corrected definitions given by Berenson et al. [1995]:

1° *Read uncommitted:* at this lowest isolation level, dirty writes are not allowed, but the transaction may do dirty or unrepeatable reads of any type.
2° *Read committed:* dirty writes and simple dirty reads are not allowed, but the transaction may do phantom dirty reads and unrepeatable reads of any type.
3° *Repeatable read:* dirty writes, simple dirty reads, and simple unrepeatable reads are not allowed, but the transaction may do phantom dirty reads and phantom unrepeatable reads.

4° *Serializable:* the transaction may not do dirty writes, dirty reads, or unrepeatable reads of any type. At this, the highest isolation level, the dirty writes, dirty reads and unrepeatable reads as we have defined above are not permitted.

The DB2 terminology for the isolation levels corresponds more closely to that used in this book. In DB2 terminology, the term "repeatable read" is used to denote the highest level, 4°, while the next highest level, 3°, is called *read stability*.

According to the SQL standard, the highest, 4° serializable, isolation level is the default, but a database management system may not be able to provide better than the next highest, 3° or repeatable-read-level isolation for the transactions. Any of the isolation levels 1°–3° can be guaranteed for a transaction, if at least level 1° is required of all other transactions. To increase concurrency, some transactions may be allowed to run at an isolation level that is weaker than normal.

## 5.6   Isolation and Transaction Rollback

Let $H$ be a history for transaction set $\{T_1, \ldots, T_n\}$ that can be run on a database $D$ to produce a database $D'$. The transaction set can include forward-rolling, committed, backward-rolling, and rolled-back transactions.

We consider a forward-rolling transaction $T_i = B\alpha_i$ in $H$. Now $T_i$ wants to abort. Then $T_i$ must be extended to the rolled-back transaction $B\alpha_i A\alpha_i^{-1} C$. A natural requirement is that this rollback must be possible in $H$. In other words, it must be possible to run the extended history $HA\alpha_i^{-1}C$ on $D$, which means that the extension sequence $A\alpha_i^{-1}C$ must be runnable on $D'$.

The following example demonstrates that rollback may not be possible if the history contains dirty writes.

*Example 5.22*  Consider the transactions

$$T_1 = BW[x, u, 1] \quad \text{and}$$
$$T_2 = BD[x, u]C$$

and their history:

$$H_1 = \begin{array}{ll} T_1: & BW[x, 0, 1] \\ T_2: & \qquad\qquad\qquad BD[x, 1]C. \end{array}$$

This history can be run on a database $D$ that contains the tuple $(x, 0)$ and will produce the database $D' = D \setminus \{(x, 0)\}$.

Now $T_1$ wishes to abort, so $BW[x, 0, 1]$ is completed to the rolled-back transaction $BW[x, 0, 1]AW^{-1}[x, 0, 1]C$. However, the respective completed history

$$H_1' = \begin{array}{ll} T_1: & BW[x, 0, 1] \qquad\qquad\qquad AW^{-1}[x, 0, 1]C \\ T_2: & \qquad\qquad\qquad BD[x, 1]C. \end{array}$$

cannot be run on $D$, because the database no longer contains the tuple $(x, 1)$ when $T_1$ wants to write it. The action $D[x, 1]$ is a dirty write by the committed transaction $T_2$.

A similar situation happens with transactions

$T_1 = BD[x, u]$ and
$T_2 = BI[x, u]C$

and the history

$$H_2 = \begin{array}{ll} T_1: & BD[x, 0] \\ T_2: & \qquad\qquad BI[x, 0]C. \end{array}$$

The completed history

$$H_2' = \begin{array}{ll} T_1: & BD[x, 0] \qquad\qquad\qquad AD^{-1}[x, 0]C \\ T_2: & \qquad\qquad BI[x, 0]C. \end{array}$$

cannot be run on any database $D$ on which $H$ can be run. The action $I[x, 0]$ by the committed transaction $T_2$ is a dirty write. □

The examples above clearly suggest that dirty writes are incompatible with transaction rollback and should therefore be prevented. The situations in which transaction rollback might still be possible in the presence of dirty writes are difficult or even impossible for the database management system to detect automatically, as is demonstrated in the following example.

*Example 5.23* Consider the following history of two forward-rolling transactions $T_1 = BW[x, u, 1]$ and $T_2 = BW[x, u, 2]$:

$$H = \begin{array}{ll} T_1: & BW[x, 0, 1] \\ T_2: & \qquad\qquad BW[x, 1, 2] \end{array}$$

Here $W[x, 1, 2]$ is a dirty write. Now both transactions wish to abort. The rollback will succeed if the undo actions of the writes are performed in reverse chronological order, that is, the rollback of $T_2$ is followed by the rollback of $T_1$:

$$H_1' = \begin{array}{ll} T_1: & BW[x, 0, 1] \qquad\qquad\qquad A \qquad\qquad\qquad W^{-1}[x, 0, 1]C \\ T_2: & \qquad\qquad BW[x, 1, 2]AW^{-1}[x, 1, 2] \qquad\qquad C \end{array}$$

(where the relative order between the two $A$ actions or between the two $C$ actions does not matter). However, rolling back the write by $T_1$ first would lead to a history that is not runnable according to our definitions:

$$H_2' = \begin{array}{ll} T_1: & BW[x, 0, 1] \qquad\qquad\qquad AW^{-1}[x, 0, 1]C \\ T_2: & \qquad\qquad BW[x, 1, 2] \qquad\qquad\qquad\qquad AW^{-1}[x, 1, 2]C \end{array}$$

Even if we defined $W[x, u, v]$ to be always runnable on a database with a tuple with key $x$ (even though its value might be different from $u$), the rollback would produce a database state that differs from the original. □

**Theorem 5.24** *Let $H$ be a history in which none of the committed or rolled-back transactions do dirty writes. The action sequences that complete the active transactions in $H$ to rolled-back transactions can be interleaved to a sequence $\gamma$ so that the completed sequence $H\gamma$ can be run on every database where $H$ can* *be run. In one such $\gamma$, the undo actions are in reverse chronological order of the corresponding forward-rolling actions of the active transactions.*

*Proof* By Theorem 4.8, the theorem holds true if the string $\gamma$ is chosen as the action sequence implied by the undo pass of the recovery algorithm presented in Chap. 4. Then, because in $H$ no dirty writes have been executed, for each undo action in $\gamma$ the database must contain the result of the corresponding forward-rolling action; so it is possible to undo that action. The undo actions performed in the undo pass are in reverse chronological order of the update actions in $H$.                             $\Box$

Theorems 5.24 and 4.8 imply that the undo pass of ARIES recovery is possible if all transactions are run at isolation level $1°$ or higher.

It makes sense to require that multiple transactions can roll back during normal transaction processing and that this can happen concurrently in any order. Thus, dirty writes must be completely prevented. The database management system must not produce any histories where a transaction would perform dirty writes. It is not feasible in practice to prevent the dirty writes of just the committed and rolled-back transactions.

In the literature, a *strict history* is one in which no committed or rolled-back transaction does dirty writes and, additionally, no transaction performs dirty reads.

**Theorem 5.25** *Let $H$ be a history in which no transactions do dirty writes. Let $\gamma$ be any interleaving of the action sequences used to complete the active transactions in $H$ to rolled-back transactions. Then we have:*

1. *The completed history $H\gamma$ can be run on any database on which $H$ can be run.*
2. *None of the transactions in $H\gamma$ do dirty writes.*
3. *None of the transactions in $H\gamma$ do dirty reads, unless one of the transactions in $H$ does a dirty read.*
4. *None of the transactions in $H\gamma$ do dirty or unrepeatable reads, unless one of the transactions in $H$ does a dirty or an unrepeatable read.*

*Proof* We leave the proving of this theorem as an exercise.                        $\Box$

## 5.7  Conflict-Serializability

We say that action $R[x, \geq z]$ *reads key range* $[z, x]$, action $R[x, > z]$ *reads key range* $(z, x]$, and actions $I[x]$, $D[x]$, and $W[x]$ *update key* $x$. The read action $R[x]$ of the read-write model does the same thing as the key-range-model action $R[x, \geq x]$. Thus, the action reads the key $x$.

Let $H$ be a history that contains no partial rollbacks, let $T_1$ and $T_2$ be two different transactions in history $H$, and let $o_1[x]$ and $o_2[y]$ be actions by $T_1$ and $T_2$, respectively, such that $o_1[x]$ precedes $o_2[y]$ in $H$. We define that $o_1[x]$ *conflicts* with $o_2[y]$ in $H$, denoted $o_1[x] < o_2[y]$, if one of the following conditions holds:

1. $x = y$ and $o_1[x]$ and $o_2[y]$ both update key $x$. This situation is called a *write-write conflict*.

2. $o_1[x]$ updates key $x$ and $o_2[y]$ reads a key range that contains $x$. This situation is called a *write-read conflict*.
3. $o_1[x]$ reads a key range and $o_2[y]$ updates a key $y$ that belongs to the range read by $o_1[x]$. This situation is called a *read-write conflict*.

Transactions $T_1$ and $T_2$ *conflict* in $H$, denoted $T_1 < T_2$, if $o_1[x] < o_2[y]$ in $H$ for some pair of actions $o_1[x]$ of $T_1$ and $o_2[y]$ of $T_2$.

*Example 5.26* Consider the transactions $T_1 = BR[x, \geq x]R[y, \geq y]$ and $T_2 = BR[x, \geq x]I[y]C$ (where $x \neq y$) and their nonserial history

$$H_1 = \begin{array}{ll} T_1: & BR[x, \geq x] \hspace{4cm} R[y, \geq y] \\ T_2: & \hspace{2cm} BR[x, \geq x]I[y]C \end{array}$$

There is only one conflict here, namely, the write-read conflict caused by the pair of actions $I_2[y]$ and $R_1[y, \geq y]$. This means that $T_2 < T_1$. The same conflict is also present in the following serial history of the same transactions:

$$H_2 = \begin{array}{ll} T_1: & \hspace{3cm} BR[x, \geq x]R[y, \geq y] \\ T_2: & BR[x, \geq x]I[y]C \end{array}$$

That is, all of $T_2$ is performed first, followed by all of $T_1$. Obviously, $H_1$ and $H_2$ are equivalent. □

Now assume that for all pairs of different transactions $T_1$ and $T_2$ in a history $H$ the following condition holds:

If $T_1 < T_2$ then $T_2 \not< T_1$.

In other words, the conflict relationship "<" between transactions of $H$ is antisymmetric (and thus the corresponding reflexive relationship "$\leq$" = "<" $\cup$ "=" is a partial order). If in addition at most one transaction in $H$ is active and $T < T'$ does not hold for the active transaction $T$ and any other transaction $T'$ of $H$, then we say that $H$ is *conflict-serializable* and that the partial order "<" is the *serializability order* of the transactions of $H$.

**Theorem 5.27** *A conflict-serializable history is equivalent to any serial history of the same transactions in which the transactions appear in an order that respects the conflict relationship "<."*

*Proof* Two histories $H_1$ and $H_2$ are clearly equivalent if $H_1$ can be transformed into $H_2$ by a series of swaps of two non-conflicting consecutive actions. This implies that a conflict-serializable history with respect to "<" can be transformed into an equivalent serial history where the transactions are ordered with respect to "<." □

*Example 5.28* For the transactions in the history $H_1$ of Example 5.26, $T_2 < T_1$ holds but $T_1 < T_2$ does not. Also, only $T_1$ is active. Thus, the history is conflict-serializable. In the serial history $H_2$ equivalent to $H_1$ the transactions appear in the order $T_2 T_1$, respecting the serializability order. □

**Theorem 5.29** *Let $H$ be a history of transactions in which at most one is forward-rolling and the other are committed and in which no transaction contains partial rollbacks. If $H$ does not contain any isolation anomalies (dirty writes, dirty reads, or unrepeatable reads), then it is conflict-serializable.*

*Proof* Assume the contrary, that is, $H$ is not conflict-serializable. Then there is a sequence of transactions $T_1, T_2, \ldots, T_n, n \geq 2$, that appear in $H$ such that

$$T_1 < T_2, \; T_2 < T_3, \ldots, \; T_n < T_1.$$

Without loss of generality we may assume that $T_1$ is committed. This means that $H$ is of the form

$$H = \begin{array}{llll} T_1 : & \ldots \quad o_1[x] & \ldots & o_1[x'] \quad \ldots \\ T_2 : & \ldots & o_2[y] & \ldots \\ & \vdots & & \\ T_n : & \ldots & & o_n[y'] \quad \ldots \end{array}$$

where $o_1[x]$ and $o_2[y]$, as well as $o_n[y']$ and $o_1[x']$ are pairs of conflicting actions. Because the commit action of $T_1$ cannot appear before $o_2[y]$, $H$ contains an isolation anomaly.                                                                 □

The converse of Theorem 5.29 does not hold.

*Example 5.30* The history

$$\begin{array}{lll} T_1 : & BW[x] & \quad\quad C \\ T_2 : & & BW[x] \quad C \end{array}$$

of two committed transactions $T_1$ and $T_2$ is conflict-serializable, with $T_1 T_2$ as the serializability order, but the write action of $T_2$ is a dirty write.                        □

Also recall that the isolation anomalies were defined for general histories that can contain any number of active transactions, besides committed and rolled-back transactions, and that transactions may contain partial rollbacks (which we excluded above when defining conflicts). Obviously, in order to be able to state a meaningful converse of Theorem 5.29, we would have to refine and extend the definition of conflicting actions, so as to observe the commit action $C$ and the undo actions contained in partial and total rollbacks. For the purposes of the approach taken in this book to concurrency control, it is not necessary to elaborate that issue any further.

## 5.8 Enforcing Isolation

As was stated earlier, it must be possible to program transactions without having to care about other concurrent transactions. The programmer of a database application needs only to ensure that each transaction preserves the consistency of the logical database when executed alone in the absence of system crashes. The *concurrency control* of the database management system enforces the actions of concurrently running transactions to be executed according to such a history in which each transaction runs at its prescribed isolation level.

Concurrency control can be termed either pessimistic or optimistic. In *pessimistic concurrency control*, an action by a transaction is allowed to be executed only when it cannot cause an isolation anomaly. Pessimistic concurrency control is typically enforced by *locking*: to execute an action, a transaction must acquire a *lock* that prevents disallowed isolation anomalies from occurring.

In *optimistic concurrency control*, transactions are allowed in their forward-rolling phase to perform dirty or unrepeatable reads and even dirty writes on data-item copies stored in private workspace. When a transaction is about to commit, that is, executing the $C$ action, the system *validates* the transaction by checking whether or not any disallowed isolation anomalies have occurred; if so, the transaction is aborted and rolled back; otherwise, the transaction is allowed to commit (and the updates stored in private workspace installed into the database).

# Problems

**5.1** We use the concept of a transaction history to model the execution order of the actions of concurrently running transactions. However, this modeling is not accurate because it does not observe what actually happens at the physical database level. For example, even if we say that an action $I[x, v]$ by transaction $T_1$ is executed before an action $R[y, w]$ by transaction $T_2$, it may very well be that the executions of these two actions are interleaved when seen as sequences of physical actions on pages, such as traversals of root-to-leaf paths in a B-tree. Explain why it still makes sense to say that $I[x, v]$ is executed before $R[y, w]$, and not vice versa.

**5.2** Explain why an unmodified page can contain dirty data items and why all the data items in a modified page can be clean (i.e., non-dirty).

**5.3** The integrity constraint on the database states that a tuple with key value $x$ can appear in the database if and only if the tuple $(y, 1)$ appears in the database. Are the transactions of the form

$$T_1 = BD[x, u]D[y, a]I[y, 0]C \quad \text{and}$$
$$T_2 = BI[x, v]D[y, b]I[y, 1]C$$

logically consistent with respect to the integrity constraint? Consider the following history:

$$T_1: \quad BD[x, u] \qquad\qquad\qquad\qquad D[y, 1]I[y, 0]C$$
$$T_2: \qquad\qquad\qquad BI[x, v]D[y, 1]I[y, 1]C$$

Give a database that satisfies the integrity constraint and on which the history can be run. What is the resulting database? What isolation anomalies (i.e., dirty writes, dirty reads, and unrepeatable reads) does the history contain?

**5.4** The integrity constraint of the database states that the value in the tuple with key 0 must be the sum of the values of two tuples with successive keys. Are the transactions of the forms

$$T_1 = BR[x, \geq 1, u]R[y, > x, v]D[0, w]I[0, u + v]C \quad \text{and}$$
$$T_2 = BD[2, 2]I[2, 2]C$$

logically consistent with respect to the integrity constraint? Consider the following history:

$$T_1: \qquad\qquad BR[1, \geq 1, 1]R[3, > 1, 3]D[0, 3]I[0, 4]C$$
$$T_2: \quad BD[2, 2] \qquad\qquad\qquad\qquad\qquad\qquad\qquad I[2, 2]C$$

Give a database that satisfies the integrity constraint and on which the history can be run. What is the resulting database? What isolation anomalies does the history contain?

**5.5** What isolation anomalies are contained in the following histories?

$$H_1 = \begin{array}{ll} T_1: & BS[P]D[x]I[x]A[P]I^{-1}[x]D^{-1}[x]C[P] \qquad\qquad\qquad\qquad I[x]C \\ T_2: & \qquad\qquad\qquad\qquad\qquad\qquad\qquad BR[x]D[x]C \end{array}$$

$$H_2 = \begin{array}{ll} T_1: & BR[x]S[P]D[x]I[x]A[P]I^{-1}[x]D^{-1}[x]C[P]R[x] \qquad\qquad D[x]C \\ T_2: & \qquad\qquad\qquad\qquad\qquad\qquad\qquad BR[x]D[x]I[x]C \end{array}$$

$$H_3 = \begin{array}{ll} T_1: & BS[P]R[x]D[x]I[x]A[P]I^{-1}[x]D^{-1}[x]C[P] \qquad\qquad\qquad I[x]C \\ T_2: & \qquad\qquad\qquad\qquad\qquad\qquad BR[x]D[x]C \end{array}$$

**5.6** The transaction, denoted $T_1$, generated from the following program fragment is run concurrently with another transaction, $T_2$, that does insertions, updates, and deletions on relation $r$.

```
exec sql set transaction isolation level L;
exec sql select avg(V) into :a from r;
exec sql select sum(V) into :s from r;
exec sql select count(∗) into :c from r;
exec sql insert into q values (:a, :s / :c);
exec sql commit.
```

How may the result produced by $T_1$ vary when (a) $L = $ **read committed**, (b) $L = $ **repeatable read**, (c) $L = $ **serializable**? We assume that all other transactions that possibly run concurrently with $T$ have isolation level **read uncommitted**, at least.

**5.7** Explain why it is not safe to run the transaction of Example 1.3 at any isolation level lower than serializable.

**5.8** Which of the following statements are true and which are false? All the transactions in history $H$ are at isolation level $1°$ (read uncommitted).

(a) A forward-rolling transaction in $H$ cannot be aborted and rolled back if it has done a dirty read.

(b) $H$ may contain two committed transactions, $T_1$ and $T_2$, where $T_1$ commits before $T_2$, although $T_2 < T_1$ (i.e., $T_2$ and $T_1$ conflict and $T_2$ does the conflicting action before $T_1$ in $H$).

(c) If $H$ is serializable, then one of the serializability orders of the committed transactions in $H$ is the *commit order* of those transactions, that is, the order in which the transactions commit in $H$.

**5.9** Consider the extended read-write transaction model in which a transaction can use the action $C[x]$ to declare the updates on $x$ done so far as committed before the transaction terminates (see Problem 1.5). An update declared as committed becomes immediately visible to other transactions. Revise the definitions of the isolation anomalies so as to observe this feature.

**5.10** Assume that the tuples $(x, y, v)$ of a relation $r(\underline{X}, Y, V)$ can be accessed besides by the unique primary key $x$ also by the (non-unique) key $y$ (see Problem 1.6). Extend the definitions of the isolation anomalies so as to cover read actions based on non-unique keys.

**5.11** Assume that our basic (single-granular) key-range transaction model is extended to a multi-granular setting with three levels: database, relation, and tuple (see Sect. 1.10). Extend accordingly the definitions of isolation anomalies. Observe, for example, that if transaction $T_1$ creates a relation $r$ and then transaction $T_2$ inserts a tuple into $r$ while $T_1$ is still active, then the action of $T_2$ must be termed a dirty write.

**5.12** In an attempt to state a converse for Theorem 5.29 in the case the history contains several active transactions, we extend the definition of a conflicting pair of actions so as to observe commit actions as follows. First, the commit actions $C_1$ and $C_2$ of two transactions $T_1$ and $T_2$ conflict if $T_1$ and $T_2$ contain a pair of conflicting actions. Second the commit action of $T_1$ conflicts with an action $o_2[y]$ of $T_2$ if $T_1$ contains an action $o_1[x]$ that conflicts with $o_2[y]$. Restate Theorem 5.29 for this extension, still assuming that the history contains no undo actions, and state a converse theorem that you can prove.

# Bibliographical Notes

The three basic isolation levels of transactions were already defined for System R [Astrahan et al. 1976], with the highest level (3) meaning repeatable read (i.e., serializable in SQL parlance). Gray et al. [1976] define four isolation levels (degrees of consistency) for transactions in a transaction model in which a transaction can declare an update as committed before the transaction terminates. A transaction $T$ sees degree 0 consistency if it does not overwrite dirty data of other transactions, degree 1 consistency if in addition it does not commit any writes before termination, degree 2 consistency in addition it does not read dirty data of other transactions, and degree 3 consistency if in addition other transactions do not dirty any data read by $T$ before $T$ terminates. In our transaction models, degree 1 is the minimum level required: each transaction commits all its writes only at the end of the transaction and dirty writes are disallowed. Dirty reads are disallowed at degree 2 and unrepeatable reads at degree 3.

Our definitions for the isolation levels follow from those defined by Berenson et al. [1995], who observe that, to ensure that transaction rollback is possible under all circumstances, the definitions of the isolation levels in the SQL 1992 standard should be corrected such that dirty writes are disallowed at all levels. Besides the basic three isolation anomalies, Berenson et al. [1995] also analyze other isolation concepts, including snapshot isolation, which is possible to enforce in systems that do transient versioning (to be discussed in Chap. 12). A thorough analysis of the isolation concepts can also be found in the textbook by Gray and Reuter [1993].

Much of the classical theory of concurrency control of transactions is based on the notion of conflict-serializability, using a read-write transaction model in which the undo actions included in transaction rollback are not modeled explicitly. This approach was also adopted in many textbooks on transaction processing, such as in the classic works by Papadimitriou [1986] and Bernstein et al. [1987]. Schek et al. [1993] and Alonso et al. [1994] develop the theory using a transaction model in which also undo actions are represented explicitly. This latter approach has been adopted in the textbook by Weikum and Vossen [2002].

# Chapter 6
# Lock-Based Concurrency Control

*Locking* is the most commonly used method for enforcing transactional isolation. Most database management systems apply some kind of locking, possibly coupled with some other mechanism (such as transient versioning). With locking-based concurrency control, transactions are required to protect their actions by acquiring appropriate locks on the parts of the database they operate on. A read action on a data item is usually protected by a shared lock on the data item, which prevents other transactions from updating the data item, and an update action is protected by an exclusive lock, which prevents other transactions from reading or updating the data item.

If a transaction requests a lock on a data item in a situation in which some other transaction holds an exclusive lock on the data item, the requesting transaction has to wait for the conflicting lock to be released. Most often this means waiting for the other transaction to commit; all the locks still held by a transaction at commit time are released when the transaction has committed.

The isolation anomalies defined in the previous chapter for our key-range transaction model can be avoided with a locking protocol called *key-range locking*. This protocol prevents phantoms by effectively locking a range between two keys with a shared lock on one key (for a read action) or with exclusive locks on two successive keys (for an insert or a delete action). We give a proof of the correctness of key-range locking in the general case in which transactions can contain partial rollbacks with the feature that commit-duration locks acquired after setting a savepoint are released after completing a partial rollback to that savepoint.

A problem with locking is that transactions waiting for locks may get into a *deadlock*, where none of the transactions can proceed. Deadlocks must be resolved by aborting and rolling back one or more of the transactions involved. With ordinary lock requests, which automatically lead to a wait if the lock cannot be immediately granted, deadlocks involving holds of and waits for latches may also occur. In this case relying on deadlock resolution is not feasible; the occurrences of deadlocks between locks and latches must be prevented. The solution is to use a *conditional*

© Springer International Publishing Switzerland 2014
S. Sippu, E. Soisalon-Soininen, *Transaction Processing*, Data-Centric Systems
and Applications, DOI 10.1007/978-3-319-12292-2_6

*lock request*, which results in granting the lock only if the lock can be granted without wait. On this line, we discuss the implementation of key-range locking for the actions in our key-range transaction model; a complete implementation is obtained by coupling the basic algorithms given in this chapter with the B-tree algorithms given in the two following chapters.


## 6.1  Locks and the Lock Table


A *lock* is a main-memory data item belonging to an active transaction that grants the transaction access to a specific part of the database. A transaction can execute an action (read, write, insert, delete) on a part of the database only if it has properly locked the relevant part of the database.

A lock includes information about its name, mode, duration, and the owning transaction. The *lock name* identifies the data item or the set of data items in the database that is the target of the lock. The name can be used to categorize locks into logical and physical locks.

The name of a *logical lock* identifies a part of the logical database, such as a single tuple in a relation, the set of tuples within a key range, or a whole relation. The name of a logical lock for tuple $(x, v)$ in relation $r(\underline{X}, V)$ is the unique key $x$ (when $r$ is the only relation in the logical database) or the pair $(r, x)$ with the relation-id $r$ and the key $x$ (if several relations exist in the database). A lock named $x$ locks the tuple with key $x$ regardless of whether or not such a tuple exists when the lock is granted. Such a lock may be used to protect actions on that tuple or, say, actions on a key range delimited by that tuple.

The name of a *physical lock* identifies a part of the physical database, such as the location of a tuple in a specific data page, or a whole page or file, or a node or a bucket in an index structure. The lock name of a physical lock on page $p$ is the page-id $p$. Such a lock can be used to lock all the record positions in page $p$ no matter if there are actual records in those positions at the time of granting the lock. The lock name of a physical lock on record position $i$ in page $p$ is the record-id $(p, i)$. Such a lock locks the record position even if it does not contain a record at the time of granting the lock.

Operating on locks is more efficient if the lock names are fixed-length short values, such as four-byte integers. Instead of variable-length or structured lock names, actual implementations usually use a hash value $h(x)$ calculated from the name $x$ using some hash function $h$. Then a lock on key $x$ locks all tuples with keys $y$ that satisfy $h(y) = h(x)$.

The *lock mode* defines what sort of actions the lock permits. For reading a data item $x$, a transaction must acquire at least a *read lock* or an *S lock* on $x$. A read lock is *shared*: multiple transactions can have a read lock on $x$ simultaneously.

For writing a data item $x$, a transaction must acquire a *write lock* or an *X lock* on $x$. Write locks are *exclusive*: when a transaction has a write lock on $x$, other transactions cannot get any locks on $x$. A write lock on $x$ allows both reading and writing $x$.

In practice, there are usually other lock modes in addition to S and X locks. For instance, there is a U lock for preparing for an update and intention locks IS, IX, and SIX for locking multi-granular data items; the meaning of those locks will be described later in Chap. 9.

The *duration* of a lock can be long or short. A lock that is held until the transaction commits or completes a total rollback is called a *long-duration lock* or *commit-duration lock*. Releasing the commit-duration locks is part of the protocol for committing or completing the rollback of a transaction, that is, among the last steps performed in the call *commit*($T$) (Algorithm 3.3), after the commit log record $\langle T, C \rangle$ has been appended to the log buffer and the log has been flushed from the log buffer onto the log disk. A *short-duration lock* is one that is released immediately after completing the database action for which the lock was acquired. Such a lock is released just after unlatching the target page of the action.

In the presence of partial rollbacks, a lock acquired for commit duration after setting a savepoint may already be released when the transaction has completed a rollback to that savepoint. Such a lock is still called a commit-duration lock, even though it is released before the transaction has committed or completed its total rollback.

The *lock owner* is always some transaction. This is in contrast to latches, which are owned by processes or process threads. The server-process thread that executes the database actions on behalf of a transaction $T$ owns the latches it acquires on database pages, while $T$ owns the locks acquired on data items read or written on data pages.

Thus, the full representation of a lock is the four-tuple

$$(T, x, m, d),$$

where $x$ is the lock name, $m$ is the lock mode (S, X, IS, IX, SIX, U, etc.), $d$ is the duration of the lock (commit duration, short duration), and $T$ is the identifier of the transaction that owns the lock.

Locks $(T, x, m, d)$ are stored in a main-memory data structure called a *lock table*, managed by the lock manager of the database management system. The lock table is created and initialized to empty when the system is starting up. The lock table only exists when an instance of the system is running: in the event of a system failure or when shutting down the system, the lock table will disappear along with the other main-memory structures of the system, such as the database and log buffers.

The lock table is organized as a hash table (or a balanced binary tree), indexed by the lock name. In this way, finding out what locks exist on a given data item is efficient. In addition, the locks held by each transaction $T$ are chained together in a two-way-linked list in the order of their acquisition. The two-way linking allows any lock to be released efficiently, given a pointer to it.

The transaction record for $T$ in the active-transaction table includes, besides the transaction identifier, the transaction state and the UNDO-NEXT-LSN value, also a pointer to the last lock acquired by $T$ in the list of all locks currently held by $T$. This pointer is denoted by LAST-LOCK($T$).

A transaction usually holds only one short-duration lock (of a given granularity) at a time, because it performs only one action at a time. When a transaction run at the highest isolation level commits, it has usually accumulated as many commit-duration locks (on data items of a given granularity) as there were different data items (of that granularity) that the transaction operated on.

## 6.2 Acquiring and Releasing Locks

For the set of lock modes used in the database management system, *compatibility* is defined as follows. Let $m_1$ and $m_2$ be lock modes. Assume that transaction $T_2$ requests an $m_2$-lock on data item $x$ on which another transaction $T_1$ already currently holds an $m_1$-lock. If the $m_2$-lock on $x$ can immediately be granted, we say that lock mode $m_2$ is *compatible* with lock mode $m_1$.

The compatibility of lock modes is often presented as a *compatibility matrix*. The compatibility matrix for the set of lock modes {S, X} is shown in Fig. 6.1. Lock mode $m_2$ is compatible with lock mode $m_1$ if and only if the the entry for $(m_1, m_2)$ in the lock-compatibility matrix has value true.

A partial order, called *exclusivity order* and denoted by $\leq$, is defined on the set of lock modes. For lock modes $m_1$ and $m_2$, $m_1 \leq m_2$ means that $m_2$ is at least as exclusive as $m_1$, so that any action permitted by a lock of mode $m_1$ is also permitted by a lock of mode $m_2$. For the set {S, X}, we have S < X, that is, S $\leq$ X and S $\neq$ X. This is because X permits more actions (reading and writing) than S (only reading).

In general, we assume that the set $M$ of lock modes used is a *lattice* so that, for any $m_1, m_2 \in M$, there exists a unique lock mode denoted sup$\{m_1, m_2\}$ (the *supremum* of $m_1$ and $m_2$) that is the least exclusive mode in $M$ that is at least as exclusive as both $m_1$ and $m_2$. We have sup{S, X} = X.

A *lock upgrade* occurs if a transaction $T$ already holds a $d$-duration $m$-mode lock on a data item $x$ and later requests a $d'$-duration $m'$-lock on $x$, where either $m < m'$ or no exclusivity order is defined between $m$ and $m'$. If no other transaction holds on $x$ a lock that is incompatible with $m'$, $T$ is granted on $x$ a $d'$-duration lock of mode sup$\{m, m'\}$.

The possible lock upgrades are presented in a *lock-upgrade matrix*. The lock-upgrade matrix for the set of lock modes {S, X} is shown in Fig. 6.2. A request to

**Fig. 6.1** Lock-compatibility matrix for the set of lock modes {S, X}

| lock requested by $T_2$ | lock held by $T_1$ | |
|---|---|---|
| | S | X |
| S | true | false |
| X | false | false |

**Fig. 6.2** Lock-upgrade
matrix for the set of lock
modes {S, X}

| lock requested by $T$ | lock held by $T$ | |
|---|---|---|
| | S | X |
| S | S | X |
| X | X | X |

upgrade an $m$ lock on $x$ to an $m'$ lock yields (when allowed) an $m''$ lock on $x$ if the
entry for $(m', m)$ in the lock-upgrade matrix contains the value $m''$.

*Example 6.1* The transaction

$$T = BR[x] R[y] W[x] R[x] C$$

requests the following locks:

1. $(T, x, S, \text{commit duration})$, granted.
2. $(T, y, S, \text{commit duration})$, granted.
3. $(T, x, X, \text{commit duration})$, X lock granted because $\sup\{S, X\} = X$.
4. $(T, x, S, \text{commit duration})$, no new lock granted because $\sup\{X, S\} = X$.

The locks are granted immediately upon request if no other transaction is holding
or waiting for an incompatible lock on $x$ or $y$. The locks are released after $T$ has
committed.                                                                              □

The call $lock(T, x, m, d)$ (Algorithm 6.1) is used to request a $d$-duration $m$ lock
on data item $x$ for transaction $T$. First, the set $M$ of modes of locks of duration $d$
that the transaction $T$ already holds on $x$ is determined. If some of these lock modes
is at least as exclusive as the requested lock mode $m$, nothing needs to be done.
For example, if $T$ requests a commit-duration S lock on $x$ when it already holds a
commit-duration S lock or X lock on $x$, no new lock needs to be granted.

Otherwise, it is checked if some transaction $T'$ other than $T$ holds a lock on
$x$ whose mode is incompatible with $m$. If such an incompatible lock exists, $T$ is
put to wait for the release of that lock. Otherwise, $T$ is granted a lock on $x$ of
mode $\sup(\{m\} \cup M)$, that is, the least exclusive mode that is at least as exclusive
as the requested lock mode $m$ and the modes of the locks of duration $d$ that $T$
already holds on $x$. For example, if $T$ requests a commit-duration X lock on $x$ while
holding a commit-duration S lock on $x$, the mode of the lock granted on $x$ will be
$\sup(\{X, S\}) = X$. In the case that $T$ requests an $m$-lock of duration $d$ on $x$ while not
holding any locks of duration $d$ on $x$, the granted lock is of mode $\sup(\{m\} \cup \emptyset) = m$.

The request serviced by the *lock* call is an *unconditional lock request* in that
either the requested lock is granted immediately or the process thread executing $T$
is forced to wait until all incompatible locks on $x$ held by other transactions are
released and the requested lock can be granted. An unconditional lock request thus
eventually leads to the granting of the lock, unless the transaction is aborted due to
a deadlock or a system failure. As will be shown later in this chapter, we also need
*conditional* lock requests that lead to the granting of the requested lock only if the
lock can be granted immediately without wait (see Sects. 6.6 and 6.7).

The *unlock*$(T, x, m, d)$ (Algorithm 6.2) is used to *release* the lock $(T, x, m, d)$. The hash-structured lock table is searched for the lock name $x$, and, given the pointer $l$ to the lock, the call *release-lock*$(T, l)$ (Algorithm 6.3) is used to release the lock and to wake up any process threads that are waiting for the release.

In order to facilitate the release of locks after completing a partial rollback to a savepoint, the setting of savepoint $P$ in the call *set-savepoint*$(T, P)$ (Algorithm 4.3) is logged with a log record,

$$n : \langle T, S[P], l, n' \rangle, \tag{6.1}$$

that also contains the pointer, $l$, to the last lock acquired by $T$ before setting the savepoint (Algorithm 6.4).

The call *complete-rollback-to-savepoint*$(T, P)$ (Algorithm 4.6) that is executed as the last step in the call *rollback-to-savepoint*$(T, P)$ (Algorithm 4.4) now takes as a third argument the pointer value $l$ logged when setting savepoint $P$. The call, given as Algorithm 6.5, releases all locks acquired by $T$ after lock $l$.

---

**Algorithm 6.1** Procedure *lock*$(T, x, m, d)$

$M \leftarrow \{m' \mid$ a lock $(T, x, m', d)$ exists in the lock table$\}$
**if** $m \not\leq m'$ for all $m' \in M$ **then**
    **while** requested lock not granted **do**
        **if** the lock table does not contain any lock $(T', x, m', d')$ where $T' \neq T$ and $m'$ is incompatible with $m$ **then**
            $m'' \leftarrow \sup(\{m\} \cup M)$
            insert $(T, x, m'', d)$ into the lock table
            put $(T, x, m'', d)$ into the two-way chain pointed to from LAST-LOCK$(T)$
            LAST-LOCK$(T) \leftarrow$ pointer to $(T, x, m'', d)$
        **else**
            wait for the conflicting locks to be released
        **end if**
    **end while**
**end if**

---

**Algorithm 6.2** Procedure *unlock*$(T, x, m, d)$

$l \leftarrow$ pointer to the lock $(T, x, m, d)$ in the lock table
*release-lock*$(T, l)$

---

**Algorithm 6.3** Procedure *release-lock*$(T, l)$

detach the lock $l$ from the two-way chain ending at LAST-LOCK$(T)$
**if** LAST-LOCK$(T) = l$ **then**
    LAST-LOCK$(T) \leftarrow$ pointer to the last record in the chain
**end if**
delete the lock $l$ from the lock table
wake up the first thread in the queue waiting for the release of lock $l$ (if any)

---

**Algorithm 6.4** Procedure *set-savepoint*($T$, $P$)

$n' \leftarrow$ UNDO-NEXT-LSN($T$)
$l \leftarrow$ LAST-LOCK($T$)
$log(n, \langle T, S[P], l', n' \rangle)$
UNDO-NEXT-LSN($T$) $\leftarrow n$

---

**Algorithm 6.5** Procedure *complete-rollback-to-savepoint*($T$, $P$, $l$)

$log(n, \langle T, C[P] \rangle)$
$l' \leftarrow$ LAST-LOCK($T$)
**while** $l' \neq$ nil **and** $l' \neq l$ **do**
    *release-lock*($T$, $l'$)
**end while**

---

## 6.3   Locking Protocol for the Read-Write Model

The programmer of a database application or a user that submits queries does not need to worry about the locking of data items: the database management system automatically invokes *lock* and *unlock* calls to acquire and release appropriate locks for the database actions, issued by transactions. The system applies a *locking protocol* that states which locks are acquired for which database actions, and when each lock is requested and released. The locking protocol must ensure transactional isolation, so that the transaction will run at the isolation levels set for it in all histories that are permitted by the locking protocol.

In a locking protocol that guarantees a high degree of isolation, most of the locks are commit-duration locks, that is, they are released only after the transaction commits or completes a partial or total rollback. Usually the programmer of a database application can affect the duration of locks only by setting appropriately the transaction boundaries, using the SQL statements **commit**, **rollback**, and **rollback to savepoint**.

For the simple read-write transaction model, we define a locking protocol, called *read-write locking* in this book. Under this protocol, every transaction must follow the following rules:

1. For a read action $R[x, v]$, a commit-duration S lock is acquired on the key $x$ of the tuple $(x, v)$ to be read.
2. For a forward-rolling write action $W[x, u, v]$, a commit-duration X lock is acquired on the key $x$ of the tuple that is to be written.
3. For an undo action $W^{-1}[x, u, v]$, no lock is requested; the action is performed under the protection of the commit-duration X lock acquired for the corresponding forward-rolling action $W[x, u, v]$.

4. After completing a partial rollback with action $C[P]$, all locks acquired after the corresponding set-savepoint action $S[P]$ are released.
5. All remaining locks are released after committing or completing a total rollback of the transaction with action $C$.

Assume that a history $H$ of transactions can be run on database $D$. We say that $H$ is *possible under a locking protocol* or *permitted by the protocol* or that the transactions in $H$ *follow the protocol* on $D$, if, when executing the history one action at a time on $D$ in the order the actions appear in $H$, all the locks requested according to the protocol can be granted immediately without wait.

*Example 6.2* Consider different histories of the transactions

$$T_1 = BR[x]R[y]W[z]C,$$
$$T_2 = BR[y]W[y]C$$

run on a database that contains tuples with keys $x$, $y$, and $z$.

The history

$$H_1 = \begin{array}{ll} T_1: & BR[x] \qquad\qquad\qquad\qquad R[y]W[z]C \\ T_2: & \qquad\quad BR[y]W[y]C \end{array}$$

is possible under the read-write locking protocol, because all the locks requested for the actions can be granted immediately:

1. For $R[x]$, a commit-duration S lock on $x$ for $T_1$.
2. For the first $R[y]$, a commit-duration S lock on $y$ for $T_2$.
3. For $W[y]$, a commit-duration X lock on $y$ for $T_2$.
4. After $C$, the locks held by $T_2$ on $y$ are released.
5. For the second $R[y]$, a commit-duration S lock on $y$ for $T_1$.
6. For $W[z]$, a commit-duration X lock on $z$ for $T_1$.
7. After $C$, the locks held by $T_1$ on $x$, $y$ and $z$ are released.

We note that $H_1$ contains no isolation anomalies.

On the other hand, the history

$$H_2 = \begin{array}{ll} T_1: & BR[x] \qquad\qquad R[y] \qquad\qquad\quad W[z]C \\ T_2: & \qquad\quad BR[y] \qquad\qquad W[y]C \end{array}$$

is not possible under the read-write locking protocol, because of the S lock held by $T_1$ on $y$ while $T_2$ requests an X lock on $y$:

1. For $R[x]$, a commit-duration S lock on $x$ for $T_1$.
2. For the first $R[y]$, a commit-duration S lock on $y$ for $T_2$.
3. For the second $R[y]$, a commit-duration S lock on $y$ for $T_1$.
4. For $W[y]$, a commit-duration X lock on $y$ for $T_2$ not granted.

Because of the lock held by $T_1$ on $y$, the exclusive lock requested by $T_2$ on $y$ cannot be granted immediately; so $T_2$ must wait for $T_1$ to commit and to release its locks, thus preventing $H_2$ from occurring. We note that in $H_2$, the read action $R[y]$ by $T_1$ is an unrepeatable read. The locking protocol enforces the following execution order for the actions of transactions $T_1$ and $T_2$:

$$H_2' = \begin{array}{llll} T_1: & BR[x] & R[y]W[z]C & \\ T_2: & BR[y] & & W[y]C \end{array}$$

with locks granted as follows:

4. For $W[z]$, a commit-duration X lock on $z$ for $T_1$.
5. After $C$, the locks held by $T_1$ on $x$, $y$ and $z$ are released.
6. For $W[y]$, a commit-duration X lock on $y$ for $T_2$.
7. After $C$, the locks held by $T_2$ on $y$ are released.

The history $H_2'$ contains no isolation anomalies.                                      ☐

Rule (4) of the read-write locking protocol states that all locks acquired after setting a savepoint $P$ are released after a completed partial rollback to $P$.

*Example 6.3* Consider the following transaction, which contains nested partial rollbacks.

$BR[x_1, u_1]W[x_1, u_1, v_1]$
$S[P_1]R[x_2, u_2]W[x_2, u_2, v_2]$
$S[P_2]W[x_2, v_2, w_2]R[x_3, u_3]$
$A[P_2]W^{-1}[x_2, v_2, w_2]C[P_2]W[x_2, v_2, w_2']$
$A[P_1]W^{-1}[x_2, v_2, w_2']W^{-1}[x_2, u_2, v_2]C[P_1]R[x_2, u_2]W[x_2, u_2, w_2'']C.$

Locks are acquired and released by the transaction as follows: (cf. Fig. 6.3):

1. For $R[x_1, u_1]$, an S lock on key $x_1$.
2. For $W[x_1, u_1, v_1]$, an X lock on key $x_1$.
3. For $R[x_2, u_2]$, an S lock on key $x_2$.
4. For $W[x_2, u_2, v_2]$, an X lock on key $x_2$.
5. For $W[x_2, v_2, w_2]$, no new lock.
6. For $R[x_3, u_3]$, an S lock on key $x_3$.
7. For $W^{-1}[x_2, v_2, w_2]$, no new lock.
8. After $C[P_2]$, the lock on $x_3$ is released.
9. For $W[x_2, v_2, w_2']$, no new lock.
10. For $W^{-1}[x_2, v_2, w_2']$, no new lock.
11. For $W^{-1}[x_2, u_2, v_2]$, no new lock.
12. After $C[P_1]$, the locks on $x_2$ are released.
13. For $R[x_2, u_2]$, an S lock on key $x_2$.
14. For $W[x_2, u_2, w_2'']$, an X lock on key $x_2$.
15. After $C$, the locks on $x_2$ and $x_1$ are released.

First we note that each of the tuples with keys $x_1$ and $x_2$ written by the transaction is kept X-locked until it is again in a committed state. Thus dirty writes are prevented. Second, the transaction holds S locks on keys $x_1$, $x_2$, and $x_3$ when reading the tuples with those keys. Thus, dirty reads are prevented. Third, the S locks are kept until the transaction commits or completes the partial rollback of an action sequence that includes the read action for which the S lock was acquired.

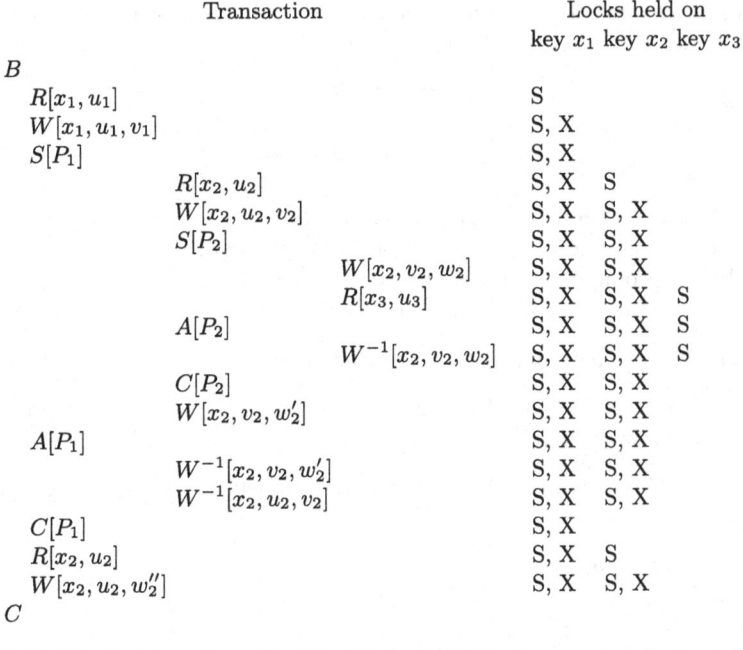

**Fig. 6.3** Locking in the presence of partial rollbacks. Rolled-back segments of the transaction are shown *indented*. The columns on the *right* show the modes of locks held by the transaction on key $x_i, i = 1, 2, 3$, at each action

We assume that the good practice of transaction programming referred to at the end of Sect. 5.4 has been followed, so that a write action following a completed partial rollback never uses information read by a read action included in the rolled-back action sequence. The write action $W[x_2, v_2, w_2']$ uses information read by $R[x_2, u_2]$, which does not belong to the partial rollback to savepoint $P_2$. The write action $W[x_2, u_2, w_2'']$ in turn uses information read by $R[x_2, u_2]$ performed after the partial rollback to savepoint $P_1$. Thus, we conclude that no unrepeatable read can occur that would violate the integrity of the database.                                           □

The following theorem states how the read-write locking protocol prevents isolation anomalies.

**Theorem 6.4** *Let H be a history of transactions in the read-write model. Then H is permitted by the read-write locking protocol if and only if H contains neither dirty writes nor dirty reads nor any unrepeatable reads by read actions not included in partially rolled-back action sequences.*

*Proof* That none of the anomalies mentioned appear in a history permitted by the read-write locking protocol follows from the following observations: (1) commit-duration X locks prevent dirty writes, (2) commit-duration X locks and short-duration S locks prevent dirty reads, and (3) short-duration X locks and commit-duration S locks prevent unrepeatable reads. These observations clearly hold true if "commit duration" is interpreted in its strictest sense, meaning a lock that, once granted, is held until the transaction has committed or completed its total rollback.

Dirty writes and dirty reads are also prevented if "commit duration" has its relaxed meaning that permits a commit-duration lock to be released when a data item updated by a transaction is returned to its committed state in a partial rollback of the transaction. This clearly is the case when the X lock has been granted for a write action that is included in the rolled-back action sequence. We also note that the S lock acquired for a read action that is not included in a rolled-back action sequence is kept until the transaction has committed or completed its total rollback.

We leave as an exercise the proving of the converse result that the absence of the anomalies implies that the history is permitted by the read-write locking protocol. □

An important property of the read-write locking protocol (also shared by all the other locking protocols presented in this book) is that an aborted transaction or a transaction doing a partial rollback does not need to acquire any new locks for the backward-rolling actions. An undo action $W^{-1}[x, u, v]$ is protected by the commit-duration X lock on key $x$ acquired by the transaction for the corresponding forward-rolling action $W[x, u, v]$. Thus, no deadlock (see Sect. 6.5) can arise that could prevent a backward-rolling transaction from completing its rollback.

In the traditional literature on concurrency control, the read-write locking protocol appears in a simplified form in which partial rollbacks (if they appear in the transaction model) involve no lock releases. That is, all locks are held until the transaction has committed or completed its total rollback. That protocol is called the *strict two-phase locking protocol*, or *strict* 2PL for short.

There is also a locking protocol called just *two-phase locking protocol* or 2PL, that is, without the "strict" attribute. In that locking protocol, a transaction must acquire an S lock on $x$ for the read action $R[x]$ and an X lock on $x$ for the write action $W[x]$, but the locks do not need to be commit duration, as long as $T$ does not acquire any new locks after releasing any one lock. Thus, the locking operations of a transaction following the 2PL protocol can be divided into two phases: (1) a phase where the transaction acquires new locks, increasing the amount of locks held by the transaction, and (2) a phase where all the locks are released. With the 2PL protocol, conflict-serializability is guaranteed for read-write-model transactions, but, because locks can be released before the commit, isolation anomalies are not prevented, so the protocol is only of theoretical interest.

## 6.4  Key-Range Locking

The simple locking protocol for the read-write model cannot be directly applied to the key-range model when full isolation is required. This is because for full isolation, also phantom dirty reads and phantom unrepeatable reads must be prevented. But if phantom reads were allowed, so that the SQL isolation level 3° would be considered sufficient, we could easily apply the read-write locking protocol to the key-range model: an insert action $I[x, v]$ or a delete action $D[x, v]$ is simply protected by a single commit-duration X lock on $x$ and a read action $R[x, \theta\ z, v]$ by a commit-duration S lock on $x$ (see Problem 6.2).

If also phantoms are to be prevented, a more sophisticated locking scheme is needed.

*Example 6.5* The following history contains a phantom dirty read but is possible under the read-write locking protocol:

$$H_1 = \begin{array}{lll} T_1: & BD[2, v] & \cdots \\ T_2: & & BR[3, >1, w] & \cdots \end{array}$$

The contents of the database in this example are at first $\{(1, u), (2, v), (3, w)\}$. Locks are acquired as follows:

1. For $D[2, v]$, a commit-duration X lock on key 2 for $T_1$.
2. For $R[3, >1, w]$, a commit-duration S lock on key 3 for $T_2$.

The locks are granted without wait.                                         □

*Example 6.6* The following history contains a phantom unrepeatable read but is possible under the read-write locking protocol:

$$H_2 = \begin{array}{lll} T_1: & BR[3, >1, w] & \cdots \\ T_2: & & BI[2, v] & \cdots \end{array}$$

The contents of the database in this example are at first $\{(1, u), (3, w)\}$. Locks are acquired as follows:

1. For $R[3, >1, w]$, a commit-duration S lock on key 3 for $T_1$.
2. For $I[2, v]$, a commit-duration X lock on key 2 for $T_2$.

The locks are granted without wait.                                         □

The basic solution to the above problem would be to S-lock the entire range that is read. In other words, when performing, for example, the read action $R[3, > 1]$, the range $(1, 3]$ should be S-locked for commit duration. The S lock on $(1, 3]$ is interpreted as S-locking all keys $x$ with $1 < x \leq 3$. Then the insertion $I[2]$ could not be performed by $T_1$ before $T_2$ commits, because $T_1$ requires an X lock on key 2 already S-locked by $T_2$ for commit duration because key 2 is in the range $(1, 3]$.

For efficiency reasons we do not want to use the scheme of locking ranges explicitly, but we implement this effect by requiring that an insert action $I[x]$ or a delete action $D[x]$ not only locks the key $x$ but also the key next to $x$ in the database. Here, as is shown in the sequel, for $I[x]$ key $x$ is locked for commit duration and the key next to $x$ only for short duration (the lock is released immediately after the action), and for $D[x]$ key $x$ is locked for short duration but the key next to $x$ for commit duration. Then the effect of range locking is obtained by requiring for a read action an S lock only on the key to be read.

*Example 6.7* The history $H_1$ of Example 6.5 is not possible under the key-range locking protocol, because locks would be requested as follows:

1. For $D[2, v]$, a short-duration X lock on key 2 and a commit-duration X lock on the next key 3 for $T_1$ (both granted).
2. After $D[2, v]$, the short-duration X lock held by $T_1$ on key 2 is released.
3. For $R[3, >1, w]$, a commit-duration S lock on key 3 for $T_2$ (not granted).

Acquiring the S lock is not possible: $T_2$ must wait for the release of the commit-duration X lock held by $T_1$, that is, for $T_1$ to commit or complete rollback.  □

*Example 6.8* The history $H_2$ of Example 6.6 is not possible under the key-range locking protocol, because locks would be requested as follows:

1. For $R[3, >1, w]$, a commit-duration S lock on key 3 for $T_1$ (granted).
2. For $I[2, v]$, a commit-duration X lock on key 2 (granted) and a short-duration X lock on the next key 3 (not granted), for $T_2$.

Acquiring the X lock on key 3 is not possible: $T_2$ must wait for the release of the S lock held by $T_1$, that is, for $T_1$ to commit or complete rollback.  □

Along these lines, we now present a protocol that ensures full isolation for transactions in our key-range transaction model. In this protocol, called the *key-range locking protocol* (or *key-value locking protocol*), every transaction obeys the following rules when acquiring and releasing locks:

1. For the read action $R[x, \theta z, v]$, a commit-duration S lock is acquired on the key $x$ of the tuple $(x, v)$ to be read.
2. For the forward-rolling insert action $I[x, v]$, a commit-duration X lock is acquired on the key $x$ of the tuple $(x, v)$ to be inserted, and a short-duration X lock is acquired on the *next key* $y$ of $x$, that is, the least key $y$ in the database with $y > x$ (if such a key exists) or the key $y = \infty$ (otherwise).
3. For the forward-rolling delete action $D[x, v]$, a short-duration X lock is acquired on the key $x$ of the tuple to be deleted, and a commit-duration X lock is acquired on the next key $y$ of $x$.
4. For an undo action $I^{-1}[x, v]$, no lock is requested; the action is performed under the protection of the commit-duration X lock on $x$ acquired by $T$ for the corresponding forward-rolling action $I[x, v]$.
5. For an undo action $D^{-1}[x, v]$, no lock is requested; the action is performed under the protection of the commit-duration X lock on the next key $y$ of $x$ acquired for the corresponding forward-rolling action $D[x, v]$.
6. After completing a partial rollback with action $C[P]$, all locks acquired after the corresponding set-savepoint action $S[P]$ are released.
7. All remaining locks are released after committing or completing a total rollback of the transaction with action $C$.

In (1), the S lock acquired on $x$ for the read action $R[x, \theta z, v]$ is interpreted as locking the entire key range $(x', x]$, where $x'$ is the greatest key less than $x$

in the database (if such a key exists) or the range $(-\infty, x]$ (otherwise). Note that $x \, \theta \, z \geq x'$.

In (2), the short-duration X lock on the next key $y$ is released when the tuple $(x, v)$ has been inserted into the data page, the insertion has been logged, and the page has been unlatched. The lock is used to check that no other transaction has locked the key range $(x', y]$, where $x'$ is the greatest key less than $y$ in the database before the insertion. After the insertion, the commit-duration X lock on $x$ is interpreted as locking the key range $(x', x]$.

In (3), the short-duration X lock on $x$ is released when the tuple $(x, v)$ has been deleted from the data page, the deletion has been logged, and the page has been unlatched. The lock is used to check that no other transaction has locked the key range $(x', x]$, where $x'$ is the greatest key less than $x$ in the database. After the deletion, the commit-duration X lock on the next key $y$ is interpreted as locking the key range $(x', y]$.

In (4) and (5) it is important to note that the undo actions indeed can be performed under the protection of the single commit-duration X lock acquired for the corresponding forward-rolling action, although for those forward-rolling actions also a short-duration lock had to be acquired besides the commit-duration lock. This is in accordance with the principle shared by all locking protocols and mentioned above in Sect. 6.3: the rollback of a transaction can never fail due to a deadlock.

*Example 6.9* Consider the database $D = \{(1, 10)\}$ and transactions

$$T_1 = BR[x, \geq 1, u] R[y, >x, v] C$$
$$T_2 = BR[x, \geq 1, u] I [2, u + 10] C$$

and their history:

$$H_3 = \begin{array}{lll} T_1: & BR[1, \geq 1, 10] & R[2, >1, 20]C \\ T_2: & & BR[1, \geq 1, 10] I [2, 20]C \end{array}$$

With the key-range locking protocol, locks are acquired and released as follows:

1. For the first $R1[1, \geq 1, 10]$, a commit-duration S lock on key 1 for $T_1$.
2. For the second $R[1, \geq 1, 10]$, a commit-duration S lock on key 1 for $T_2$.
3. For $I [2, 20]$, a commit-duration X lock on key 2 and a short-duration X lock on the next key $\infty$ for $T_2$.
4. After $I [2, 20]$, the short-duration lock on key $\infty$ is released.
5. After the first $C$, the commit-duration locks held by $T_2$ on keys 1 and 2 are released.
6. For $R[2, >1, 20]$, a commit-duration S lock on key 2 for $T_1$.
7. After the second $C$, the commit-duration locks held by $T_1$ on keys 1 and 2 are released.

In this case, all locks can be granted immediately in the order in which they are requested. Thus the history $H_3$ is possible under the key-range locking protocol. □

A major task in proving the correctness of key-range locking is to show that locking the key $y$ next to $x$ implies that at all later points in the history the (possibly different) next key of $x$ will be locked. For example, it is not immediately obvious that the history

$T_1$:   ...   $D[x]$   ...
$T_2$:            ...   $I[x]$   ...

(exhibiting a dirty write) is prevented by the rules (2) and (3) of the key-range locking protocol. This is because, though the key next to $x$ is X-locked by $T_1$ for commit duration when performing the delete action $D[x]$, the key next to $x$ may have changed by the time $T_2$ wants to execute its insert action $I[x]$. In Lemmas 6.10 and 6.11, we prove that in such a case $T_1$ always holds an X lock on the current next key of $x$.

**Lemma 6.10** *Assume that a history $H$ is permitted by the rules (2)–(7) of the key-range locking protocol on a given database. Then for any undo-delete action $D^{-1}[x]$ in $H$ the key next to $x$ in the database at the time the action is performed is the same as the key next to $x$ in the database at the time the corresponding forward-rolling delete action $D[x]$ was performed.*

*Proof* The proof is by induction on the number of undo-delete actions in $H$. The claim holds trivially in the base case in which $H$ contains no undo-delete actions.

In the induction step we consider the last undo-delete action in $H$ and write $H$ as

$$H = \alpha D_i[x]\beta D_i^{-1}[x]\gamma,$$

where $D_i[x]$ is the delete action by transaction $T_i$ corresponding to the undo-delete action $D_i^{-1}[x]$ by $T_i$ and $\gamma$ contains no undo-delete actions. Let $y$ be the key next to $x$ in the database at $\alpha D_i[x]$ and $y'$ the key next to $x$ in the database at $\alpha D_i[x]\beta D_i^{-1}[x]$. We have to prove that $y' = y$. By rule (3) of the key-range locking protocol, $T_i$ holds a commit-duration X lock on $y$.

First we note that if the transaction $T_i$ itself changes the key next to $x$ between the database states at $\alpha D_i[x]$ and at $\alpha D_i[x]\beta D_i^{-1}[x]$, then those changes are undone before performing $D_i^{-1}[x]$. Thus if $y' \neq y$, some actions in the sequence $\beta$ by transactions other than $T_i$ are responsible for the change, although those actions may be intermixed with actions of $T_i$ in $\beta$.

Changing $y$ to $y'$ by a sequence of one or more actions can only be accomplished by deleting $y$ or by inserting some key $z$ with $x < z < y$. Because of the commit-duration X lock held by $T_i$ on $y$, the sequence $\beta$ cannot contain any delete action $D_j[y]$ or insert action $I_j[z]$ by a transaction $T_j$ other than $T_i$ with $x \leq z < y$ (by rules (2) and (3)). Also an undo-insert action $I_j^{-1}[y]$ in $\beta$ is impossible because $T_j$ cannot hold a commit-duration X lock on $y$.

There remains the case in which $\beta$ contains an undo-delete action $D_j^{-1}[z]$ that inserts a key $z$ between $x$ and $y$. The history $\alpha D_i[x]\beta$ has one undo-delete action less than $H$. By the induction hypothesis, the key next to $z$ in the database at the time the corresponding forward-rolling delete action $D_j[z]$ was performed is equal to the key next to $z$ in the database at the time $D_j^{-1}[z]$ is performed, that is, $y$. But then, by rule (3), $T_j$ would hold a commit-duration X lock on $y$ at some timepoint between $D_i[x]$ and $D_i^{-1}[x]$, which is impossible because of the X lock on $y$ held by $T_i$. Thus $y' = y$, as claimed.                                                                                  □

Lemma 6.10 implies that when performing an undo-delete action $D^{-1}[x]$, the transaction holds a commit-duration X lock on the current next key $y$ of the deleted key $x$.

**Lemma 6.11** *Assume that the history*

$$H = \alpha o_1[x]\beta$$

*is permitted by the rules (2) to (7) of the key-range locking protocol on a given database, transaction $T_1$ is active at $H$, $o_1[x]$ is either an insert action $I[x]$ or a delete action $D[x]$, and the action sequence $\beta$ does not contain the undo action for $o_1[x]$. Then $T_1$ holds a commit-duration X lock on the least key $y \geq x$ in the database at the end of the history.*

*Proof* First we note that if the action sequence $\beta$ contains some undo actions by $T_1$, then the corresponding forward-rolling actions must also be included in $\beta$, because otherwise also the action $o_1[x]$ would be undone in $\beta$, contradicting the assumption. Let $\beta'$ be the sequence $\beta$ stripped off all actions involved in partial rollbacks by $T_1$. The history $H' = \alpha o_1[x]\beta'$ then produces exactly the same database as $H$. Because by rule (6), upon completing a partial rollback to a savepoint, exactly those locks are released that were acquired after setting the savepoint, the set of locks held by $T_1$ at $H'$ is the same as at $H$.

The proof of the lemma is now by induction on the number of actions in $\beta'$ that change the least key $\geq x$ in the database at $\alpha o_1[x]$ through $\alpha o_1[x]\beta'$.

In the base case, there are no actions in $\beta'$ that change the least key $\geq x$ in the database. Thus, the least key $y \geq x$ in the database at $\alpha o_1[x]\beta'$ is the same as that in the database at $\alpha o_1[x]$. If $o_1[x]$ is an insert action $I[x]$, $y = x$ and $T_1$ holds a commit-duration X lock on $x$, by rule (2). If $o_1[x]$ is a delete action $D[x]$, $y$ is the key next to $x$ and $T_1$ holds a commit-duration X lock on $y$, by rule (3). This proves the base case.

For the induction step, we assume that $\beta'$ contains $n > 0$ actions that change the least key $\geq x$ in the database and assume that the lemma holds for prefixes of $\beta'$ that contain $n - 1$ such actions. Let $o_2[y']$ be the last such action in $\beta'$. Then $\beta'$ can be written as $\beta_1 o_2[y']\beta_2$, where the least key $\geq x$ remains the same through $\beta_2$. As an induction hypothesis, we assume that $T_1$ holds a commit-duration X lock on the least key $y \geq x$ in the database at $\alpha o_1[x]\beta_1$. We have to prove that $T_1$ also holds such a lock at

$$H' = \alpha o_1[x]\beta_1 o_2[y']\beta_2 = \alpha o_1[x]\beta'.$$

Because of the commit-duration X lock held by $T_1$ on the least key $y \geq x$ in the database at $\alpha o_1[x]\beta_1$, the action $o_2[y']$ that changes this situation cannot be an insert action $I_2[y']$ with $y > y' \geq x$ or a delete action $D_2[y]$ or an undo-insert action $I_2^{-1}[y]$, by any transaction $T_2$ other than $T_1$. Note that, by rules (2) and (3), such a transaction $T_2$ would have to hold a short-duration X lock on the key $y$ next to the inserted key $y'$ for $I_2[y']$ or a short-duration X lock on the deleted key $y$ for $D_2[y]$ or a commit-duration X lock on the key $y$ to be restored to the database for $I_2^{-1}[y]$. Also, $o_2[y']$ cannot be an undo-delete action $D_2^{-1}[y']$ with $y > y' \geq x$ by a transaction $T_2 \neq T_1$, because then the key next to $y'$ would be $y$ and so Lemma 6.10 would imply that $T_2$ must hold a commit-duration X lock on $y$. Thus we conclude that $o_2[y']$ can only be an action by $T_1$ itself.

As $\beta'$ contains no undo actions by $T_1$, the action $o_2[y']$ can only be an insert action $I[y']$ or a delete action $D[y]$ by $T_1$. In the case of an insert action $I[y']$ with $y > y' \geq x$, $T_1$ holds a commit-duration X lock on the least key $y' \geq x$, by rule (2). In the case of a delete action $D[y]$, $T_1$ holds, by rule (3), a commit-duration X lock on the key next to $y$, which is then also the key next to $x$. Because the least key $\geq x$ remains the same through $\beta_2$, we conclude that in each case $T_1$ holds a commit-duration X lock on the least key $\geq x$ in the database at $H'$ and hence also at $H$. □

**Lemma 6.12** *Assume that a history $H$ is permitted by the rules (2)–(7) of the key-range locking protocol on a given database. Then $H$ contains no dirty writes.*

*Proof* For the sake of contradiction, assume that $H$ contains a dirty write. By Lemma 5.13, we can write $H$ as

$$H = \alpha o_1[x]\beta o_2[x]\gamma = H'o_2[x]\gamma,$$

where $o_1[x]$ is a forward-rolling insert action $I_1[x]$ or delete action $D_1[x]$ by some transaction $T_1$ active at $H'$, $o_2[x]$ is a forward-rolling insert action $I_2[x]$ or delete action $D_2[x]$ by some transaction $T_2$ other than $T_1$, and the action sequence $\beta$ does not contain the corresponding undo action for $o_1[x]$.

By Lemma 6.11, $T_1$ holds a commit-duration X lock on the least key $y \geq x$ at $H'$, that is, on $x$ if $x$ is in the database at $H'$ and on the key next to $x$ if $x$ is not in the database at $H'$. Moreover, in the case $o_1[x] = I_1[x]$, $T_1$ holds a commit-duration X lock on $x$ at $H'$, by rule (2) of the key-range locking protocol. Note that rule (6) does not apply for $T_1$ and $o_1[x]$, because $T_1$ does not undo $o_1[x]$ in $\beta$.

In the case $o_2[x] = I_2[x]$, $x$ is not in the database at $H'$, $y$ is the key next to $x$, and $T_2$ must acquire at $H'$ a commit-duration X lock on $x$ and a short-duration X lock on $y$, by rule (2). In the case $o_2[x] = D_2[x]$, $x$ is in the database at $H'$, $y = x$, and $T_2$ must acquire at $H'$ a short-duration X lock on $x$ (and a commit-duration X lock on the key next to $x$), by rule (3).

We conclude the impossibility of all the following cases:

Case (a) $o_1[x] = I_1[x]$ and $o_2[x] = I_2[x]$. Then $T_1$ holds a commit-duration X lock on $x$ at $H'$, and $T_2$ should acquire a commit-duration X lock on $x$.

Case (b) $o_1[x] = I_1[x]$ and $o_2[x] = D_2[x]$. Then $T_1$ holds a commit-duration X lock on $x$ at $H'$, and $T_2$ should acquire a short-duration X lock on $x$.

Case (c) $o_1[x] = D_1[x]$ and $o_2[x] = I_2[x]$. Then $x$ is not in the database at $H'$, $T_1$ holds a commit-duration X lock on the key $y$ next to $x$ at $H'$, and $T_2$ should acquire a short-duration X lock on $y$.

Case (d)   $o_1[x] = D_1[x]$ and $o_2[x] = D_2[x]$. Then $x$ is in the database at $H'$, $T_1$ holds a commit-duration X lock on $y = x$ at $H'$, and $T_2$ should acquire a short-duration X lock on $x$.

Thus, no dirty write can appear in $H$.                                          □

**Lemma 6.13** *Assume that a history $H$ is permitted, on a given database, by the rules (2)–(7) and the following relaxed version of the rule (1) of the key-range locking protocol:*

*(1') For the read action $R[x, \theta z, v]$, a short-duration S lock is acquired on the key $x$ of the tuple $(x, v)$ to be read.*

*Then $H$ contains no dirty reads.*

*Proof* For the sake of contradiction, assume that $H$ contains a dirty read. By Lemma 5.14, we can write $H$ as

$$H = \alpha o_1[y]\beta R_2[x, \theta z]\gamma = H' R_2[x, \theta z]\gamma,$$

where $x \geq y \ \theta \ z$, $o_1[y]$ is a forward-rolling insert action $I_1[y]$ or delete action $D_1[y]$ by some transaction $T_1$ active at $H'$, $R_2[x, \theta z, v]$ is a read action by some transaction $T_2$ other than $T_1$, and the action sequence $\beta$ does not contain the corresponding undo action for $o_1[y]$.

By Lemma 6.11, $T_1$ holds a commit-duration X lock on the least key $y' \geq y$ in the database at $H'$. If $o_1[y]$ is an insert action $I_1[y]$, $T_1$ also holds a commit-duration X lock on $y$ at $H'$, by rule (2). Because of the S lock acquired by $T_2$ on $x$, $y \neq x$. Thus $x > y \ \theta \ z$ and $y$ cannot be in the database at $H'$. Since, by the definition of the read action $R_2[x, \theta z]$, $x$ is the least key in the database at $H'$ with $x \ \theta \ z$, we conclude that $x = y'$, which however is not possible because of the X lock held on $y'$ by $T_1$ at $H'$.

We are left with the case that $o_1[y]$ is a delete action $D_1[y]$. If $y$ is in the database at $H'$, the condition $x \geq y \ \theta \ z$ and the fact that $x$ is the least key with $x \ \theta \ z$ and $y'$ the least key $\geq y$ in the database at $H'$ imply that $y' = y = x$, which however is impossible because of the X lock held by $T_1$ on $y'$ at $H'$ and the S lock acquired by $T_2$ on $x$. On the other hand, if $y$ is not in the database at $H'$, we conclude that $y' = x$, which is also impossible for the same reasons.                    □

In order to prove that unrepeatable reads are not permitted, we need to show that the key-range locking protocol does not permit a history

$T_1$:    ...    $R[x, \theta z]$   ...
$T_2$:                    ...    $o[y]$   ...

where $o[y]$ is a forward-rolling insert or delete action, $x \geq y \ \theta \ z$, and $T_1$ has not committed or performed a partial rollback or initialized a total rollback that includes $R[x, \theta z]$ at the time $T_2$ performs the action $o[y]$. Here we cannot conclude that in the case $x > y$, the key $x$ would be the key next to $y$, because after performing $R[x, \theta z]$ transaction $T_1$ may have performed actions that have changed the key next to $z$. However, we have:

**Lemma 6.14**  *Assume that the history*

$$H = \alpha R_1[y, \theta z]\beta$$

*is permitted by the key-range locking protocol on a given database, and transaction $T_1$ is active and has not performed a partial rollback or initialized a total rollback that includes the action $R_1[y, \theta z]$. Then $T_1$ holds commit-duration locks on all keys $y''$ with $y' \geq y'' \theta z$ in the database at the end of the history, where $y'$ is the least key with $y' \geq y$ in that database.*

*Proof*  As in the proof of Lemma 6.11, we consider the history

$$H' = \alpha R_1[y, \theta z]\beta',$$

where the action sequence $\beta'$ is $\beta$ stripped off all actions by $T_1$ involved in partial rollbacks. We use induction on the number of insert, delete, undo-insert, and undo-delete actions in $\beta'$.

In the base case, there are no such actions in $\beta'$. Thus $y' = y$ is the least key with $y' \geq y$ and $y'' = y$ the only key with $y' \geq y'' \theta z$ in the database at $\alpha R_1[y, \theta z]$ as well as in the database at $\alpha R_1[y, \theta z]\beta'$ and hence in the database at $H$. By rule (1) of the key-range locking protocol, the key $y$ is S-locked by $T_1$ for commit duration. This proves the base case.

In the induction step, we assume that $\beta'$ contains $n > 0$ insert, delete, undo-insert, or undo-delete actions and that the lemma holds for the longest prefix $\beta_1$ of $\beta'$ that does not include the last action, $o_2[y'']$, of interest. Then $H'$ can be written as

$$H' = \alpha R_1[y, \theta z]\beta' = \alpha R_1[y, \theta z]\beta_1 o_2[y'']\beta_2.$$

As an induction hypothesis, we assume that $T_1$ holds commit-duration locks on all keys $y''$ with $y' \geq y'' \theta z$ in the database at $\alpha R_1[y, \theta z]\beta_1$ where $y'$ is the least key $\geq y$ in that database.

Because of the commit-duration locks held by $T_1$ on all keys $y''$ with $y' \geq y'' \theta z$ in the database at $\alpha R_1[y, \theta z]\beta_1$, the action $o_2[y'']$ cannot be an insert action $I_2[y'']$ or a delete action $D_2[y'']$ or an undo-insert action $I_2^{-1}[y'']$, by any transaction $T_2$ other than $T_1$, if $y' \geq y'' \theta z$. Note that, by rules (2) and (3), such a transaction $T_2$ should hold a commit-duration X lock on $y''$ for $I_2[y'']$ or for $I_2^{-1}[y'']$, or a short-duration X lock on $y''$ for $D_2[y'']$. Also, if $o_2[y'']$ were an undo-delete action $D_2^{-1}[y'']$, then, by Lemma 6.10, the key next to $y''$ would be the same as in the database at the time the corresponding forward-rolling action $D_2[y'']$ was performed and $T_2$ would hold a commit-duration X lock on that next key at $\alpha R_1[y, \theta z]\beta_1$. However, this is not possible with $y' \geq y'' \theta z$. Note that $y''$ cannot be $y'$ (locked by $T_1$); hence $y'' < y'$ and the key next to $y''$ is among those locked for commit duration by $T_1$. Thus we conclude that $o_2[y'']$, with $y' \geq y'' \theta z$, can only be an action by $T_1$ itself.

As $\beta'$ contains no undo actions by $T_1$, the action $o_2[y'']$ with $y' \geq y'' \theta z$ can only be an insert action $I[y'']$ or a delete action $D[y'']$ by $T_1$. In the case of an insert action $I[y'']$, $T_1$ holds a commit-duration X lock on $y''$, by rule (2), as desired. The case of a delete action $D[y'']$ is interesting only in the case $y''$ is the least key $\geq y$ in the database at $\alpha R_1[y, \theta z]\beta_1$. Then the key $y'$ next to the deleted key is required to be X-locked for commit duration, as it is, by rule (3). Because the set of keys $y''$ with $y' \geq y'' \, \theta \, z$ remains the same thru $\beta_2$, we conclude the lemma. $\qquad\square$

**Lemma 6.15** *Assume that a history $H$ is permitted, on a given database, by the key-range locking protocol. Then $H$ contains no unrepeatable reads, except possibly ones included in partially rolled-back action sequences.*

*Proof* For the sake of contradiction, assume that $H$ contains an unrepeatable read. By Lemmas 5.15 and 6.13, we can write $H$ as

$$H = \alpha R_1[x, \theta z]\beta o_2[y]\gamma = H'o_2[y]\gamma,$$

where $x \geq y \, \theta \, z$, $R_1[x, \theta z]$ is a read action by a transaction $T_1$ forward-rolling at $H'$, and $o_2[y]$ is a forward-rolling insert action $I_2[y]$ or delete action $D_2[y]$ by some transaction $T_2$ other than $T_1$.

If $T_1$ has not performed a partial rollback or initialized a total rollback in $H'$ that includes $R_1[x, \theta z]$, then, by Lemma 6.14, $T_1$ holds commit-duration locks on all keys $y''$ with $y' \geq y'' \, \theta \, z$ in the database at $H'$, where $y'$ is the least key with $y' \geq y$ in that database. If $o_2[y]$ is an insert action $I_2[y]$, $y$ is not in the database at $H'$, and $T_2$ must acquire a commit-duration X lock on $y$. But then $y$ would be one of the keys locked by $T_1$ for commit duration. On the other hand, if $o_2[y]$ is a delete action $D_2[y]$, $y$ is in the database at $H'$, and $T_2$ must acquire a commit-duration X lock on the key next to $y$ at $H'$. But that next key is also among those keys locked by $T_1$ for commit duration. $\qquad\square$

Lemmas 6.12, 6.13 and 6.15 imply:

**Theorem 6.16** *Let $H$ be a history of transactions in the key-range model that is permitted, on a given database, by the key-range locking protocol. Then $H$ contains no dirty writes nor dirty reads nor any unrepeatable reads by read actions not included in partially rolled-back action sequences.* $\qquad\square$

## 6.5   Deadlocks

When using a concurrency-control scheme based on locking, *deadlocks* can occur: in a set of transactions, each transaction is waiting for a lock to a data item locked by some other transaction in the set, so that none of the transactions can proceed.

*Example 6.17* Let $T_1$ and $T_2$ be active transactions.

1. First $T_1$ acquires an S lock on key $x$ and reads the tuple with key $x$.
2. Then $T_2$ acquires an S lock on key $y$ and reads the tuple with key $y$.
3. Next $T_1$ wants to update the tuple with key $y$. Therefore it requests an X lock on $y$. Because of the S lock held by $T_2$ on $y$, the requested X lock cannot be granted immediately, and so $T_1$ is put to sleep, waiting for $T_2$ to release its lock.
4. Now $T_2$ wants to update the tuple with key $x$. Therefore it requests an X lock on $x$. Because of the S lock held by $T_1$ on $x$, the requested X lock cannot be granted immediately, and so $T_2$ is put to sleep, waiting for $T_1$ to release its lock.

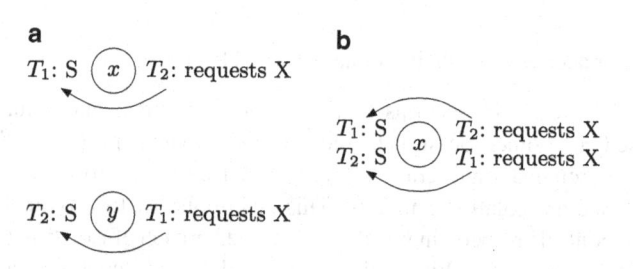

**Fig. 6.4** Deadlock between two transactions: (a) locking two data items in different orders; (b) upgrading locks on a single data item. The arcs indicate wait-for relationships between transactions waiting for locks

Now both transactions need to wait for each other forever (see Fig. 6.4a).  □

In the above example, the deadlock results from two transactions operating on two data items in different orders and locking the keys with incompatible locks. A lock upgrade from a read lock to a write lock can also lead to a deadlock.

*Example 6.18* Let $T_1$ and $T_2$ be active transactions.

1. $T_1$ acquires an S lock on key $x$ and reads the tuple with key $x$.
2. $T_2$ acquires an S lock on key $x$ and reads the tuple with key $x$. Recall that S locks are compatible.
3. $T_1$ wants to update the tuple with key $x$. Therefore it requests an X lock on $x$, that is, a lock upgrade from its S lock to an X lock. Because of the S lock held by $T_2$ on $x$, the upgrade cannot be granted immediately, and so $T_1$ is put to sleep to wait for $T_2$ to release its lock.
4. $T_2$ wants to update the tuple with key $x$. Therefore it requests an X lock on $x$, that is, a lock upgrade from its S lock to an X lock. Because of the S lock held by $T_1$ on $x$, the upgrade cannot be granted immediately, and so $T_2$ is put to sleep to wait for $T_1$ to release its lock.

Again both transactions need to wait for each other forever (Fig. 6.4b).  □

Actually, in practice lock upgrades are the most important reason for deadlocks. This is because the following operation sequence is very common in transactions:

1. Acquire an S lock on key $x$ and read the tuple with key $x$.
2. Find out if the tuple needs to be updated.
3. If an update is needed, upgrade the S lock on $x$ to an X lock and perform the update.

If the tuple is a *hotspot tuple*, one that is updated very often from different transactions, the possibility of a deadlock is high.

In general: the database system is in *deadlock*, if the following holds for some active transactions $T_1, \ldots, T_n$:

$T_1$ waits for access to a data item that is locked by $T_2$.
$T_2$ waits for access to a data item that is locked by $T_3$.

$$\cdots$$

$T_n$ waits for access to a data item that is locked by $T_1$.

This kind of cycle of transactions that wait for each other can occur only with locking-based and other pessimistic concurrency-control protocols that ensure isolation by forcing a transaction to wait if it tries to perform an action that would introduce an isolation anomaly. This is in contrast to a typical optimistic concurrency-control protocol, in which an attempted breach of isolation is prevented by aborting the transaction. With such a protocol, deadlocks cannot occur.

With lock-based concurrency control, there are two basic approaches for deadlock management: (1) deadlock prevention protocols and (2) deadlock detection and resolution schemes. First we note that, in general, it is not possible to enforce *deadlock prevention* without forcing some transactions to abort and roll back. This is because transactions are formed by performing one action at a time, resulting from SQL requests coming from the application process; a transaction when it starts does not know beforehand what actions it will eventually perform or in what order it will access the data items or whether or not it will later update a data item it has read earlier.

Thus, every database management system with lock-based concurrency control must include a *deadlock detection* and *resolution* scheme. Deadlocks are allowed to occur, and when so happens, one of the transactions involved in the deadlock is aborted. That transaction is called the *victim*.

The basic solution is to maintain a *wait-for graph* of transactions waiting for each other. A directed edge $T \longrightarrow T'$ is added to the wait-for graph when transaction $T$ requests a lock on a data item locked by transaction $T'$ and needs to wait. The edge $T \longrightarrow T'$ is removed when $T$ no longer waits for access to any data item locked by $T'$.

The system is in deadlock if and only if there is a cycle in the wait-for graph. To detect deadlocks, an algorithm that looks for cycles in the wait-for graph is run periodically. A deadlock is resolved by repeatedly aborting a transaction that is a part of a cycle, until there are no more cycles.

Besides the locks acquired by transactions on data items in the database, the latches acquired by server-process threads on database pages can be the cause for a deadlock.

*Example 6.19* Assume that a server-process thread executing a transaction $T_1$ is running concurrently with a server-process thread executing a transaction $T_2$.

1. The thread executing $T_1$ read-latches page $p$.
2. The thread executing $T_2$ read-latches page $q$.
3. The thread executing $T_1$ requests a write latch on page $q$. Because of the latch held by the other thread, the requested latch cannot be granted immediately, and so the requester has to wait.
4. The thread executing $T_2$ requests a write latch on page $p$. Because of the latch held by the other thread, the requested latch cannot be granted immediately, and so the requester has to wait.

Both threads have to wait for each other.                                                    □

Latching and unlatching must be as efficient as possible. This is why it is normal practice not to keep a wait-for graph of the latches acquired by processes to notice deadlocks. Thus, the latching protocol applied by all processes must prevent deadlocks.

If a process or thread must keep more than one page latched simultaneously, such as in a B-tree a parent page and one or two child pages, the latches must always be acquired in a predefined order, such as in the order first-parent-then-child, first-older-sibling-then-younger-sibling. Upgrading of a read-latch to a write latch is forbidden altogether.

## 6.6  Conditional Lock Requests

Deadlocks can also occur between locks and latches.

*Example 6.20*  Assume two forward-rolling transactions $T_1$ and $T_2$.

1. $T_1$ performs the insert action $I[x, v]$ on data page $p$. Therefore the key $x$ is X-locked by $T_1$ for commit duration, page $p$ is write-latched, the insertion is performed and logged, and page $p$ is unlatched.
2. $T_2$ wants to perform the read action $R[x > z, v]$. Therefore page $p$ is read-latched, the key $x$ satisfying the search condition is found, and an S lock on $x$ for $T_2$ is requested. Because $x$ is currently X-locked by $T_1$, the requested S lock cannot be granted immediately, and so $T_2$ must wait.
3. $T_1$ is still active and wants to perform the insert action $I[y, w]$ on page $p$ (which also covers key $y$). Therefore a write latch on page $p$ is requested. Because $p$ is currently read-latched by the thread executing $T_2$, the requested latch cannot be granted immediately, and so $T_1$ must wait.

Now the thread executing $T_1$ is waiting for a write latch on page $p$ that is currently read-latched by the thread executing $T_2$ and waiting for an S lock on key $x$ (Fig. 6.5).
□

Deadlocks between latches and locks are prevented by requiring that a process or thread must never hold a latch on any page when it is waiting for a lock. Thus, going to sleep to wait for a lock is only allowed when the thread holds no latches.

**Fig. 6.5**  Deadlock between locks and latches. Transaction $T_2$ is holding page $p$ latched while waiting for an S lock on key $x$ currently X-locked by $T_1$, which now wants to write-latch $p$ for $I[y, w]$

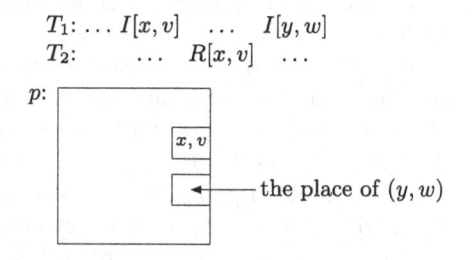

Enforcing this restriction requires that the lock manager must be able to process *conditional lock requests* in addition to normal unconditional lock requests (that can lead to waiting).

A $d$-duration $m$ lock for transaction $T$ on data items $x$ can be requested conditionally with the call

*conditionally-lock*$(T, x, m, d)$

to the lock manager. The call returns with the lock granted if the lock could be granted immediately (i.e., no incompatible locks were held by other transactions) or with the response "lock not granted" otherwise.

The use of unconditional lock requests leads to the following sequence of operations:

1. Locate and latch the target data page $p$.
2. Conditionally lock the target key $x$ covered by page $p$.
3. If the lock was granted, go to step 6. Otherwise, save the PAGE-LSN of page $p$, unlatch $p$, and unconditionally lock key $x$.
4. When the lock on $x$ is granted, relatch page $p$ and examine its PAGE-LSN.
5. If the PAGE-LSN is not changed from the saved value (i.e., the page has not been updated in between) or if it can been seen from the current page contents that page $p$ is still the correct target page covering key $x$, go to step 6; otherwise, unlatch $p$ and go to step 1.
6. Perform the action on the tuple with key $x$ on page $p$.

Examples of the need of this kind of operation sequences are given in the next section.

## 6.7   Algorithms for Read, Insert, and Delete Actions

In the implementation of the key-range locking protocol for the logical database actions $R[x, \theta z, v]$, $I[x, v]$, and $D[x, v]$, we apply a conditional lock request while holding the target data page latched. For the read action $R[x, \theta z, v]$ (Algorithm 6.6), the S lock on $x$ is requested conditionally after the page holding the least key $x$ with $x \, \theta \, z$ has been located and latched. For the insert and delete actions $I[x, v]$ and $D[x, v]$ (Algorithms 6.7 and 6.8), the X lock on the key $y$ next to $x$ is requested conditionally after the page holding that key has been located and latched.

We have to make some assumptions about the underlying physical database structure that stores the tuples of the relation $r(\underline{X}, V)$. The data pages of the physical database allocated for $r$ must partition the entire key space $[-\infty, \infty)$ into disjoint key ranges $[l_i, h_i)$, $i = 1, \ldots, n$, one for each allocated data page $p_i$. More specifically, each page $p_i$ *covers* its key range $[l_i, h_i)$, that is, contains only tuples $(x, v)$ with $l_i \leq x < h_i$, and the key ranges satisfy the following conditions: (1) $l_1 = -\infty$, (2) $l_i < h_i = l_{i+1}$ for $i = 1, \ldots, n - 1$, and (3) $h_n = \infty$ (Fig. 6.6). We

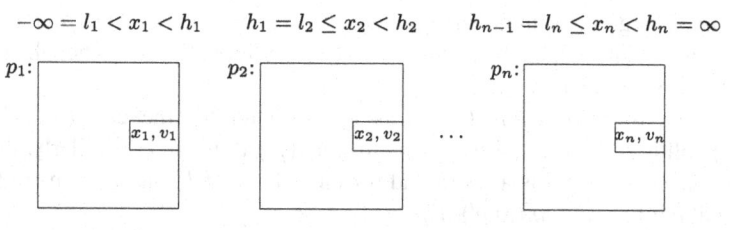

$$-\infty = l_1 < x_1 < h_1 \qquad h_1 = l_2 \le x_2 < h_2 \qquad h_{n-1} = l_n \le x_n < h_n = \infty$$

**Fig. 6.6** Data pages $p_1, p_2, \dots, p_n$ covering the key space $[-\infty, \infty)$ partitioned into disjoint key ranges $(-\infty, h_1), [h_1, h_2), \dots, [h_{n-1}, \infty)$

---

**Algorithm 6.6** Procedure $read(T, x, \theta z, v)$

---

**while** true **do**
   *find-page-for-read*$(p, p', x, \theta z)$
   *conditionally-lock*$(T, x, S, \text{commit-duration})$
   **if** the lock on $x$ was granted **then**
      exit the **while** loop
   **else**
      $m \leftarrow \text{PAGE-LSN}(p)$
      $m' \leftarrow \text{PAGE-LSN}(p')$
      *unlatch-and-unfix*$(p, p')$
      *lock*$(T, x, S, \text{commit-duration})$
      *fix-and-read-latch*$(p, p')$
      **if** PAGE-LSN$(p) = m$ **and** PAGE-LSN$(p') = m'$ **or** $p = p'$ **and** $p$ is a data page **and** $x$
      is the least key in page $p$ with $x \, \theta \, z \ge z_1$ where $z_1$ is the least key in page $p$ **then**
         exit the **while** loop
      **else**
         *unlatch-and-unfix*$(p, p')$
         *unlock*$(T, x, S, \text{commit-duration})$
      **end if**
   **end if**
**end while**
**if** $x = \infty$ **then**
   $v \leftarrow 0$
**else**
   $v \leftarrow$ the value of the tuple with key $x$
**end if**
*unlatch-and-unfix*$(p, p')$

---

assume that the key range covered by page $p_i$ can change only if the page itself is modified and hence its PAGE-LSN is advanced.

The read action $R[x, \theta z, v]$ is implemented by the procedure $read(T, x, \theta z, v)$ (Algorithm 6.6). In this procedure, the call *find-page-for-read*$(p, p', x, \theta z)$ is used to locate and read-latch the data page $p$ that covers key $z$ and the data page $p'$ that holds the key to be read, that is, the least key $x$ in the database with $x \, \theta \, z$. If $p' \ne p$, the key ranges $[l', h')$ and $[l, h)$ covered by these pages are two successive subranges of the key space, so that $l' < h' = l < h$.

In Sect. 7.3 we give an implementation for the procedure call *find-page-for-read*$(p, p', x, \theta z)$ in the case the physical database structure is a sparse B-tree. Then the page $p'$ is either $p$ or the page next to $p$ at the leaf level of the B-tree. If the lock on $x$ cannot be granted immediately, the PAGE-LSNs of pages $p$ and $p'$ are saved, the pages are unlatched, the lock is requested unconditionally, and, when the lock is granted, the pages are relatched.

The call *fix-and-read-latch*$(p, p')$ is a shorthand for the call *fix-and-read-latch*$(p)$ followed by the call *fix-and-read-latch*$(p')$ if $p' \neq p$. Similarly, the call *unlatch-and-unfix*$(p, p')$ is a shorthand for the call *unlatch-and-unfix*$(p)$ followed by the call *unlatch-and-unfix*$(p')$ if $p' \neq p$.

After relatching the pages $p$ and $p'$, the page contents are inspected so as to find out whether or not they can still be used to satisfy the read action. The read action can be satisfied if one of the following two conditions holds:

1. PAGE-LSN$(p)$ and PAGE-LSN$(p')$ have not advanced from the saved values, that is, the pages have not been modified in the interim.
2. $p = p'$ and PAGE-LSN$(p)$ has advanced from the saved value, but page $p$ is still a data page, and $x$ is the least key in page $p$ with $x\ \theta\ z \geq z_1$, where $z_1$ is the least key in page $p$.

In case (1) it is clear that the read action can be satisfied from pages $p$ and $p'$. Recall that the pages $p'$ and $p$, determined by the call *find-page-for-read*$(p, p', x, \theta z)$, cover two successive subranges of the key space and that the key range covered by a page can change only if the page is modified and hence its PAGE-LSN is advanced.

Also in case (2), even if one of the pages $p$ and $p'$ has been modified in the interim, so that the PAGE-LSN value of one of these pages is greater than the saved value, it may still be possible to satisfy the read action, provided that we can be sure that the pages $p$ and $p'$ together still cover the key range from $z$ to the least key $x$ with $x\ \theta\ z$. This is true at least in the case that $p = p'$ and $x$ is the least key in page $p$ with $x\ \theta\ z \geq z_1$, where $z_1$ is the least key in page $p$. Note that if $p \neq p'$ and one of the pages has been modified in the interim, it may no longer be the case that these are the pages that together cover the range from $z$ to the least key $x$ with $x\ \theta\ z$.

If neither (1) nor (2) holds, we cannot be sure that pages $p$ and $p'$ can be used to satisfy the read action; hence we must unlatch pages $p$ and $p'$, unlock key $x$, and repeat the search for the correct target pages and the correct target key by performing again the call *find-page-for-read*$(p, p', x, \theta z)$, locking the found key $x$ conditionally, and so on.

The insert action $I[x, v]$ is implemented by the procedure *insert*$(T, x, v)$ (Algorithm 6.7). First, the key $x$ is unconditionally X-locked for commit duration. Then the call *find-page-for-insert*$(p, p', x, v, y)$ is used to locate and write-latch the data page $p$ that covers key $x$ and to locate and read-latch the data page $p'$ that contains the key $y$ next to $x$ if that key is not found in page $p$ (and *high-key*$(p) < \infty$). If $p \neq p'$, the pages $p$ and $p'$ cover two successive subranges of the key space. The call also ensures (by appropriate structure modifications) that page $p$ does not overflow by the insertion, that is, there is room for the tuple $(x, v)$ in page $p$. In Sect. 7.4 we give an implementation for the call *find-page-for-insert*$(p, p', x, v, y)$ in the case of a sparse B-tree.

---

**Algorithm 6.7** Procedure *insert*$(T, x, v)$

---

$lock(T, x, X, \text{commit-duration})$
**while** true **do**
    *find-page-for-insert*$(p, p', x, v, y)$
    *conditionally-lock*$(T, y, X, \text{short-duration})$
    **if** the lock on $y$ was granted **then**
        exit the **while** loop
    **else**
        $m \leftarrow \text{PAGE-LSN}(p)$
        $m' \leftarrow \text{PAGE-LSN}(p')$
        *unlatch-and-unfix*$(p, p')$
        $lock(T, y, X, \text{short-duration})$
        *fix-and-write/read-latch*$(p, p')$
        **if** $\text{PAGE-LSN}(p) = m$ **and** $\text{PAGE-LSN}(p') = m'$ **or** $p = p'$ **and** $p$ is a data page with
        $x_1 \leq x \leq x_2$ where $x_1$ is the least and $x_2$ the greatest key in page $p$ **and** $p$ has room for
        the tuple $(x, v)$ **and** $y$ is the least key in $p$ greater than $x$ **then**
            exit the **while** loop
        **else**
            *unlatch-and-unfix*$(p, p')$
            *unlock*$(T, y, X, \text{short-duration})$
        **end if**
    **end if**
**end while**
**if** page $p$ contains a tuple with key $x$ **then**
    *unlatch-and-unfix*$(p, p')$
    return with error
**else**
    *insert-into-page*$(T, p, x, v)$
    *unlatch-and-unfix*$(p, p')$
    *unlock*$(T, y, X, \text{short-duration})$
**end if**

---

The found next key $y$ is conditionally X-locked for short duration. If the lock is granted, the call *insert-into-page*$(T, p, x, v)$ (Algorithm 3.1) given in Sect. 3.6 is used to perform and log the insertion of $(x, v)$ into page $p$, after which the pages $p$ and $p'$ are unlatched and the lock on $y$ released. Otherwise, the PAGE-LSNs of pages $p$ and $p'$ are saved, the pages are unlatched, the lock on $y$ is requested unconditionally, and, when the lock is granted, the pages are relatched. The call *fix-and-write/read-latch*$(p, p')$ is a shorthand for the call *fix-and-write-latch*$(p)$ followed by the call *fix-and-read-latch*$(p')$ if $p' \neq p$.

After relatching the pages $p$ and $p'$, the page contents are inspected so as to find out whether or not they can still be used to satisfy the insert action. The insert action can be satisfied if one of the following two conditions holds:

1. $\text{PAGE-LSN}(p)$ and $\text{PAGE-LSN}(p')$ have not advanced from the saved values.
2. $p = p'$ and $\text{PAGE-LSN}(p)$ has advanced from the saved value, but page $p$ is still a data page with $x_1 \leq x \leq x_2$ where $x_1$ is the least and $x_2$ the greatest key in page $p$, page $p$ has room for the tuple $(x, v)$, and $y$ is the least key in page $p$ greater than $x$.

Otherwise, we must unlatch pages $p$ and $p'$, unlock key $y$, and repeat the search for the pages and the next key by performing again the call *find-page-for-insert*$(p, p', x, v, y)$, locking the found key $y$ conditionally, and so on.

The delete action $D[x, v]$ is implemented analogously by the procedure *delete*$(T, x, v)$ (Algorithm 6.8). First, the key $x$ is unconditionally X-locked for short duration. Then the call *find-page-for-delete*$(p, p', x, y)$ is used to locate and write-latch the data page $p$ that covers key $x$ and to locate and read-latch the data page $p'$ that contains the key $y$ next to $x$ if that key is not found in page $p$ (and *high-key*$(p) < \infty$). Again, if $p \neq p'$, the pages $p$ and $p'$ cover two successive subranges of the key space. The call also ensures (by appropriate structure modifications) that page $p$ will not underflow by the deletion of a single tuple (if the physical database structure requires that each data page should have at least a specified minimum number of tuples). In Sect. 7.5 we give an implementation for the call *find-page-for-delete*$(p, p', x, y)$ in the case of a sparse B-tree.

---

**Algorithm 6.8** Procedure *delete*$(T, x, v)$

---

   *lock*$(T, x, X,$ short-duration$)$
   **while** true **do**
      *find-page-for-delete*$(p, p', x, y)$
      *conditionally-lock*$(T, y, X,$ commit-duration$)$
      **if** the lock on $y$ was granted **then**
         exit the **while** loop
      **else**
         $m \leftarrow$ PAGE-LSN$(p)$
         $m' \leftarrow$ PAGE-LSN$(p')$
         *unlatch-and-unfix*$(p, p')$
         *lock*$(T, y, X,$ commit-duration$)$
         *fix-and-write/read-latch*$(p, p')$
         **if** PAGE-LSN$(p) = m$ **and** PAGE-LSN$(p') = m'$ **or** $p = p'$ **and** $p$ is a data page with
         $x_1 \leq x \leq x_2$ where $x_1$ is the least and $x_2$ the greatest key in page $p$ **and** $p$ does not
         underflow by the deletion of a tuple **and** $y$ is the least key in $p$ greater than $x$ **then**
            exit the **while** loop
         **else**
            *unlatch-and-unfix*$(p, p')$
            *unlock*$(T, y, X,$ commit-duration$)$
         **end if**
      **end if**
   **end while**
   **if** page $p$ does not contain a tuple with key $x$ **then**
      *unlatch-and-unfix*$(p, p')$
      return with error
   **else**
      *delete-from-page*$(T, p, x)$
      *unlatch-and-unfix*$(p, p')$
      *unlock*$(T, x, X,$ short-duration$)$
   **end if**

---

The found next key $y$ is conditionally X-locked for commit duration. If the lock is granted, the call *delete-from-page*$(T, p, x)$ (Algorithm 3.2) given in Sect. 3.6 is used to perform and log the deletion of the tuple with key $x$ from page $p$, after which the pages $p$ and $p'$ are unlatched and the lock on $x$ released. Otherwise, the PAGE-LSNs of pages $p$ and $p'$ are saved, the pages are unlatched, the lock on $y$ is requested unconditionally, and, when the lock is granted, the pages are relatched.

After relatching the pages $p$ and $p'$, the page contents are inspected so as to find out whether or not they can still be used to satisfy the delete action. The delete action can be satisfied if one of the following two conditions holds:

1. PAGE-LSN$(p)$ and PAGE-LSN$(p')$ have not advanced from the saved values.
2. $p = p'$ and PAGE-LSN$(p)$ has advanced from the saved value, but page $p$ is still a data page with $x_1 \leq x \leq x_2$ where $x_1$ is the least and $x_2$ the greatest key in page $p$, page $p$ does not underflow by the deletion of a single tuple, and $y$ is the least key in page $p$ greater than $x$.

Otherwise, we must unlatch the pages $p$ and $p'$, unlock key $y$, and repeat the search for the pages and the next key by performing again the call *find-page-for-delete*$(p, p', x, y)$, locking the found key $y$ conditionally, and so on.

Obviously, because of the use of conditional lock calls, no deadlocks can be caused by the interplay of locks and latches in Algorithms 6.6–6.8. In Chap. 7 we give deadlock-free algorithms for the procedures *find-page-for-read*, *find-page-for-insert*, and *find-page-for-delete* used in Algorithms 6.6–6.8. This means that for any set of concurrent transactions, where each transaction consists of only a single action (a read, an insert, or a delete action), no deadlock can occur. Note that the two keys locked in the insert and delete algorithms are locked in ascending key order, and no lock is ever upgraded.

Unfortunately, there is no upper limit on the number of iterations of the **while** loop in Algorithms 6.6–6.8. Thus, with key-range locking, there is no upper limit on the access-path-length of read, insert, and delete actions: *starvation* may occur. We will discuss this problem later.

## 6.8 Predicate Locking

Assume that a transaction reads a set of tuples from $r(\underline{X}, V)$ using an SQL query

**select** $*$ **from** $r$ **where** $P$,

where $P$ is a predicate over $X$ and $V$. If $P$ is more complicated than a single primary-key range or a union of primary-key ranges, then in our key-range transaction model, the query must be mapped to the action sequence

$$R[x_1, > -\infty, v_1], R[x_2, > x_1, v_2] \ldots R[x_n, > x_{n-1}, v_n] R[\infty, > x_n, 0]$$

that reads all the tuples from $r$. The program generating the transaction then checks at each tuple $(x_i, v_i)$ whether or not it qualifies for $P$.

With key-range locking, all the tuples in $r$ are S-locked, meaning that the entire key space $(-\infty, \infty]$ is S-locked, even though only a small subset qualifies for $P$. With multi-granular locking (presented in Sect. 9.2), a single S lock would be used to lock the entire relation $r$.

We would like to lock only those tuples that qualify for the query predicate $P$. In *predicate locking*, the lock names are predicates over the attributes of the logical database, such as

$$x < X < y \wedge X \neq z \wedge V \geq v.$$

A query with such a predicate $P$ as the query predicate would be protected by a predicate lock of mode $S$ and name $P$. An update action $W[x, u, v]$, generated from

**update** $r$ **set** $V = v$ **where** $X = x$,

would be protected by a predicate lock of mode X and name

$$X = x \wedge (V = u \vee V = v).$$

Two predicate locks with names $P_1$ and $P_2$ are incompatible if their modes are incompatible and the conjunction $P_1 \wedge P_2$ is satisfiable.

The main problem with general predicate locks is the complexity of testing their incompatibility. Recall that satisfiability of general boolean predicates is an NP-complete problem. The incompatibility test cannot be reduced to an equality test between two simple hash values computed from the predicates. For this reason, general predicate locking is not applied in concurrency control of transactions.

The key-range locking protocol can be regarded as an efficient implementation of predicate locking that uses a very simple form of predicates, namely:

$$
\begin{array}{ll}
\text{``}x \geq X > z\text{,''} & \text{for } R[x, > z, v]. \\
\text{``}x \geq X \geq z\text{,''} & \text{for } R[x, \geq z, v]. \\
\text{``}X = x\text{,''} & \text{for } I[x, v] \text{ or } D[x, v].
\end{array}
$$

The implementation by key-range locking is conservative in that more is locked than is implied by the above predicate locks. Assume that $x$ is a key which is or is not in the current database and let $x'$ be the greatest key less than $x$ in the database (if such a key exists) or $-\infty$ (otherwise). The S lock on $x$ acquired for $R[x, > z, v]$ or $R[x, \geq z, v]$ actually locks the predicate "$x \geq X > x'$," even if $z > x'$. The commit-duration X lock acquired for $I[x, v]$ on $x$ actually X-locks the predicate "$x \geq X > x'$." The commit-duration X lock acquired for $D[x, v]$ on the next key $y$ of $x$ actually X-locks the predicate "$y \geq X > x'$."

*Example 6.21* The following histories, although free of anomalies, are not possible under the key-range locking protocol when run on the database $\{(2, u)\}$:

1. $B_1 R_1[2, > 1, u] B_2 I_2[1, v]$.
2. $B_1 I_1[3, v] B_2 D_2[2, u]$.
3. $B_1 I_1[4, v] B_2 I_2[3, w]$.
4. $B_1 D_1[2, u] B_2 I_2[1, v]$.
5. $B_1 D_1[2, u] B_2 R_2[\infty, > 2, 0]$.

However, all are possible under predicate locking:

1. $T_1$ S-locks "$1 < X \leq 2$" and $T_2$ X-locks "$X = 1$."
2. $T_1$ X-locks "$X = 3$" and $T_2$ X-locks "$X = 2$."
3. $T_1$ X-locks "$X = 4$" and $T_2$ X-locks "$X = 3$."
4. $T_1$ X-locks "$X = 2$" and $T_2$ X-locks "$X = 1$."
5. $T_1$ X-locks "$X = 2$" and $T_2$ S-locks "$2 < X \leq \infty$."

The histories are permitted because in each case the conjunction of the names of the locks acquired by $T_1$ and $T_2$ is not satisfiable. □

Another drawback with key-range locking is that the name of one of the locks needed for an insert action $I[x, v]$ or a delete action $D[x, v]$, namely, the lock on the next key, can only be determined after accessing the page that holds the next key. Moreover, as shown above, implementing this in a deadlock-free manner requires the use of conditional lock requests. With predicate locking, only one lock is needed for an insert or a delete action, and that lock can be acquired before accessing the target page.

A question arises whether there still might exist a reasonably efficient implementation for the simple case of predicate locks "$x \geq X > z$," "$x \geq X \geq z$," and "$X = x$." The lock table then stores locks of the form $(T, r, m, d)$, where $r$, the lock name, is a key range of one of the forms $(z, x]$, $[z, x]$, and $[x, x]$.

The lock table can be organized as follows. The X locks $[x, x]$ for updates are stored in a balanced search tree indexed by keys $x$. The S locks that protect ranges of the forms $(z, x]$ or $[z, x]$ are stored in another balance search tree indexed by the pairs $(x, z)$ of upper and lower bounds of the ranges. Obviously, locked ranges with different upper bounds never overlap. Thus the question whether or not a given key $x'$ belongs to some S-locked range can be answered in time logarithmic in the number of locked ranges stored in the tree. Similarly, inserting new locked ranges and deleting existing locked ranges can be done in logarithmic time.

This organization is still less efficient than searching a hash-table bucket for occurrences of the four-byte hash value computed from a given key, but it may nonetheless be feasible in some cases. The problem with long keys of varying lengths can be alleviated by truncating the keys from the tail to a prefix of fixed maximum length, with, of course, the penalty of sacrificing in the accuracy of locking.

# Problems

**6.1** Are the following histories possible under the read-write locking protocol?

(a) $B_1 R_1[x]S_1[P]R_1[y]W_1[x]A_1[P]W_1^{-1}[x]C_1[P]W_1[y]B_2 R_2[x]W_2[x]C_2 W_1[x]C_1$.

(b) $B_1 R_1[y]S_1[P]R_1[x]W_1[x]A_1[P]W_1^{-1}[x]C_1[P]B_2 R_2[x]W_2[x]C_2 R_1[x]W_1[y]C_1$.

What if transaction $T_1$ is run at SQL isolation level $2°$ (read committed)?

**6.2** Consider applying the read-write locking protocol to the key-range transaction model, so that for the read action $R[x, \theta z, v]$ the key $x$ is S-locked for commit duration as in key-range locking, but for the insert and delete actions $I[x, v]$ and $D[x, v]$, only the key $x$ is X-locked for commit duration. Show that this protocol prevents dirty writes, simple dirty reads and simple unrepeatable reads. Also show that the SQL isolation level 2° is achieved by keeping the S locks only for short duration.

**6.3** We apply the key-range locking protocol. What locks are acquired by the transactions in the following histories and when are those locks released?

(a) $B_1 I_1[1] B_2 I_2[2] B_3 I_3[3] A_2 I_2^{-1}[2] C_2 C_3 I_1[2]$.
(b) $B_1 D_1[0] A_1 D_1^{-1}[0] B_2 R_2[0] C_2 C_1$.
(c) $B_1 D_1[0] I_1[0] D_1[0] A_1 D_1^{-1}[0] B_2 R_2[0] C_2 I_1^{-1}[0] D_1^{-1}[0] C_1$.

Which of the histories are possible under the locking protocol? In (a), the database is empty initially, and in (b) and (c), the database contains initially a tuple with key value 0. If a history is possible, what is the serializability order of the transactions? If a history is not possible under the key-range locking protocol, what isolation anomalies does it possibly contain?

**6.4** The physical structure of relation $r(\underline{X}, V)$ is a sparse (primary) B-tree indexed by key $X$. With the key-range locking protocol, what locks must be acquired by the transaction generated by the following application program fragment, assuming the transaction is run at the serializable isolation level?

> **exec sql select min($X$) $-$ 1 into** : x **from** $r$;
> **exec sql select max($X$) $+$ 1 into** : y **from** $r$;
> **exec sql insert into** $r$ **values** ( : x, : u);
> **exec sql insert into** $r$ **values** ( : y, : v);
> **exec sql commit**.

**6.5** The key-range locking protocol applied in Algorithms 6.6–6.8 ensures the highest (i.e., serializable) isolation level for the transaction. Consider relaxing the isolation level. How do you modify the algorithms if the transaction is only required to run at (a) SQL isolation level 3°, (b) SQL isolation level 2°, or (c) SQL isolation level 1°.

**6.6** The proof of Theorem 6.16, stating the correctness of key-range locking, assumes the basic key-range transaction model in which the forward-rolling update actions are insert actions $I[x, v]$ and delete actions $D[x, v]$. Modify the proof to cover also write actions $W[x, u, v]$.

**6.7** In Algorithms 6.6–6.8, there is a (tiny) possibility that the **while** loop is iterated indefinitely, so that the read, insert, or delete action will never be completed. In other words, the transaction *starves* in its repeated attempts at locking the target keys and latching the target pages. Can this problem be avoided?

**6.8** Consider the extended read-write transaction model in which a transaction can use the action $C[x]$ to declare the updates on $x$ done so far as committed before the transaction terminates (see Problems 1.5 and 5.9). Revise the read-write locking protocol so as to observe this feature.

**6.9** Assume that the tuples $(x, y, v)$ of a relation $r(\underline{X}, Y, V)$ can be accessed besides by the unique primary key $x$ also by the (non-unique) key $y$ (see Problems 1.6 and 5.10). Extend the key-range locking protocol so as to cover read actions based on non-unique keys.

**6.10** Show that predicate locking with predicates "$z < X \leq x$," "$z \leq X \leq x$," and "$X = x$" prevents all the isolation anomalies defined for the key-range transaction model.

## Bibliographical Notes

Eswaran et al. [1974, 1976] argue that a transaction needs to lock a logical rather than a physical subset of the database; they present the idea of general predicate locks. The locking protocols needed to achieve the different isolation levels for transactions were well understood by the authors of System R [Astrahan et al. 1976, Eswaran et al. 1974, 1976, Gray et al. 1975, 1976, 1981]. Fine-granular (tuple-level) locking was already applied in the logical locking protocol implemented in System R [Astrahan et al., 1976]. This locking protocol also observed that when a transaction does a partial rollback to a savepoint, all locks acquired since that savepoint can be released [Astrahan et al. 1976, Gray et al. 1981]. System R is said to have been the first database management system to do key-range locking and guarantee serializability [Mohan, 1996a].

The key-range locking protocol described in this chapter is a simplified version of the ARIES/KVL locking protocol presented by Mohan [1990a] (revisited by Mohan [1996a]). More specifically, our key-range locking protocol assumes that the keys used as lock names are unique and that only the two basic lock modes S and X are used, while ARIES/KVL includes locking protocols for both unique and non-unique indexes and uses besides S and X also the intention lock mode IX (to be discussed in Chap. 9), thus achieving a locking protocol that permits more concurrent histories than our simple key-range locking. Also, ARIES/KVL is defined observing the underlying physical database structure (B-tree index); in particular, it is observed that a short-duration lock requested on a data item in a data page needs only be of "instant duration," so that the lock is not actually inserted into the lock table but only checked if it can be granted, provided that the data page is kept latched during the entire action (as it is in the algorithms in Sect. 6.7).

For reducing the number of logical locks ($2n + 1$ locks for an insert action on a relation with $n$ indexes) needed in ARIES/KVL, Mohan and Levine [1992] defined the ARIES/IM (ARIES for index management) locking protocol. In the "data-only locking" variant of ARIES/IM, the lock names are tuple identifiers, and in the "index-specific locking" variant, the lock names are record identifiers of index records pointing to tuples. The original ARIES/IM protocol used the basic lock modes S and

X; an enhanced protocol, obtained by adding some features from ARIES/KVL such as the use of IX locks, is presented by Mohan [1996a]. Enhanced key-range locking protocols for improved concurrency are presented by Lomet [1993].

The authors of System R observed that transaction rollback cannot tolerate a deadlock, implying that it must be able to perform undo actions without acquiring any new locks [Gray et al. 1981]. In System R, deadlocks were resolved by selecting from a cycle of the wait-for graph the youngest transaction whose lock is of short duration, or just the youngest one if none of the locks in the cycle is of short duration, and then rolling that transaction back to the savepoint preceding the offending lock request [Astrahan et al. 1976]. The need to use conditional lock requests in order to prevent deadlocks that may arise between latches and locks is discussed by Mohan [1990a, 1996a], Mohan and Levine [1992], and Mohan et al. [1990, 1992a].

The classical theory of concurrency control based on the read-write transaction model is extensively treated in the textbooks by Papadimitriou [1986], Bernstein et al. [1987], and Weikum and Vossen [2002]. Gray and Reuter [1993] present in detail how locking is implemented; their textbook also includes a detailed history of the development of lock-based and other concurrency-control mechanisms.

# Chapter 7
# B-Tree Traversals

The *B-tree*, or, more specifically, the $B^+$-tree, is the most widely used physical database structure for primary and secondary indexes on database relations. Because of its balance conditions that must be maintained under all circumstances, the B-tree is a highly dynamic structure in which records are often moved from one page to another in structure modifications such as page splits caused by insertions and page merges caused by deletions. In a transaction-processing environment based on fine-grained concurrency control, this means that a data page can hold uncommitted updates by several transactions at the same time and an updated tuple can migrate from a page to another while the updating transaction is still active.

In this chapter we show how the B-tree is traversed when performing a read, insert, or delete action under the key-range locking protocol described in Sect. 6.4. The traversal algorithms to be presented augment the read, insert, and delete algorithms presented in Sect. 6.7. The latching protocol applied in the B-tree traversals is deadlock-free, ensuring that no deadlocks can occur between latches held by different process threads and that no deadlock can occur between a latch and a lock. Repeated work done for the traversals is minimized by heavy use of *saved paths*, which often make it possible to start a new traversal by the same thread at some lower-level page in the previously traversed path, rather than at the root page. Saved paths thus provide a simple means of shortening the access-path-length of database actions.

The B-tree structure we consider is the very basic one in which no sideways links are maintained and in which deletions are handled uniformly with insertions. The management of this basic B-tree structure seems to be the most challenging as compared to some B-tree variants such as those with sideways links maintained at the leaf level or to those in which the balance conditions are relaxed by allowing pages to become empty until they are detached from the B-tree and deallocated. One motivation of our approach is that for write-optimized B-trees (to be discussed in Chap. 15), sideways-linking is not feasible at all.

© Springer International Publishing Switzerland 2014
S. Sippu, E. Soisalon-Soininen, *Transaction Processing*, Data-Centric Systems and Applications, DOI 10.1007/978-3-319-12292-2_7

## 7.1   Sparse B-Tree Indexes

We assume that the physical database used to implement our logical database, that is, the relation $r(\underline{X}, V)$, is a *sparse B-tree* defined as follows:

1. The sparse B-tree is a tree whose nodes are database pages and whose root-to-leaf paths are all of the same length; this length is called the *height* of the B-tree.
2. Each B-tree page $p$ stores records with keys $x$ satisfying

   $$low\text{-}key(p) \leq x < high\text{-}key(p),$$

   where $low\text{-}key(p)$ and $high\text{-}key(p)$ are fixed keys, called the *low key* and *high key*, respectively, of page $p$. The low and high keys are not stored in the page; the low key may appear as the least key in the records stored in the page, but this need not be the case: the low key is just a lower bound on the keys stored in the page. We say that page $p$ *covers* the half-open key range $[low\text{-}key(p), high\text{-}key(p)]$.
3. The *level* of a B-tree page is one for a leaf page and one plus the level of its child pages for a non-leaf page. For each level $l = 1, \ldots, h$, where $h$ is the height of the B-tree, the key ranges covered by the sequence of pages $p_1, \ldots, p_n$ at level $l$ form a disjoint partition of the entire key space $[-\infty, \infty)$, that is,

   $$low\text{-}key(p_1) = -\infty,$$
   $$low\text{-}key(p_i) < high\text{-}key(p_i) = low\text{-}key(p_{i+1}), \text{ for } i = 1, \ldots, n-1,$$
   $$\text{and } high\text{-}key(p_n) = \infty.$$

4. The leaf pages of the B-tree are data pages and store the tuples of $r$.
5. The non-leaf pages $p$ of the B-tree are index pages and store *index records* of the form $(low\text{-}key(q), q)$, where $q$ is the page-id of a child page of page $p$. There is one index record for each child page of page $p$. The child pages of a B-tree page are ordered in ascending order by the low keys of the child pages.

We say that such a tree structure $b$ is a *sparse B-tree index on* relation $r$ and that $r$ is the *logical database indexed by* B-tree $b$, denoted $r = db(b)$.

*Example 7.1* Figure 7.1 shows (a part of) a sparse B-tree index on relation $r(X, V)$ with tuples with keys 1, 3, 5, 6, 7, 9, 10, 11, 12, 14, 15, 17, 18, 19, 20, 21, 24, 25, 26, 27, etc. We assume (unrealistically) that only six tuples fit in a page. Pages $p_4$ to $p_9$ are leaf pages, $p_2$ and $p_3$ are index pages of height 2, and $p_1$ (of height 3) is the root page.                                                                    □

We call a B-tree as defined above a *consistent B-tree*. This basic definition states how the pages of the B-tree are organized as a tree structure and what key ranges are covered by each page. The following lemma states the most important properties of a consistent B-tree:

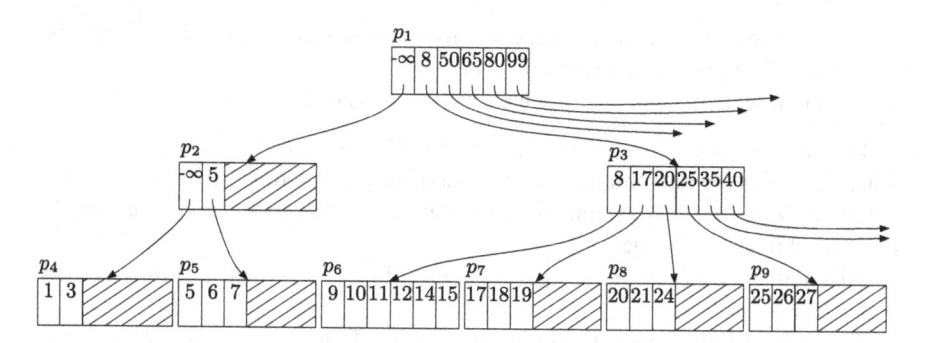

**Fig. 7.1** Part of a B-tree of height 3

**Lemma 7.2** *In a consistent B-tree, the following holds:*

(a) *For each non-leaf page p in the tree, the key range covered by p is the union of the key ranges covered by the child pages of p; in particular, low-key(p) = low-key($q_1$) and high-key(p) = high-key($q_2$), where $q_1$ is the eldest and $q_2$ the youngest child of p.*

(b) *The root page is the only page of the B-tree that covers the entire key space $[-\infty, \infty)$.*

(c) *For any non-root page q, low-key(q) is stored in the parent of q; for any non-leaf page q, low-key(q) is stored in q.*

(d) *For any non-root page q, either high-key(q) $= \infty$ and q is the last page at its level or high-key(q) is stored in the nearest ancestor p of q with high-key(p) > high-key(q).*

□

The above lemma readily implies the following method for accessing the data page that covers a given key $z$:

> $p \leftarrow$ the page-id of the root page
> *fix-and-latch*($p$)
> **while** page $p$ is a non-leaf page **do**
> $\quad q \leftarrow$ the page-id of the child page of $p$ that covers key $z$
> $\quad$ *fix-and-latch*($q$)
> $\quad$ *unlatch-and-unfix*($p$)
> $\quad p \leftarrow q$
> **end while**

The child page $q$ that covers $z$ is the one for which page $p$ contains an index record $(x, q)$ where $x$ is the greatest key in page $p$ with $x \leq z$. Also note that we use latch-coupling in traversing the path from the root page down to the leaf page: we latch the child first and only then release the latch on the parent.

The B-tree definition above says nothing of how many records should be stored in a page of a consistent B-tree. The definition even allows pages to be empty. We say that a consistent B-tree is *balanced* if it satisfies the following *balance conditions*:

1. If the tree has more than one page, the root contains at least two index records, and each leaf page contains at least $e$ tuples.
2. Each non-leaf non-root page contains at least $d$ records.

Here $e$ and $d$ are prespecified constants with $2 \leq e < E/2$ and $2 \leq d < D/2$, where $E \geq 5$ is the maximum number of maximum-length tuples that fit in a single page and $D \geq 5$ is the maximum number of index records with a maximum-length key that fit in a single page.

For example, the (part of the) B-tree shown in Fig. 7.1 is balanced with $e = d = 2$ and $E = D = 6$.

That at least five records must fit in a page comes from the requirement that if the records in two sibling pages do not fit in a single page, then it must be possible to distribute the records between the two pages in a way that both pages receive more than the minimum number of records required in a page. We observe that if two pages together have six records, which thus do not fit in a single page, then after distributing the records equally between the two pages results in three records in both pages, which is one more than the required minimum of two records per page.

The balance conditions ensure that the height of a balanced B-tree is always logarithmic in the number of tuples in its leaf pages:

**Lemma 7.3** *The height of a balanced sparse B-tree index on relation $r$ is $O(\log N)$, where $N$ is the number of tuples in $r$.*                                □

Thus, also the time complexity of accessing the data page covering a given key by the above method is $O(\log N)$.

The algorithms presented in this chapter all assume that the page-id of the root page of the B-tree does not change. Accordingly, the B-tree structure modifications given in Chap. 8 all retain the page-id of the root page. The page-id of the root page can only change in a reorganization operation, which must then be protected with a lock of coarser granularity (an X lock on the relation) that prevents simultaneous access by user transactions.

## 7.2  Saved Paths

When a server-process thread accesses a B-tree index for the first time while servicing a forward-rolling action by a transaction, it must begin the search at the root page of the B-tree. As further searches by the same thread often traverse a path that is the same as or near to the path traversed last, it is possible to accelerate the further searches by remembering the latest path traversed, so that a new search can be started at some page lower down than the root, thus saving in latching (Fig. 7.2).

We assume that the root-to-leaf path traversed by a thread is *saved* in a main-memory array *path* maintained by the thread in its private workspace. The array *path* is indexed by the height of the page in the latest path traversed, and the entry for index $i$ contains the page-id, PAGE-LSN, the low and high keys of the page, the

**Fig. 7.2** Utilizing the saved
path when searching the
B-tree for key $x$, then for key
$x'$, and then for key $x''$

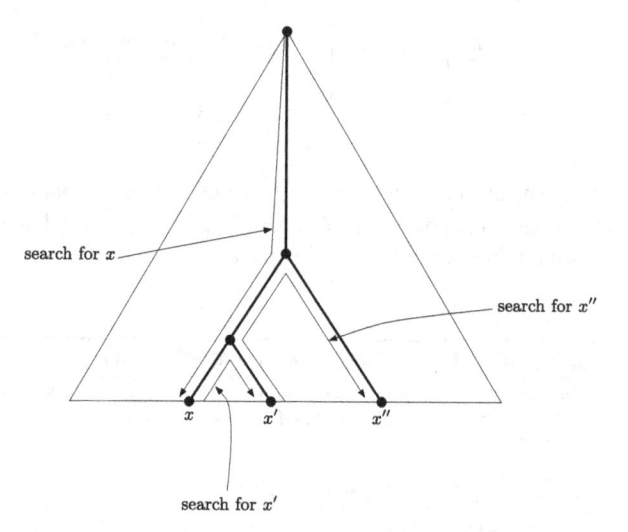

maximum key in the page, the number of records in the page, and the space left (in bytes or words) in the record area of the page. Except for the low and high keys and the maximum key, the values of these attributes can be obtained directly from corresponding fields of the header of the page; the maximum key can be found in the record area of the page and so can the low key in the case of a non-leaf page, while the low key of a leaf page comes from the parent page and the high key of any page from some ancestor page (see Lemma 7.2).

The array *path* is initialized with the call *initialize-saved-path*() (Algorithm 7.1), which saves a path with the property that, for any key $x$, the least index $i$ with $path[i].low\text{-}key \leq x < path[i].high\text{-}key$ is 2, and $path[i].page\text{-}id$ is the page-id of the root. When such a saved path is used, the root is latched for the initial traversal in read mode rather than in write mode (which in turn is used in latching a leaf page for an initial traversal for an update action).

The *path* entry for the latched root page $p$ is updated by the call *save-root*($p$) (Algorithm 7.2), and the *path* entry for a latched child page $q$ of a latched parent page $p$ is updated by the call *save-child*($p, q$) (Algorithm 7.3). Observe that the call *save-child*($p, q$) sets the high-key value of the child page $q$ from the high-key value saved for the parent page $p$ when $q$ is the youngest child. Thus, this call can only be used in the case when the high-key value saved for the parent can be trusted.

*Example 7.4* Assume that we wish to read the tuples with keys 3 and 7 from the relation indexed by the B-tree of Fig. 7.1. The initial contents of the array *path* are:

|     | page-id | page-lsn | low-key   | high-key | max-key  | #records | space-left |
| --- | ------- | -------- | --------- | -------- | -------- | -------- | ---------- |
| 1:  | $p_1$   | . . .    | $\infty$  | $\infty$ | $\infty$ | 0        | 0          |
| 2:  | $p_1$   | . . .    | $-\infty$ | $\infty$ | $\infty$ | 0        | 0          |

Thus the search for key 3 starts from the root and the path $p_1 \rightarrow p_2 \rightarrow p_4$ is traversed and saved:

|    | page-id | page-lsn | low-key | high-key | max-key | #records | space-left |
|----|---------|----------|---------|----------|---------|----------|------------|
| 1: | $p_4$   | . . .    | $-\infty$ | 5      | 3       | 2        | . . .      |
| 2: | $p_2$   | . . .    | $-\infty$ | 8      | 5       | 2        | . . .      |
| 3: | $p_1$   | . . .    | $-\infty$ | $\infty$ | 99    | 6        | 0          |

Then, when searching for key 7, we use the saved path, observing that the lowest-level page that covers key 7 is page $p_2$. Thus, provided that $p_2$ still covers key 7, we can start the search for key 7 at page $p_2$. □

---

**Algorithm 7.1** Procedure *initialize-saved-path*()

$path[1].page\text{-}id \leftarrow path[2].page\text{-}id \leftarrow$ the page-id of the root page
$path[1].page\text{-}lsn \leftarrow path[2].page\text{-}lsn \leftarrow 0$
$path[1].low\text{-}key \leftarrow \infty$
$path[2].low\text{-}key \leftarrow -\infty$
$path[1].high\text{-}key \leftarrow path[2].high\text{-}key \leftarrow \infty$
$path[1].max\text{-}key \leftarrow path[2].max\text{-}key \leftarrow \infty$
$path[1].\#records \leftarrow path[2].\#records \leftarrow 0$
$path[1].space\text{-}left \leftarrow path[2].space\text{-}left \leftarrow 0$

---

**Algorithm 7.2** Procedure *save-root*($p$)

$i \leftarrow height(p)$
$path[i].page\text{-}id \leftarrow p$
$path[i].page\text{-}lsn \leftarrow$ PAGE-LSN($p$)
$path[i].low\text{-}key \leftarrow -\infty$
$path[i].high\text{-}key \leftarrow \infty$
$path[i].max\text{-}key \leftarrow$ maximum key of the records in page $p$
$path[i].\#records \leftarrow$ number of records in page $p$
$path[i].space\text{-}left \leftarrow$ space left in page $p$

---

**Algorithm 7.3** Procedure *save-child*($p, q$)

$i \leftarrow height(q)$
$path[i].page\text{-}id \leftarrow q$
$path[i].page\text{-}lsn \leftarrow$ PAGE-LSN($q$)
$path[i].low\text{-}key \leftarrow$ the key in the index record of child page $q$ in page $p$
**if** $q$ is the youngest child of $p$ **then**
    $path[i].high\text{-}key \leftarrow path[i+1].high\text{-}key$
**else**
    $path[i].high\text{-}key \leftarrow$ the key in the index record of next younger sibling of page $q$ in page $p$
**end if**
$path[i].max\text{-}key \leftarrow$ maximum key of the records in page $q$
$path[i].\#records \leftarrow$ number of records in page $q$
$path[i].space\text{-}left \leftarrow$ space left in page $q$

In general, in search for a key $x$, the saved path is used as follows. First the least index $i$ with $path[i].low\text{-}key \leq x < path[i].high\text{-}key$ is determined, and the page $p = path[i].page\text{-}id$ is latched. Then the contents of page $p$ are examined so as to check whether or not this page is still a correct page for starting the traversal. If PAGE-LSN$(p)$ has not advanced from the saved value $path[i].page\text{-}lsn$, we can be sure that the saved information is still valid, and we can search for $x$ in the subtree rooted at $p$. Otherwise, if PAGE-LSN$(p)$ has advanced, we can no longer trust the saved information. However, inspecting the current contents of page $p$ may still reveal that it is safe to start at $p$. For example, if we see from the page header that page $p$ is still part of the B-tree index of our relation $r$ and if $x_1 \leq x \leq x_2$ where $x_1$ is the least and $x_2$ the greatest key currently in page $p$, then certainly $p$ is a correct place to start at. Otherwise, we have to unlatch $p$ and try the page at the next higher level on the saved path, and so on, in the worst case arriving at the root.

## 7.3  Traversals for Read Actions

In this section and in the sections that follow, we present in detail how the B-tree is traversed when executing the forward-rolling actions $R[x]$, $I[x]$ and $D[x]$ or the backward-rolling actions $I^{-1}[x]$ and $D^{-1}[x]$ of our transaction model. To make the traversals to work correctly and efficiently, several problems have to be solved. First we must take care that the traversals are deadlock-free. This means that B-tree pages must be latched in a prespecified order, which is the following:

1. If both a parent page and a child page must be kept latched simultaneously, the parent is latched first.
2. If two sibling or cousin pages, that is, two pages at the same level of the tree, must be kept latched simultaneously, the elder one is latched first.

With traversals for read actions $R[x, \theta z, v]$, we have the problem that the tuple $(x, v)$ to be read may not reside in the page $p$ that covers the key $z$ given as the input argument of the action: it may reside in the leaf page $p'$ next to page $p$ (Fig. 7.3b). (Because, by the balance conditions, there are no non-root empty pages, we know however for sure that $(x, v)$ must reside in $p'$ if it does not reside in $p$.) In this case we must begin the traversal at a page that is an ancestor to both $p$ and $p'$ and the traversal must end with both pages being latched.

As it is however less common that the tuple $(x, v)$ to be read does not reside in the page that covers the key $z$ given as the input argument, we always start the search for $x$ optimistically assuming that $(x, v)$ resides in the page covering $z$, and only in the case that the assumption turns out to be false, we perform a pessimistic traversal that locates both the page that covers $z$ and the one that covers $x$.

The procedure call $read(T, x, \theta z, v)$ (Algorithm 6.6) implements the read action $R[x, \theta z, v]$ of transaction $T$. The call $find\text{-}page\text{-}for\text{-}read(p, p', x, \theta z)$ is used to locate and read-latch the leaf page $p$ that covers key $z$ and the leaf page $p'$ that contains the least key $x$ with $x \theta z$ if such a key exists. The procedure $find\text{-}page\text{-}$

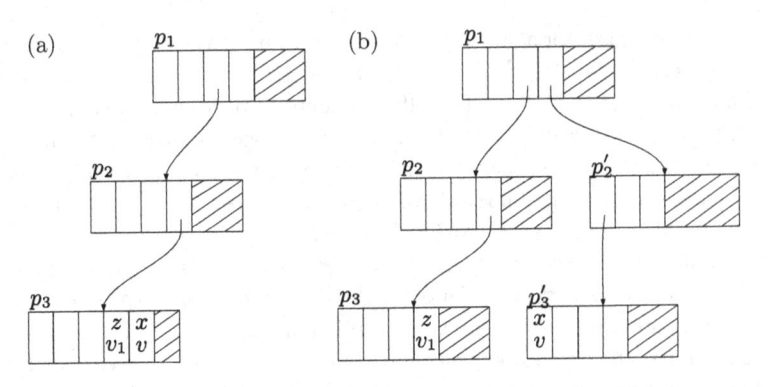

**Fig. 7.3** Traversal performed for $R[x, > z, v]$ **(a)** in the case when tuples with keys $z$ and $x$ reside in the same leaf page, **(b)** in the case when they reside in different leaf pages

*for-read* is given as Algorithm 7.6. It uses an auxiliary procedure *find-page-for-fetch*$(p, p', \theta z, both)$ (Algorithm 7.4), where the last parameter is a boolean value indicating whether an optimistic traversal (*both* = false) or a pessimistic traversal (*both* = true) is to be performed. In an optimistic traversal, only the page $p$ is determined, while in a pessimistic traversal both $p$ and $p'$ are determined.

The procedure *find-page-for-fetch*$(p, p', \theta z, both)$ is first called with *both* = false, thus assuming optimistically that the key $x$ to be read resides in page $p$, the page that covers key $z$. If it turns out that page $p$ does not contain any key $x$ with $x \theta z$, even if it is known that *high-key*$(p) < \infty$, page $p$ is unlatched and *find-page-for-fetch*$(p, p', \theta z, both)$ is called with *both* = true. Page $p$ must be unlatched before starting the new traversal, because latching an ancestor of a latched page could lead to a deadlock. The new traversal must search both for the page that covers $z$ and for the page that covers the least key $x$ with $x \theta z$, because after unlatching $p$ both pages may change in the interim.

The procedure *find-page-for-fetch*$(p, p', \theta z, both)$ uses the saved path to determine the lowest-level page $p$ at which to start the traversal for the search for the leaf page covering key $z$. The first approximation for $p$ is determined using only the information stored in the saved path. That page $p$ is latched and then, if different from the root, checked if it satisfies one of the two conditions below; if not, the page is unlatched, and the page at the next higher level on the saved path is probed. In the following, $i$ is the index of the page on the saved path.

1. PAGE-LSN$(p) = path[i].page\text{-}lsn$ and $path[i].low\text{-}key \leq z < path[i].high\text{-}key$ and, in addition, either $path[i].high\text{-}key = \infty$ or $path[i].max\text{-}key \ \theta \ z$.
2. PAGE-LSN$(p) > path[i].page\text{-}lsn$ but $p$ is a B-tree page with $x_2 \ \theta \ z \geq x_1$ for the keys $x_2$ and $x_1$ of some records in page $p$.

Both of these conditions ensure that the least key $x$ with $x \ \theta \ z$, if such a key exists, is found in the subtree rooted at page $p$. Both conditions also ensure that when the path down from $p$ is traversed and the path is saved using the *save-child*$(p, q)$ call, the high key for $q$ is correctly determined. Note that the high key can come from the saved high key of the parent $p$ only in case (1), in which case it can be trusted because PAGE-LSN$(p) = path[i].page\text{-}lsn$. The starting page $p$ is determined as a result of the first **while** loop of Algorithm 7.4.

Next the path from the starting page down to the leaf page that covers key $z$ is traversed using latch-coupling with read latches and saving the path so traversed. This is done in the second **while** loop of Algorithm 7.4. If *both* = true, then at each page $p$ visited, we also keep track of the root $p'$ of the subtree that may contain the least key $x$ with $x\ \theta\ z$ should the key not reside in the subtree rooted at $p$ (that covers key $z$). The page $p'$, if distinct from $p$, is always the page next to $p$ at the same level. At the start of the traversal, $p' = p$. When advancing from $p$ to its child page $q$, the call *find-next-page*$(p, q, \theta z, p', q')$ (Algorithm 7.5) is used to determine the page $q'$ whose subtree may contain the searched key $x$ should the key not reside in the subtree rooted at $q$ (Fig. 7.4). If *both* = false, the call *find-page-for-fetch*$(p, p', \theta z, both)$ returns with $p' = p$.

---

**Algorithm 7.4** Procedure *find-page-for-fetch*$(p, p', \theta z, both)$

---

$i \leftarrow$ the least index with $path[i].low\text{-}key \leq z < path[i].high\text{-}key$ and with either $path[i].high\text{-}key = \infty$ or $path[i].max\text{-}key\ \theta\ z$
$p \leftarrow path[i].page\text{-}id$
**while** $p$ is not the page-id of the root page **do**
    *fix-and-read-latch*$(p)$
    **if** PAGE-LSN$(p) = path[i].page\text{-}lsn$ **and** $path[i].low\text{-}key \leq z < path[i].high\text{-}key$ **and** either $path[i].high\text{-}key = \infty$ or $path[i].max\text{-}key\ \theta\ z$ **or** $p$ is a B-tree page with $x_2\ \theta\ z \geq x_1$ for the keys $x_2$ and $x_1$ of some records in page $p$ **then**
        exit the **while** loop
    **else**
        *unlatch-and-unfix*$(p)$
        $i \leftarrow i + 1$
        $p \leftarrow path[i].page\text{-}id$
    **end if**
**end while**
$p' \leftarrow p$
**if** page $p$ is the root **then**
    *save-root*$(p)$
**end if**
**while** page $p$ is a non-leaf page **do**
    $q \leftarrow$ the page-id of the child page of $p$ that covers key $z$
    *fix-and-read-latch*$(q)$
    *save-child*$(p, q)$
    **if** *both* **then**
        *find-next-page*$(p, q, \theta z, p', q')$
        $p' \leftarrow q'$
    **else**
        $p' \leftarrow q$
    **end if**
    *unlatch-and-unfix*$(p)$
    $p \leftarrow q$
**end while**

---

---

**Algorithm 7.5** Procedure *find-next-page*$(p, q, \theta z, p', q')$

---

**if** page $q$ contains a key $x$ with $x \; \theta \; z$ **then**
  $q' \leftarrow q$
**else if** $q$ is the youngest child of $p$ **then**
  **if** $p' = p$ **then** {case (a)}
    $q' \leftarrow q$
  **else** {case (b)}
    $q' \leftarrow$ the page-id of the eldest child of $p'$
    *fix-and-read-latch*$(q')$
  **end if**
**else** {case (c)}
  $q' \leftarrow$ the page-id of the next younger sibling of $q$
  *fix-and-read-latch*$(q')$
**end if**
**if** $p' \neq p$ **then**
  *unlatch-and-unfix*$(p')$
**end if**

---

**Fig. 7.4** Different situations
for the call
*find-next-page*$(p, q, \theta z, p', q')$.
(**a**) $p' = p$ and $q$ is the
youngest child of page $p$;
(**b**) $p' \neq p$ and page $q$ is the
youngest child of page $p$;
(**c**) page $q$ is not the youngest
child of page $p$

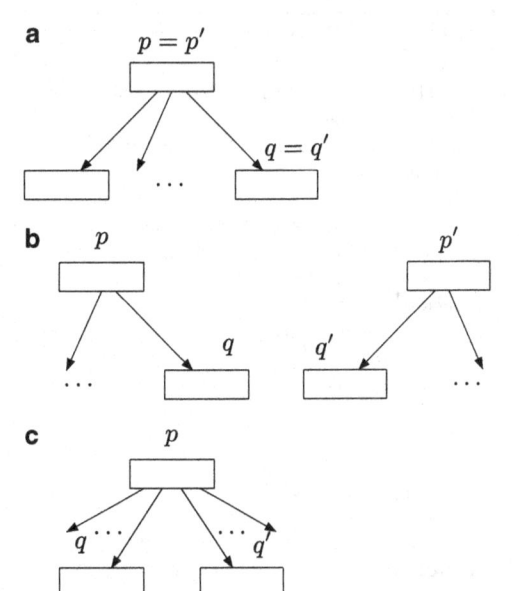

*Example 7.5* Consider executing the read action $R[x, > \; 7, v]$ by the call
*read*$(T, x, >7, v)$ (Algorithm 6.6) on the database indexed by the B-tree of Fig. 7.1.
Assume that the saved path has the initial value resulting from the call *initialize-
saved-path*(). In the procedure call *find-page-for-read*$(p, p', x, >7)$, an optimistic
traversal is first performed with the call *find-page-for-fetch*$(p, p', >7, \text{false})$, starting
at the root page $p_1$ and going down to the leaf page, $p = p_5$, that covers key 7.
Latch-coupling with read latches is used, retaining a read latch on page $p_5$. The
traversed path $p_1 \rightarrow p_2 \rightarrow p_5$ (shown in Fig. 7.5a) is saved:

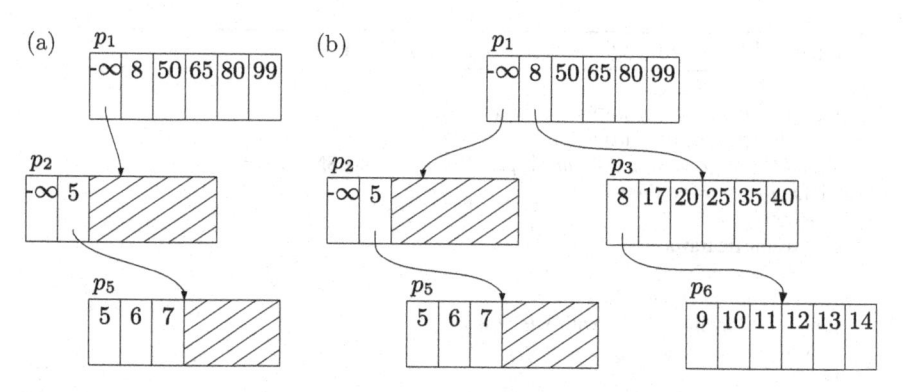

**Fig. 7.5** Performing the read action $R[x, >7, v]$ on the B-tree of Fig. 7.1. The pages visited (**a**) in the optimistic traversal and (**b**) in the pessimistic traversal that not only determines the page $p_5$ that covers key 7 but also page $p_6$ that covers the key 9 next to key 7

| | page-id | page-lsn | low-key | high-key | max-key | #records | space-left |
|---|---|---|---|---|---|---|---|
| 1: | $p_5$ | . . . | 5 | 8 | 7 | 2 | . . . |
| 2: | $p_2$ | . . . | $-\infty$ | 8 | 5 | 2 | . . . |
| 3: | $p_1$ | . . . | $-\infty$ | $\infty$ | 99 | 6 | 0 |

Because page $p_5$ does not contain a key greater than 7, the page is unlatched and the pessimistic call $find\text{-}page\text{-}for\text{-}fetch(p, p', > 7, \text{true})$ is performed.

Using the path saved from the previous traversal, we determine the first page to be latched. It is page $p_1 = path[3].page\text{-}id$, because $i = 3$ is the least index with $path[i].low\text{-}key \leq 7 < path[i].high\text{-}key$ and with either $path[i].high\text{-}key = \infty$ or $path[i].max\text{-}key > 7$. Thus, the second top-down traversal is started at the root.

Both the page $p$ that covers key 7 and the page $p'$ that contains the key next to 7 are determined. The procedure $find\text{-}next\text{-}page(p, q, > 7, p', q')$ is first called with arguments $p = p_1$, $q = p_2$ and $p' = p_1$, returning $q' = p_3$, and then with arguments $p = p_2$, $q = p_5$ and $p' = p_3$, returning $q' = p_6$. Thus the call $find\text{-}page\text{-}for\text{-}fetch(p, p', > 7, \text{true})$ returns with $p = p_5$ and $p' = p_6$, meaning that the call $find\text{-}page\text{-}for\text{-}read(p, p', x, > 7)$ returns with $p = p_5$ and $p' = p_6$ and $x = 9$. Here pages $p_5$ and $p_6$ are read-latched. See Fig. 7.5b. The tuple with key 9 can now be read. □

**Lemma 7.6** *The latching protocol applied in the B-tree traversals for read actions is deadlock-free. At most four pages are kept latched simultaneously. For a B-tree of height h, at most 2h latchings on B-tree pages are performed in all for a read action $R[x, \theta z, v]$ implemented by the call $read(T, x, \theta z, v)$ with an initialized saved path, provided that the S lock on key x is granted immediately as a result of the conditional lock call.*

*Proof* In Algorithms 7.4–7.6, only read latches are acquired, and while a page is kept latched, no latch is requested on its parent or elder sibling. Also note that in the call $find\text{-}next\text{-}page(p, q, \theta z, p', q')$ (Algorithm 7.5), the page $q$ is already latched

---

**Algorithm 7.6** Procedure *find-page-for-read*($p$, $p'$, $x$, $\theta z$)

---

*find-page-for-fetch*($p$, $p'$, $\theta z$, false)
**if** page $p$ contains a key $x$ with $x \, \theta \, z$ **then**
    $x \leftarrow$ the least such key in page $p$
**else if** *path*[1].*page-id* $= p$ and *path*[1].*page-lsn* $=$ PAGE-LSN($p$) and
*path*[1].*high-key* $= \infty$ **then**
    $x \leftarrow \infty$
**else** {pessimistic traversal}
    *unlatch-and-unfix*($p$)
    *find-page-for-fetch*($p$, $p'$, $\theta z$, true)
    **if** page $p'$ contains a key $x$ with $x \, \theta \, z$ **then**
        $x \leftarrow$ the least such key in page $p'$
    **else**
        $x \leftarrow \infty$
    **end if**
**end if**

---

when its younger cousin $q'$ is latched, thus respecting the rule that when two siblings or cousins are to be kept latched, the elder one is latched first. The worst case of four simultaneous read latches occurs when page $p$, its child $q$, a cousin $p'$ of $p$, and a child of $p'$ must all be latched simultaneously in the procedure *find-next-page*. By Lemma 7.3, this makes at most $2h$ latchings in all, when the traversal is started, using the initialized saved path, from the root page of the B-tree.     □

If the saved path has a value different from the initial value at the start of executing a read action $R[x, \theta z, v]$, then the total number of latchings needed may be less or greater than the number stated in the above lemma. In the worst case, the saved path dates from a time when the B-tree was higher that it is at the time the read action starts and the entire path has to be climbed bottom-up just to end up starting the traversal from the root page.

## 7.4  Traversals for Insert Actions

The procedure call *insert*($T$, $x$, $v$) (Algorithm 6.7) that implements the insert action $I[x, v]$ of transaction $T$ uses the call *find-page-for-insert*($p$, $p'$, $x$, $v$, $y$) to locate and write-latch the leaf page $p$ that covers key $x$ and to locate and read-latch the page $p'$ that contains the key $y$ next to $x$ if that key is not found in page $p$ (when *high-key*($p$) $< \infty$). The procedure *find-page-for-insert*, given as Algorithm 7.8, also arranges that page $p$ has sufficient room for the tuple $(x, v)$ to be inserted. To accomplish this arrangement, pages may need to be split, and the tree height may have to be increased, in a top-down fashion, as will be explained in Chap. 8.

In the same way as the procedure *find-page-for-read* uses the auxiliary procedure *find-page-for-fetch*, the procedure *find-page-for-insert* uses an auxiliary procedure

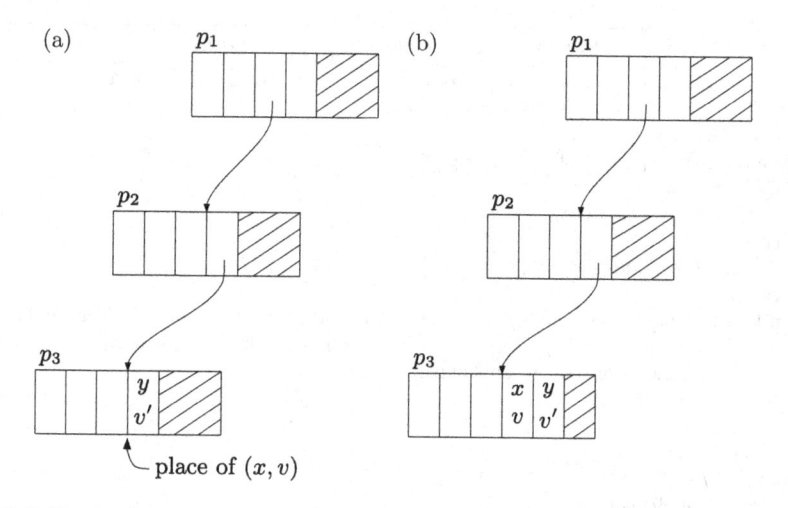

**Fig. 7.6** The simple case of insertion $I[x, v]$. The page to which the tuple $(x, v)$ belongs has room for the tuple, and the tuple $(y, v')$ with the least key $y$ greater than $x$ resides in the same page. (**a**) Before insertion (**b**) After insertion

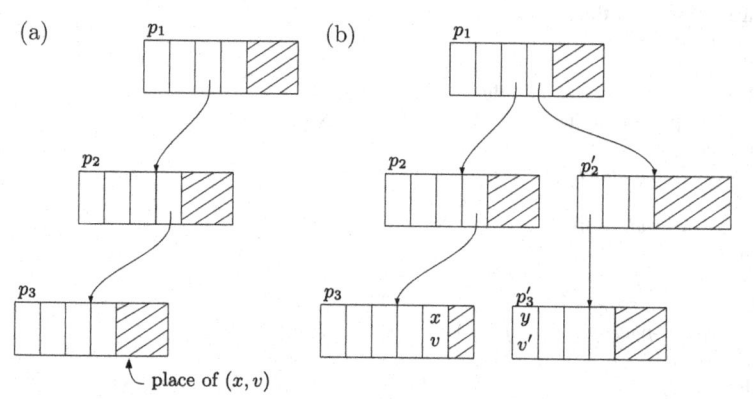

**Fig. 7.7** A more complicated case of insertion $I[x, v]$. The tuple $(x, v)$ to be inserted fits into the covering leaf page, but the tuple with the least key $y$ greater than $x$ does not reside in the same page. A pessimistic traversal is needed to locate $y$. (**a**) Before insertion (**b**) After insertion

*find-page-for-update*$(p, p', x, both)$ (Algorithm 7.7), where the last parameter is a boolean value indicating an optimistic traversal where only the page $p$ is determined ($both = $ false) or a pessimistic traversal where both $p$ and $p'$ are determined ($both = $ true). The procedure is first called with $both = $ false, thus assuming optimistically that the next key $y$ resides in page $p$, the page that covers key $x$. Figure 7.6 represents the simple case in which only the optimistic traversal is needed in order to accomplish the insertion: the next key resides in the page that covers the key to be inserted, and the page has room for the tuple to be inserted. Figure 7.7 represents a more complicated case.

**Algorithm 7.7** Procedure *find-page-for-update*($p, p', x, both$)

$i \leftarrow$ the least index with *path*[$i$].*low-key* $\leq x <$ *path*[$i$].*high-key* and with either
*path*[$i$].*high-key* $= \infty$ or *path*[$i$].*max-key* $> x$
$p \leftarrow$ *path*[$i$].*page-id*
**while** $p$ is not the page-id of the root page **do**
    **if** $i = 1$ **then**
        *fix-and-write-latch*($p$)
    **else**
        *fix-and-read-latch*($p$)
    **end if**
    **if** PAGE-LSN($p$) $=$ *path*[$i$].*page-lsn* **and** *path*[$i$].*low-key* $\leq x <$ *path*[$i$].*high-key* **and** either
    *path*[$i$].*high-key* $= \infty$ or *path*[$i$].*max-key* $> x$ **or** $p$ is a B-tree page with $x_1 \leq x < x_2$ for
    the keys $x_1$ and $x_2$ of some records in page $p$ **then**
        exit the **while** loop
    **else**
        *unlatch-and-unfix*($p$)
        $i \leftarrow i + 1$
        $p \leftarrow$ *path*[$i$].*page-id*
    **end if**
**end while**
$p' \leftarrow p$
**if** page $p$ is the root **then**
    *save-root*($p$)
**end if**
**while** page $p$ is a non-leaf page **do**
    $q \leftarrow$ the page-id of the child page of $p$ that covers key $x$
    **if** *height*($p$) $= 2$ **then**
        *fix-and-write-latch*($q$)
    **else**
        *fix-and-read-latch*($q$)
    **end if**
    *save-child*($p, q$)
    **if** *both* **then**
        *find-next-page*($p, q, > x, p', q'$)
        $p' \leftarrow q'$
    **else**
        $p' \leftarrow q$
    **end if**
    *unlatch-and-unfix*($p$)
    $p \leftarrow q$
**end while**

---

**Algorithm 7.8** Procedure *find-page-for-insert*($p, p', x, v, y$)

---

*find-page-for-update*($p, p', x$, false)
**while** true **do**
  **if** page $p$ has no room for $(x, v)$ **then**
    *unlatch-and-unfix*($p, p'$)
    *start-for-page-splits*($p, x, v$)
    *top-down-page-splits*($p, x, v$)
  **end if**
  **if** page $p$ contains a key greater than $x$ **then**
    $y \leftarrow$ the least key greater than $x$ in page $p$
    $p' \leftarrow p$
    exit the **while** loop
  **else if** $path[1].page\text{-}id = p$ and $path[1].page\text{-}lsn = \text{PAGE-LSN}(p)$ and
  $path[1].high\text{-}key = \infty$ **then**
    $y \leftarrow \infty$
    $p' \leftarrow p$
    exit the **while** loop
  **else** {pessimistic traversal}
    *unlatch-and-unfix*($p, p'$)
    *find-page-for-update*($p, p', x$, true)
    **if** page $p'$ contains a key greater than $x$ **then**
      $y \leftarrow$ the least key greater than $x$ in page $p'$
    **else**
      $y \leftarrow \infty$
    **end if**
    **if** page $p$ has room for $(x, v)$ **then**
      exit the **while** loop
    **end if**
  **end if**
**end while**

---

After the optimistic traversal, if we find that page $p$ has no room for the tuple $(x, v)$, page $p$ is unlatched, and the procedure call *start-for-page-splits*($p, x, v$) (Algorithm 8.6) is issued so as to locate and write-latch the highest-level ancestor page $p$ of the leaf page covering $x$ that may need a structure modification to accommodate the insertion of $(x, v)$. Then the procedure call *top-down-page-splits*($p, x, v$) (Algorithm 8.7) is used to perform a sequence of page splits, possibly preceded by a tree-height increase, from page $p$ down to the leaf page covering $x$. The call *top-down-page-splits*($p, x, v$) returns with a write-latched leaf page $p$ that covers $x$ and has room for $(x, v)$.

If it now turns out that page $p$ does not contain any key greater than $x$, even if it is known that $high\text{-}key(p) < \infty$, page $p$ is unlatched and *find-page-for-update*($p, p', x, both$) is called with $both =$ true. The key $y$ next to $x$ can now be determined. Because page $p$ or its contents may have changed in the interim, we must again check if there is room for $(x, v)$. If not, pages $p$ and $p'$ must be unlatched, and the process be repeated, calling again *start-for-page-splits* and *top-down-page-splits* and maybe also *find-page-for-update*, hence the "**while** true **do**" loop in Algorithm 7.8.

The procedure *find-page-for-update*$(p, p', x, both)$ (Algorithm 7.7) is similar to
the procedure *find-page-for-fetch*$(p, p', \theta z, both)$ (Algorithm 7.4). First, the saved
path is used to determine the lowest-level page $p$ at which to start the traversal for
the search for the leaf page covering key $x$. This page $p$, if different from the root,
must satisfy one of the following two conditions, where $i$ is the index of the page
on the saved path:

1. PAGE-LSN$(p) = path[i].page\text{-}lsn$ and $path[i].low\text{-}key \leq x < path[i].high\text{-}key$
   and either $path[i].high\text{-}key = \infty$ or $path[i].max\text{-}key > x$.
2. PAGE-LSN$(p) > path[i].page\text{-}lsn$, but $p$ is a B-tree page with $x_1 \leq x < x_2$ for
   the keys $x_1$ and $x_2$ of some records in page $p$.

Both of these conditions ensure that the least key $y$ with $y > x$, if such a
key exists, is found in the subtree rooted at page $p$. Both conditions also ensure
that when the path down from $p$ is traversed and the path is saved using the *save-
child*$(p, q)$ call, the high key for $q$ is correctly determined.

Next the path from the starting page down to the leaf page that covers key $x$
is traversed using latch-coupling with read latches on non-leaf pages and a write-
latch on the leaf page. This is done in the second **while** loop of Algorithm 7.7.
If *both* = true, then at each page $p$ visited, we also keep track of the root
$p'$ of the subtree that may contain the key next to $x$ should that key not reside
in the subtree rooted at $p$ (that covers key $x$). The page $p'$, if distinct from $p$,
is always the page next to $p$ at the same level. At the start of the traversal,
$p' = p$. When advancing from $p$ to its child page $q$, the call *find-next-page*$(p, q, >
x, p', q')$ (Algorithm 7.5) is used to determine the page $q'$ whose subtree may
contain the searched key $x$ should the key not reside in the subtree rooted
at $q$. If *both* = false, the call *find-page-for-update*$(p, p', x, both)$ returns with
$p' = p$.

*Example 7.7*  Consider executing the insert action $I[4, v']$ by the call *insert*$(T, 4, v')$
(Algorithm 6.7) on the database indexed by the B-tree of Fig. 7.1. Assume that
the saved path has the initial value resulting from the call *initialize-saved-path*$()$.
In the procedure call *find-page-for-insert*$(p, p', 4, v', y)$, an optimistic traversal
performed by the call *find-page-for-update*$(p, p', 4, false)$ starts at the root page
$p_1$ and goes down to the leaf page, $p = p_4$, that covers key 4. Latch-coupling
is used, with read latches acquired on the non-leaf pages $p_1$ and $p_2$ and a write
latch on the leaf page $p_4$. The traversed path $p_1 \rightarrow p_2 \rightarrow p_4$ is saved (see
Example 7.4).

Page $p_4$ has room for the tuple $(4, v')$ to be inserted, but does not contain a
key greater than 4 (Fig. 7.8a). Hence the write latch on the page is released, and
the call *find-page-for-update*$(p, p', 4, true)$ is performed. Using the path saved from
the previous traversal, we determine the first page to be latched. It is page $p_2 =
path[2].page\text{-}id$, because $i = 2$ is the least index for which $path[i].low\text{-}key \leq 4 <
path[i].high\text{-}key$ and with either $path[i].high\text{-}key = \infty$ or $path[i].max\text{-}key > 4$.
Thus, the top-down traversal can be started at $p_2$.

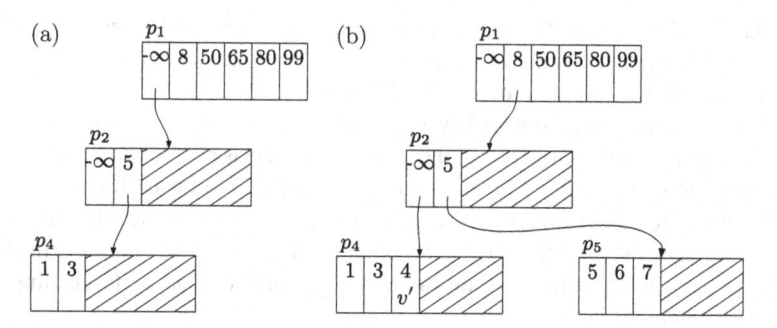

**Fig. 7.8** Insert action $I[4, v']$ on the B-tree of Fig. 7.1. (**a**) In the optimistic traversal, the path $p_1 \rightarrow p_2 \rightarrow p_4$ is traversed, leaving page $p_4$ write-latched. (**b**) As page $p_4$ does not contain the least key greater than key 4, a pessimistic traversal is performed from the lowest-level page, $p_2$, that covers 4 and whose subtree contains a key greater than 4, yielding pages $p_4$ and $p_5$; finally the tuple $(4, v')$ is inserted into $p_4$

The child page $p_4$ of page $p_2$ is write-latched and the call *find-next-page*$(p_2, p_4, > 4, p_2, q')$ is performed, returning with $q' = p_5$. Thus, the call *find-page-for-update*$(p, p', 4, \text{true})$ returns with page $p = p_4$ write-latched and page $p' = p_5$ read-latched, and the call *find-page-for-insert*$(p, p', 4, v', y)$ returns with $y = 5$. The next key $y$ can now be locked and the tuple $(4, v')$ be inserted into page $p = p_4$ (Fig. 7.8b).

In a more complicated situation, the leaf page covering the key of the tuple to be inserted does not have room for the tuple. This happens with the insert action $I[16, v']$. In this case, the leaf page, $p_6$, covering the key has no room for the tuple, besides that the next key resides in the next page. This situation will be considered in detail in Example 8.7.                                                                                    □

**Lemma 7.8** *The latching protocol applied in the B-tree traversals for insert actions is deadlock-free. If the leaf page covering key x has room for tuple $(x, v)$, then in the traversal for insert action $I[x, v]$, at most four read latches or one write latch and three read latches are ever held simultaneously. If the X lock on the key next to x is granted immediately as a result of the conditional lock call in the procedure insert (Algorithm 6.7) and no updates by other transactions or structure modifications on the B-tree occur during the insertion and the leaf page covering key x has room for the tuple $(x, v)$ and the saved path has its initial value, then in the worst case two write-latchings and $3h - 2$ read-latchings on B-tree pages in all are performed, where h is the height of the B-tree. In the best case, only one latching (write-latching) is performed.*

*Proof* In Algorithms 7.5, 7.7, and 7.8, no read latch is ever upgraded to a write latch, and while a page is kept latched, no latch is requested on its parent or elder sibling or elder cousin. By Lemma 8.9, the latching protocol applied in the procedures *start-for-page-splits* and *top-down-page-splits* is deadlock-free. The worst case of one write latch and three read latches held simultaneously occurs when page $p$ is read-latched, its child, the leaf page covering $x$ is write-latched, and a cousin of $p$ and its child containing the key next to $x$ are read-latched.

When the X lock on the next key is granted immediately, only one call of the procedure *find-page-for-insert* from the procedure *insert* is performed. In *find-page-for-insert*, the call *find-page-for-update*$(p, p', x,$ false$)$ is first performed. As the saved path has its initial value, the root page is read-latched, and then the path down to the leaf page covering key $x$ is latch-coupled with read latches on the remaining non-leaf pages and a write latch on the leaf page, making one write-latching and $h - 1$ read-latchings in all.

When the leaf page covering $x$ has room for $(x, v)$, the procedures *start-for-page-splits* and *top-down-page-merges* are not called from *find-page-for-insert*. However, if the leaf page $p$ covering $x$ does not contain the key next to $x$ (and *high-key*$(p) <$ ∞), then a call *find-page-for-update*$(p, p', x,$ true$)$ is performed. This call uses the path saved from the call *find-page-for-update*$(p, p', x,$ false$)$. As we assumed that no other updates or structure modifications occur simultaneously, the saved path is valid and the starting page for the top-down traversal is thus determined correctly by the first assignment statement in Algorithm 7.7, so no climbing up the path occurs.

In the worst case, the top-down traversal starts from the root page, because the next key may be located in a subtree of the root different from the subtree covering key $x$. Because of the calls of *find-next-page*, one write-latching and and $2h - 1$ read-latchings are performed by the traversal in the worst case.                  □

## 7.5  Traversals for Delete Actions

The procedure call *delete*$(T, x, v)$ (Algorithm 6.8) that implements the delete action $D[x, v]$ of transaction $T$ uses the call *find-page-for-delete*$(p, p', x, y)$ to locate and write-latch the leaf page $p$ that contains the tuple with key $x$ and to locate and read-latch the page $p'$ that contains the key $y$ next to $x$ if that key is not found in page $p$ (when *high-key*$(p) <$ ∞). The procedure *find-page-for-delete*, given as Algorithm 7.9, also ensures that page $p$ does not underflow by the deletion. The procedure is analogous to the procedure *find-page-for-insert* (Algorithm 7.8). In the same way it uses the auxiliary procedure *find-page-for-update*$(p, p', x, both)$ (Algorithm 7.7), which is first called optimistically with $both =$ false.

Figure 7.9 represents the simple case in which only the optimistic traversal is needed in order to accomplish the deletion: the tuple to be deleted and the next key reside in the same page, and the page does not underflow by the deletion. Figure 7.10 represents a more complicated case.

---

**Algorithm 7.9** Procedure *find-page-for-delete*($p, p', x, y$)

*find-page-for-update*($p, p', x$, false)
**while** true **do**
  **if** page $p$ would underflow by the deletion of a single tuple **then**
    *unlatch-and-unfix*($p, p'$)
    *start-for-page-merges*($p, x$)
    *top-down-page-merges*($p, x$)
  **end if**
  **if** page $p$ contains a key greater than $x$ **then**
    $y \leftarrow$ the least key greater than $x$ in page $p$
    $p' \leftarrow p$
    exit the **while** loop
  **else if** *path*[1]*.page-id* $= p$ and *path*[1]*.page-lsn* $=$ PAGE-LSN($p$) and
  *path*[1]*.high-key* $= \infty$ **then**
    $y \leftarrow \infty$
    $p' \leftarrow p$
    exit the **while** loop
  **else** {pessimistic traversal}
    *unlatch-and-unfix*($p, p'$)
    *find-page-for-update*($p, p', x$, true)
    **if** page $p'$ contains a key greater than $x$ **then**
      $y \leftarrow$ the least key greater than $x$ in page $p'$
    **else**
      $y \leftarrow \infty$
    **end if**
    **if** page $p$ will not underflow by the deletion of a single tuple **then**
      exit the **while** loop
    **end if**
  **end if**
**end while**

---

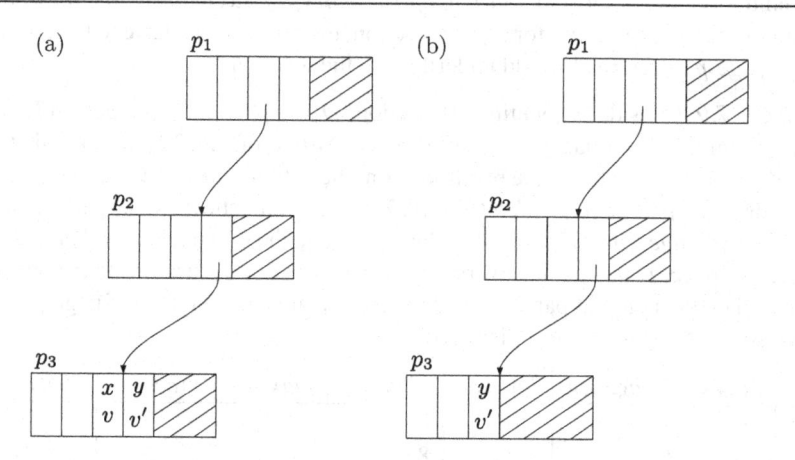

**Fig. 7.9** The simple case of deletion $D[x, v]$. The leaf page containing $(x, v)$ does not underflow by the deletion, and the tuple with the least key $y$ greater than $x$ resides in the same page. **(a)** Before deletion **(b)** After deletion

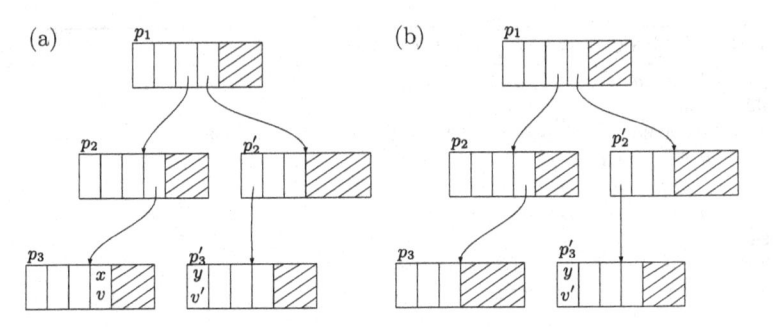

**Fig. 7.10** A case of deletion $D[x, v]$ where a pessimistic traversal is needed. The leaf page containing $(x, v)$ does not underflow by the deletion, but the tuple with the least key $y$ greater than $x$ does not reside in the same page. (**a**) Before deletion (**b**) After deletion

If the leaf page $p$ determined by the optimistic traversal would underflow by the deletion of a tuple, page $p$ is unlatched, the procedure call *start-for-page-merges*$(p, x)$ (Algorithm 8.8) is issued so as to locate and write-latch the highest-level ancestor page $p$ of the page covering $x$ that may need a structure modification to accommodate the deletion. Then the procedure call *top-down-page-merges*$(p, x)$ (Algorithm 8.9) is used to perform a sequence of page merges or records redistributes, possibly preceded by a tree-height decrease, from page $p$ down to the leaf page covering $x$. The call *top-down-page-merges*$(p, x)$ returns with a write-latched leaf page $p$ that covers $x$ and does not underflow by the deletion of a tuple.

Then the key $y$ next to $x$ is determined, possibly needing the unlatching of page $p$ and the issuing of the call *find-page-for-update*$(p, p', x, both)$ with $both = $ true. As in the case of traversing for an insert action, the process may have to be repeated, if the page $p$ so returned would underflow by the deletion.

*Example 7.9* Consider executing the action $D[7, v']$ by the call *delete*$(T, 7, v')$ (Algorithm 6.8) on a database indexed by the B-tree of Fig. 7.1. Assume that the saved path has the initial value resulting from the call *initialize-saved-path*(). In the procedure call *find-page-for-delete*$(p, p', 7, y)$, an optimistic traversal performed by the call *find-page-for-update*$(p, p', 7, $ false$)$ starts at the root page $p_1$ and goes down to the leaf page, $p = p_5$, that covers key 7. Latch-coupling is used, with read latches acquired on the non-leaf pages $p_1$ and $p_2$ and a write latch on the leaf page $p_5$. The traversed path $p_1 \rightarrow p_2 \rightarrow p_5$ is saved:

|     | page-id | page-lsn | low-key | high-key | max-key | #records | space-left |
|-----|---------|----------|---------|----------|---------|----------|------------|
| 1:  | $p_5$   | . . .    | 5       | 8        | 7       | 3        | . . .      |
| 2:  | $p_2$   | . . .    | $-\infty$ | 8      | 5       | 2        | . . .      |
| 3:  | $p_1$   | . . .    | $-\infty$ | $\infty$ | 99    | 6        | 0          |

Page $p_5$ can accommodate the deletion but does not contain a key greater than 7 (Fig. 7.11a). Hence the write latch on the page is released and the call *find-page-for-update*$(p, p', 7, $ true$)$ is performed. Using the above-saved path, we determine

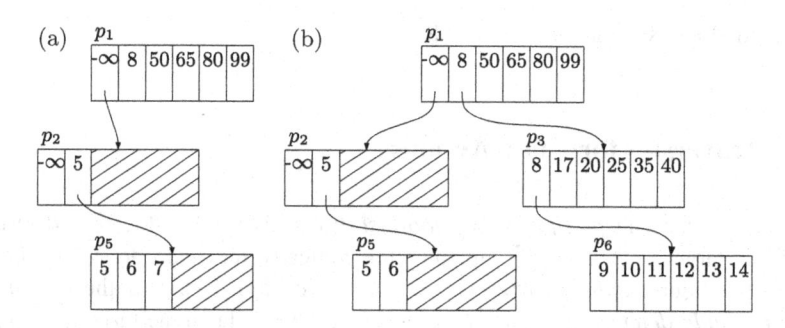

**Fig. 7.11** Delete action $D[7, v']$ on the B-tree of Fig. 7.1. (**a**) In the optimistic traversal, the path $p_1 \rightarrow p_2 \rightarrow p_5$ is traversed. (**b**) As page $p_5$ does not contain the least key greater than key 7, a pessimistic traversal is performed from the lowest-level page, $p_1$, whose subtree contains both key 7 and its next key, yielding pages $p_5$ and $p_6$; finally the tuple $(7, v')$ is deleted from $p_5$

the first page to be latched. It is page $p_1 = path[3].page\text{-}id$, because $i = 3$ is the least index for which $path[i].low\text{-}key \leq 7 < path[i].high\text{-}key$ and with either $path[i].high\text{-}key = \infty$ or $path[i].max\text{-}key > 7$. Thus the top-down traversal is started at the root page $p_1$.

The child page $p_2$ of page $p_1$ is read-latched, and the call *find-next-page*$(p_1, p_2, > 7, p_1, q')$ is performed, returning with $q' = p_3$. Next the child page $p_5$ of page $p_2$ is write-latched and the call *find-next-page*$(p_2, p_5, > 7, p_3, q')$ is performed, returning with $q' = p_6$. Thus, the call *find-page-for-update*$(p, p', 7, \text{true})$ returns with page $p = p_5$ write-latched and page $p' = p_6$ read-latched, and the call *find-page-for-delete*$(p, p', 7, y)$ returns with $y = 9$. The next key $y$ can now be locked and the tuple with key 7 be deleted from page $p = p_5$ (Fig. 7.11b).

In a more complicated situation, the leaf page containing the tuple to be deleted is about to underflow. This happens with the delete action $D[3, v']$. In this case, the leaf page, $p_4$, containing the tuple has only the minimum number of tuples, besides that the next key resides in the next page. This situation will be considered in detail in Example 8.8.                                                                                     □

**Lemma 7.10** *The latching protocol applied in the B-tree traversals for delete actions is deadlock-free. If the leaf page containing tuple $(x, v)$ does not underflow by the deletion, then in the traversal for delete action $D[x, v]$, at most four read latches or one write latch and three read latches are ever held simultaneously. If the X lock on the key next to $x$ is granted immediately as a result of the conditional lock call in the procedure delete (Algorithm 6.8) and no updates by other transactions or structure modifications on the B-tree occur during the deletion and the leaf page containing the tuple $(x, v)$ does not underflow by the deletion and the saved path has its initial value, then in the worst case two write-latchings and $3h - 2$ read-latchings on B-tree pages in all are performed, where $h$ is the height of the B-tree. In the best case, only one latching (write-latching) is performed.*

*Proof* Analogous to that of Lemma 7.8.                                    □


## 7.6   Traversals for Undo Actions

In Sect. 4.3, the call *find-page-for-undo-insert*$(q, x)$ is used to locate and write-latch the target data page in implementing logically the undo action $I^{-1}[x, v]$ in the procedure *undo-insert*$(n, T, p, x, n')$ (Algorithm 4.7), and the call *find-page-for-undo-delete*$(q, x, v)$ is used to locate and write-latch the target data page in implementing logically the undo action $D^{-1}[x, v]$ in the procedure *undo-delete*$(n, T, p, x, v, n')$ (Algorithm 4.9). Both calls are assumed to perform any structure modifications that are needed to accommodate the undo action.

As the undo actions are performed under the protection of the commit-duration X locks obtained for the corresponding forward-rolling actions, no next keys need be determined. Thus, *find-page-for-undo-insert*$(p, x)$ and *find-page-for-undo-delete*$(p, x, v)$ are just *find-page-for-delete*$(p, p', x, y)$ and *find-page-for-insert*$(p, p', x, v, y)$, respectively, simplified so that the next page $p'$ and next key $y$ are not determined. Accordingly, the call *find-page-for-update*$(p, p', x, both)$ is only used with *both* = false.

*Example 7.11* Assume that page $p_8$ of the B-tree of Fig. 7.1 was created in a split of page $p_7$ in the following way. Page $p_7$ contained tuples with keys 17, 18, 19, 21, and 22 at the time when transaction $T_1$ inserted into it a tuple with key 20, thus making the page full. Then transaction $T_2$ inserted a tuple with key 24, causing a page split in which the tuples with keys 20, 21, and 22 were moved to the new sibling page $p_8$, into which the tuple with key 24 went. Then $T_2$ deleted the tuple with key 22 and committed. The pages $p_7$ and $p_8$ thus have the contents as shown in Fig. 7.1.

Now $T_1$ is still active and wants to abort and roll back. In undoing the insertion of the tuple with key 20, the attempt at physical undo fails because the tuple no longer resides in the page, $p_7$, mentioned in the log record for the insertion. Thus, a logical undo must be performed. To that end, page $p_7$ is unlatched, and the call *find-page-for-undo-insert*$(p, 20)$ is performed, returning $p = p_8$, which can accommodate the undoing.

The entire history of the two transactions is

$$T_1: \quad BI[20] \qquad\qquad\qquad AI^{-1}[20]C$$
$$T_2: \qquad\qquad BI[24]D[22]C$$

where the action $I[20]$ is performed on page $p_7$ and the actions $I[24]$, $D[22]$ and $I^{-1}[20]$ on page $p_8$.                                    □

As the procedure calls *find-page-for-undo-insert*$(p, x)$ and *find-page-for-undo-delete*$(p, x, v)$ are only used in logical undo actions, that is, to perform an undo action in the case in which an attempt at a physical undo fails, it is likely that the leaf-level page on the saved path is not a correct page to start the top-down traversal. This was the case in the example above. Thus, we suggest that in this case, the leaf-level page on the saved path is ignored in determining the starting page of the top-down traversal in the procedure *find-page-for-update*.

**Lemma 7.12** *The latching protocol applied in the B-tree traversals for undo-insert and undo-delete actions is deadlock-free. If no structure modifications are needed, then in a traversal for an undo-insert or undo-delete action, at most two read latches or one write latch and one read latch are ever held simultaneously. Even if updates by other transactions and structure modifications on the B-tree occur during the undoing, then in the worst case, assuming that the saved path reflects the current height of the B-tree, one write-latching and $2h - 2$ read-latchings on B-tree pages in all are performed, where $h$ is the height of the B-tree. In the best case, only one read-latching and one write-latching are performed.*

*Proof* Follows from the proofs of Lemmas 7.8 and 7.10, observing that the *find-page-for-update*$(p, p', x, both)$ is only called with $both = $ false. Climbing the saved path from level 2 up to the root and traversing the path again down to the leaf page means $2h - 2$ read-latchings and one write-latching. The best case of one read latch and one write latch in all occurs when the level-two page on the saved path is the lowest-level non-leaf page that covers the key in question.                                   □

# Problems

**7.1** Consider executing the transaction

$$BR[x_1, > 10, v_1] R[x_2, > x_1, v_2] R[x_3, > x_2, v_3] R[x_4, > x_3, v_4] C$$

on the database indexed by the B-tree of Fig. 7.1. Assuming that the saved path has its initial value set by the call *initialize-saved-path*() and that no other transactions are in progress simultaneously, what page latchings and unlatchings of B-tree pages occur during the execution of the transaction?

**7.2** Consider executing the transaction

$$BI[22] I [23] D [24] C$$

on the database indexed by the B-tree of Fig. 7.1. Assuming that the saved path has its initial value set by the call *initialize-saved-path*() and that no other transactions are in progress simultaneously, what page latchings and unlatchings of B-tree pages occur during the execution of the transaction?

**7.3** Consider executing the transaction

$$BD[9] D[10] D[20] A D^{-1}[20] D^{-1}[10] D^{-1}[9] C$$

on the database indexed by the B-tree of Fig. 7.1. Assuming that the saved path has its initial value set by the call *initialize-saved-path*() and that no other transactions are in progress simultaneously, what page latchings and unlatchings of B-tree pages occur during the execution of the transaction?

**7.4** We add to our key-range transaction model a read action $R[y, z, V]$ that reads from the logical database $r$ all tuples whose keys are in the range $[y, z]$. More specifically, given keys $y$ and $z$ with $y \leq z$, the action $R[y, z, V]$ returns the set

$$V = \{(x, v) \mid (x, v) \in r, \ y \leq x \leq z\}.$$

Outline an implementation for this action, keeping the number of page latchings as small as possible.

**7.5** In the B-tree structure considered in this chapter, B-tree pages do not carry their high keys, and leaf pages do not carry their low keys, that is, these keys are not stored explicitly in the page and hence cannot be deduced by looking only into the page contents. Assume that a B-tree page is latched, its PAGE-LSN is saved, after which the page is unlatched. Later the page is latched again, and its PAGE-LSN is inspected. If it is found to be equal to the saved value, we can be sure that the page contents have not changed in the interim. Moreover, we can be sure that the key range covered by the page remains unchanged, that is, the low and high keys of the page are still the same. Explain why. Point out where this fact is made use of in the algorithms of this chapter.

**7.6** Many presentations of the B-tree index structure use the minor space optimization that the key in the index record of the eldest child is omitted, meaning that in the search algorithm the missing key is regarded as the least possible key ($-\infty$). Work out this modification for the algorithms in this chapter.

**7.7** Consider a modification of the B-tree structure in which the high key of every B-tree page is stored explicitly in the page. How can B-tree traversals gain from this modification? Work out this modification for the algorithms in this chapter. What further advantages can be gained if each leaf page also carries its low key?

## Bibliographical Notes

The B-tree was already used as an index structure in early database management systems such as System R [Astrahan et al. 1976]. The original B-tree index structure was introduced by Bayer and McCreight [1972]. The B$^+$-tree version considered in this book (and called simply the B-tree) was introduced by Knuth [1973]. Different B-tree variants and techniques are surveyed by Comer [1979] and Srinivasan and Carey [1993] and, more recently, by Graefe [2010, 2011].

Controlling concurrent access to a B-tree or one of its variants has been studied by several authors, including Samadi [1976], Bayer and Schkolnick [1977], Lehman and Yao [1981], Kwong and Wood [1982], Mond and Raz [1985], and Sagiv [1986]. Lock-coupling—the analog of latch-coupling but short-duration page locks substituted for latches—is used to traverse the B-tree in the algorithms by Samadi [1976], Bayer and Schkolnick [1977], Kwong and Wood [1982], and Mond and Raz [1985], while in algorithms designed for a variant of the B-tree called the B-link tree (to be briefly discussed in Sect. 8.7), lock-coupling is avoided at the expense of relaxing the balance conditions of the B-tree [Lehman and Yao, 1981, Sagiv, 1986]. In these early studies the focus was on synchronizing B-tree traversals for single read, insert and delete actions, and structure modifications triggered by single actions rather than ensuring isolation across entire multi-action transactions. Nor was transaction rollback or recovery from failures discussed in these studies.

In the algorithms of Samadi [1976] and Bayer and Schkolnick [1977], all the pages along an insertion or a deletion path that need to be modified are kept exclusively locked during the modification. To decrease the number of exclusive locks to be acquired when lock-coupling from the root down to the target leaf page, Bayer and Schkolnick [1977] suggest that the search for the target leaf page be started with an optimistic traversal that locks non-leaf pages in shared mode, and if it turns out that structure modifications are needed, the locks are released, and a new, pessimistic traversal is performed with exclusive locks.

The first industrial-strength algorithms for recoverable and concurrent B-trees were published by Mohan [1990a] (ARIES/KVL), Mohan and Levine [1992] (ARIES/IM), Lomet and Salzberg [1992, 1997], Gray and Reuter [1993], and Mohan [1996a] (ARIES/KVL and ARIES/IM revisited). These studies show in detail how a B-tree is traversed, pages are latched, operations are logged, and keys are locked when ensuring repeatable-read-level isolation for transactions in an environment that allows a page to contain uncommitted updates by several active transactions at the same time. The idea of maintaining a saved path so as to avoid repeated traversals is implicit in the algorithms by Mohan [1990a, 1996a] and Mohan and Levine [1992] and is discussed explicitly by Lomet and Salzberg [1992, 1997] and by Gray and Reuter [1993].

For accelerating key-range searches and key-range locking, many B-tree traversal algorithms (including those for ARIES/KVL and ARIES/IM) assume that the leaf pages are sideways linked. In the algorithms in this book, we have deliberately omitted that feature. Graefe [2004, 2011] and Graefe et al. [2012] advocate B-trees in which no page has more than one incoming link and in which each page stores explicitly its low and high keys. The motivation for this are the write-optimized B-tree of Graefe [2004, 2011] and the Foster B-tree of Graefe et al. [2012] designed to enable large sequential writes and efficient defragmentation, and wear leveling on flash devices, issues to be discussed briefly in Chap. 10.

# Chapter 8
# B-Tree Structure Modifications

A B-tree structure modification is an update operation that changes the tree structure of the B-tree, so that at least one index record (a parent-to-child link) is inserted, deleted, or updated. The structure modifications are encompassed by the following five types of primitive modifications: page split, page merge, records redistribute, tree-height increase, and tree-height decrease. Each of these primitive modifications modifies three B-tree pages, namely, a parent page and two child pages.

Structure modifications are triggered by insert or delete actions on leaf pages that cannot accommodate the action: an attempt to insert into a leaf page that has no room for the tuple to be inserted or an attempt to delete from a leaf page that would underflow by the deletion. In such cases, to make the insertion possible, a sequence of one or more page splits, possibly preceded by a tree-height increase, is needed, and to make the deletion possible, a sequence of one or more page merges or records redistributes, possibly preceded by a tree-height decrease, is needed.

In this chapter we show how B-tree structure modifications are managed in the ARIES-based transaction-processing environment developed in the previous chapters, using the traversal algorithms presented in Chap. 7. Following the principle of redo-only structure modifications outlined in Sects. 3.5 and 4.11, we give algorithms for the five primitive structure modifications. We also show how sequences of these modifications, when performed in a top-down fashion, retain the B-tree in a consistent and balanced state during normal transaction processing with any number of concurrent forward-rolling and backward-rolling transactions and that in the event of a process failure or a system crash the B-tree is brought back to a consistent and balanced state in the redo pass of ARIES recovery.

© Springer International Publishing Switzerland 2014
S. Sippu, E. Soisalon-Soininen, *Transaction Processing*, Data-Centric Systems and Applications, DOI 10.1007/978-3-319-12292-2_8

## 8.1  Top-Down Redo-Only Modifications

In Sects. 7.4 and 7.5, when discussing B-tree traversals for insertions and deletions, we presented the procedures *find-page-for-insert*$(p, p', x, v, y)$ and *find-page-for-delete*$(p, p', x, y)$ (Algorithms 7.8 and 7.9) that locate the leaf page $p$ covering key $x$ and the leaf page $p'$ containing the next key $y$ and also arrange that page $p$ can accommodate the insertion or deletion of the tuple with key $x$.

In the case of an insertion, if it is found that the leaf page covering key $x$ does not have room for the tuple $(x, v)$, the calls

*start-for-page-splits*$(p, x, v)$
*top-down-page-splits*$(p, x, v)$

are performed in order to make room for the tuple. The former call returns the page-id $p$ of the highest-level page on the root-to-leaf path to the leaf page covering $x$ that needs modification (page split or tree-height increase). The latter call then performs the splits, in top-down order, starting at the child of page $p$ on the path.

*Example 8.1*  In the case of Fig. 8.1a, the call *start-for-page-splits*$(p, x, v)$ returns with $p = p_2$, because pages $p_3$ and $p_4$ have to be split. Figure 8.1b shows the situation after calling *top-down-page-splits*$(p_2, x, v)$ and inserting the tuple into page $p_4$. First, page $p_3$ was split by allocating a new page $p_3'$, moving the upper half of the records in $p_3$ to $p_3'$ and linking $p_3'$ as a child of $p_2$. Second, page $p_4$ was split by allocating a new page $p_4'$, moving the upper half of the records in $p_4$ to $p_4'$ and linking $p_4'$ as a child of $p_3$. Third, the tuple $(x, v)$ was inserted into page $p_4$ that now has room for the tuple.                                                    □

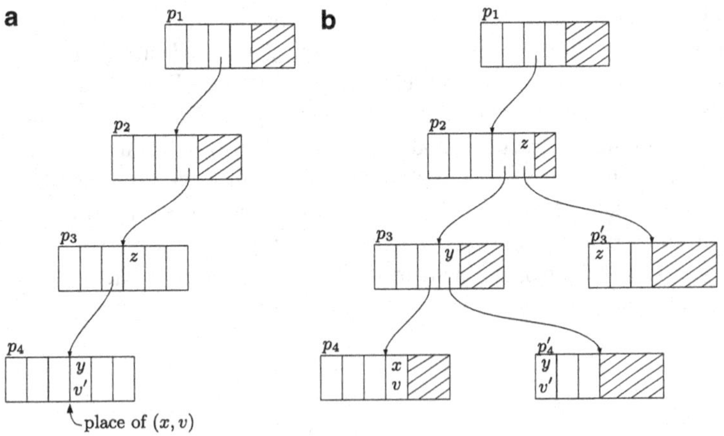

**Fig. 8.1**  Page splits are needed to accommodate the insertion $I[x, v]$ because the tuple $(x, v)$ does not fit into the leaf page that covers key $x$. (**a**) Before insertion (**b**) After insertion

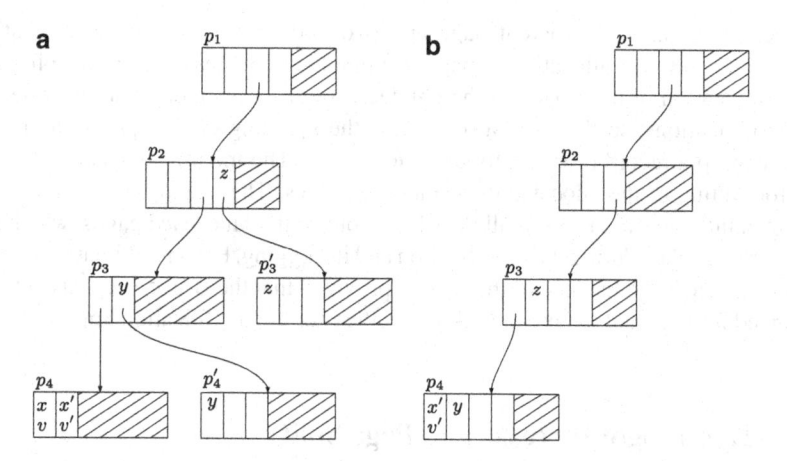

**Fig. 8.2** Page merges are needed to accommodate the deletion $D[x, v]$ because the page containing the tuple $(x, v)$ to be deleted is about to underflow. (**a**) Before deletion (**b**) After deletion

Similarly, in the case of a deletion, if it is found that the leaf page containing the tuple $(x, v)$ to be deleted would underflow by the deletion, the calls

*start-for-page-merges*$(p, x)$
*top-down-page-merges*$(p, x)$

are performed in order to make the leaf page not about to underflow. The former call returns the page-id $p$ of the highest-level page on the root-to-leaf path to the leaf page covering $x$ that needs modification (page merge, records redistribute, or tree-height decrease). The latter call then performs the merges, in top-down order, starting at the child of page $p$ on the path.

*Example 8.2* In the case of Fig. 8.2a, the call *start-for-page-merges*$(p, x)$ returns with $p = p_2$, because pages $p_3$ and $p_4$ have to be merged with their sibling pages $p_3'$ and $p_4'$, respectively. Figure 8.2b shows the situation after calling *top-down-page-merges*$(p_2, x, v)$ and deleting the tuple $(x, v)$ from page $p_4$. First, pages $p_3$ and $p_3'$ were merged by moving the records in $p_3'$ to $p_3$, detaching $p_3'$ as a child $p_2$, and deallocating $p_3'$. Second, pages $p_4$ and $p_4'$ were merged by moving the tuples in $p_4'$ to $p_4$, detaching $p_4'$ as a child of $p_3$, and deallocating $p_4'$. Third, the tuple $(x, v)$ was deleted from page $p_4$ that now does not underflow by the deletion. ☐

The page splits and merges are thus performed in a top-down order, each retaining the consistency and balance of the B-tree. Each split or merge can thus be made a redo-only atomic unit of work, logging it with a single redo-only log record, following the principle discussed in Sects. 2.8, 3.5, and 4.11. In the case of Example 8.1, two redo-only page-split log records are generated, and in the case of Example 8.2, two redo-only page-merge log records are generated.

Each of the five B-tree structure modifications to be defined (tree-height increase, page split, tree-height decrease, page merge, and records redistribute) modifies three

B-tree pages, namely, a parent page and two child pages. In the basic solution presented below, the allocation of pages (in tree-height increase and page split) and the deallocation of pages (in tree-height decrease and page merge) are included in the modifications, so that they also involve the updating of the space-map pages that keep up a record of currently allocated pages. The modifications also include the formatting of the allocated or deallocated pages. The LSN of the log record is stamped into the PAGE-LSNs of all the (three, four or five) modified pages, which are kept write-latched during the modification and its logging. Each modification retains the consistency and balance of the B-tree, provided that the B-tree is consistent and balanced initially and the precondition specified for the modification holds.

## 8.2  Tree-Height Increase and Page Split

A *tree-height increase* may be needed when the root page $p$ of a B-tree is full, so that it has no room for a record of maximum length. Then two pages, $q$ and $q'$, are allocated and formatted as B-tree pages of the height of $p$, the records in page $p$ are distributed between $q$ and $q'$ such that the records with keys greater than or equal to some key $x'$ go to page $q'$, the height of $p$ is incremented, and the index records $(-\infty, q)$ and $(x', q')$ are inserted to page $p$ thus making $q$ and $q'$ the only children of $p$, as illustrated in Fig. 8.3.

The tree-height-increase modification is performed by the procedure call *tree-height-increase*$(p, q, x', q')$, where the page-id $p$ is given as an input parameter (Algorithm 8.1). The precondition for this modification is that the root page $p$ is write-latched and contains enough records so that when these are distributed equally between the two children, neither of them will underflow. The modification is logged with the redo-only log record

$$n : \langle tree\text{-}height\text{-}increase, p, q, x', q', s, V, V', h \rangle, \qquad (8.1)$$

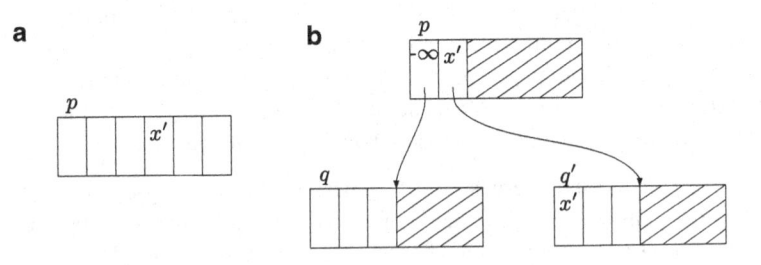

**Fig. 8.3** Increasing the height of the B-tree. The records in the root page $p$ are evenly distributed at key $x'$ between two new pages $q$ and $q'$, which become the only children of $p$. (**a**) Root before height increase (**b**) After height increase

where $s$ is the page-id of the space-map page that records the allocation of $q$ and $q'$, $V$ is the set of records moved to $q$, $V'$ is the set of records moved to $q'$, and $h$ is the height of pages $q$ and $q'$. The LSN $n$ of the log record is stamped in the PAGE-LSN fields of all the affected pages: $p$, $q$, $q'$, and $s$. At the end, page $s$ is unlatched, but pages $p$, $q$, and $q'$ remain write-latched.

---

**Algorithm 8.1** Procedure *tree-height-increase*$(p, q, x', q')$

---

$s \leftarrow$ the page-id of some space-map page used to allocate pages for the B-tree
*fix-and-write-latch*$(s)$
$q \leftarrow$ the page-id of some page designated as free in $s$
mark $q$ as allocated in $s$
$q' \leftarrow$ the page-id of some page designated as free in $s$
mark $q'$ as allocated in $s$
*fix-and-write-latch*$(q)$
*fix-and-write-latch*$(q')$
format $q$ and $q'$ as empty B-tree pages
$height(q) \leftarrow height(q') \leftarrow height(p)$
move the lower half $V$ of the records in page $p$ to page $q$
move the upper half $V'$ of the records in page $p$ to page $q'$
$x' \leftarrow$ the least key of the records in page $q'$
insert the index records $(-\infty, q)$ and $(x', q')$ into page $p$
$height(p) \leftarrow height(p) + 1$
$log(n, \langle tree\text{-}height\text{-}increase, p, q, x', q', s, V, V', h\rangle)$, where $h = height(q)$
PAGE-LSN$(p) \leftarrow$ PAGE-LSN$(q) \leftarrow$ PAGE-LSN$(q') \leftarrow$ PAGE-LSN$(s) \leftarrow n$
*unlatch-and-unfix*$(s)$

---

In the *page split* of a full child page $q$ of non-full parent page $p$, a page $q'$ is allocated and write-latched and formatted as an empty B-tree page of the same height as $q$, records from the upper half of page $q$ starting from some key $x'$ are moved to page $q'$, and the index record $(x', q')$ is inserted into the parent page $p$, thus making $q'$ the next younger sibling of $q$ (Algorithm 8.2), as illustrated in Fig. 8.4.

The page-split modification is performed by the procedure call *page-split*$(p, q, x', q')$, where the page-ids $p$ and $q$ are given as input parameters

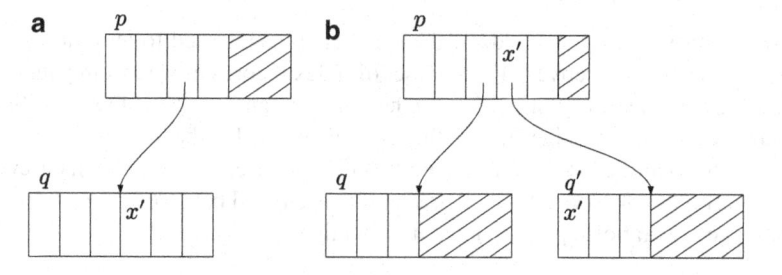

**Fig. 8.4** Splitting a full page. The records in page $q$ are evenly distributed at key $x'$ between $q$ and a new sibling page $q'$. (**a**) Before split (**b**) After split

(Algorithm 8.2). The precondition for this modification is that the parent page $p$ and its child page $q$ are write-latched, page $p$ has room for an index record with a maximum-length key, and page $q$ contains records enough so that distributing them equally between two pages makes neither of them underflow. The modification is logged with the redo-only log record

$$n : \langle page\text{-}split, p, q, x', q', s', V', h\rangle, \tag{8.2}$$

where $s'$ is the page-id of the space-map page that records the allocation of page $q'$, $V'$ is the set of records that were moved from page $q$ to page $q'$, and $h$ is the height of pages $q$ and $q'$. The LSN $n$ of the log record is stamped in the PAGE-LSN fields of pages $p, q, q'$ and $s'$. At the end, page $s'$ is unlatched, but pages $p, q$, and $q'$ remain write-latched.

---

**Algorithm 8.2** Procedure *page-split*$(p, q, x', q')$

$s' \leftarrow$ the page-id of some space-map page used to allocate pages for the B-tree
*fix-and-write-latch*$(s')$
$q' \leftarrow$ the page-id of some page designated as free in $s'$
mark $q'$ as allocated in $s'$
*fix-and-write-latch*$(q')$
format $q'$ as an empty B-tree page
$height(q') \leftarrow height(q)$
move the upper half $V'$ of the records in page $q$ to page $q'$
$x' \leftarrow$ the least key of the records in page $q'$
insert the index record $(x', q')$ into the parent page $p$
$log(n, \langle page\text{-}split, p, q, x', q', s', V', h\rangle)$, where $h = height(q')$
PAGE-LSN$(p) \leftarrow$ PAGE-LSN$(q) \leftarrow$ PAGE-LSN$(q') \leftarrow$ PAGE-LSN$(s') \leftarrow n$
*unlatch-and-unfix*$(s)$

---

*Example 8.3* In the case of Example 8.1 (Fig. 8.1), the following log records are written:

$n_1 : \langle page\text{-}split, p_2, p_3, z, p'_3, s'_3, V'_3, 2\rangle,$
$n_2 : \langle page\text{-}split, p_3, p_4, y, p'_4, s'_4, V'_4, 1\rangle,$
$n_3 : \langle T, I, p_4, x, v, n'\rangle,$

where $s'_3$ and $s'_4$ are the page-ids of the space-map pages used to allocate the new pages $p'_3$ and $p'_4$, respectively $V'_3$ is the set of index records moved from page $p_3$ to page $p'_3$; $z$ is the least key in $p'_3$; 2 is the height of $p_3$ and $p'_3$; $V'_4$ is the set of tuples moved from page $p_4$ to page $p'_4$; $y$ is the least key in $p'_4$; 1 is the height of $p_4$ and $p'_4$, and $T$ is the identifier of the transaction that did the insertion and $n'$ is its previous UNDO-NEXT-LSN. Observe that the B-tree is consistent and balanced after each of the three logged operations, provided that it is so initially.                                      □

## 8.3   Tree-Height Decrease, Page Merge, and Records Redistribute

In the *tree-height decrease* of a B-tree rooted at page $p$ with two child pages, $q$ and $q'$, the index records pointing to these child pages are deleted from the parent page $p$, the records from pages $q$ and $q'$ are moved to page $p$, the height of $p$ is decremented, and the pages $q$ and $q'$ are deallocated (Algorithm 8.3), as illustrated in Fig. 8.5.

The tree-height-decrease modification is performed by the procedure call *tree-height-decrease*$(p, q, x', q')$, where the page-ids $p$, $q$, and $q'$ are given as input parameters, and the least key in page $q'$ is returned in the output parameter $x'$ (Algorithm 8.3). The precondition for this modification is that the root page $p$ has exactly two child pages $q$ and $q'$, all three pages are write-latched, and the records in $q$ and $q'$ all fit into a single page. The modification is logged with the redo-only log record

$$n : \langle tree\text{-}height\text{-}decrease, p, q, x', q', s, s', V \rangle, \tag{8.3}$$

where $s$ and $s'$ are the page-ids of the space-map pages that record the allocation of $q$ and $q'$, and $V$ is the set of records moved to $p$. The LSN $n$ of the log record is stamped in the PAGE-LSN fields of pages $p$, $q$, $q'$, $s$, and $s'$. At the end, pages $s$, $s'$, $q$ and $q'$ are unlatched, but page $p$ remains write-latched.

In the *page merge* of sibling pages $q$ and $q'$, one of which is about to underflow, the index record pointing to $q'$ is deleted from the parent page $p$, the records in page $q'$ are moved to page $q$, and page $q'$ is deallocated (Algorithm 8.4), as illustrated in Fig. 8.6.

---

**Algorithm 8.3** Procedure *tree-height-decrease*$(p, q, x', q')$

---

delete the index records $(-\infty, q)$ and $(x', q')$ from the root page $p$
move the records in the child pages $q$ and $q'$ to the parent page $p$
$V \leftarrow$ the set of records moved
format $q$ and $q'$ as free pages
$height(p) \leftarrow height(p) - 1$
$s \leftarrow$ the page-id of the space-map page that records $q$
$s' \leftarrow$ the page-id of the space-map page that records $q'$
*fix-and-write-latch*$(s, s')$
mark $q$ as free in $s$ and $q'$ as free in $s'$
$log(n, \langle tree\text{-}height\text{-}decrease, p, q, x', q', s, V \rangle)$
PAGE-LSN$(p) \leftarrow$ PAGE-LSN$(q) \leftarrow$ PAGE-LSN$(q') \leftarrow n$
PAGE-LSN$(s) \leftarrow$ PAGE-LSN$(s') \leftarrow n$
*unlatch-and-unfix*$(s, s')$
*unlatch-and-unfix*$(q)$
*unlatch-and-unfix*$(q')$

---

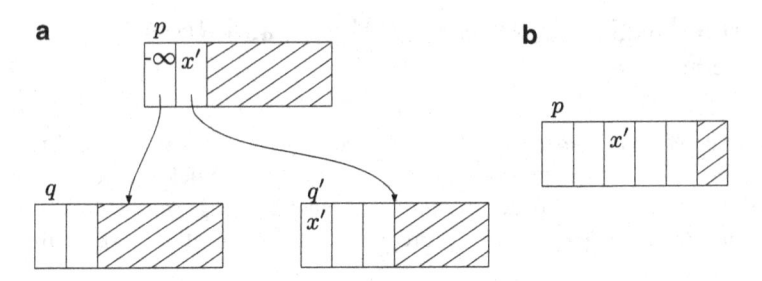

**Fig. 8.5** Decreasing the height of the B-tree. The pages $q$ and $q'$ are detached from the B-tree, their records are moved to the root page $p$, and $q$ and $q'$ are deallocated. (**a**) Before height decrease (**b**) After height decrease

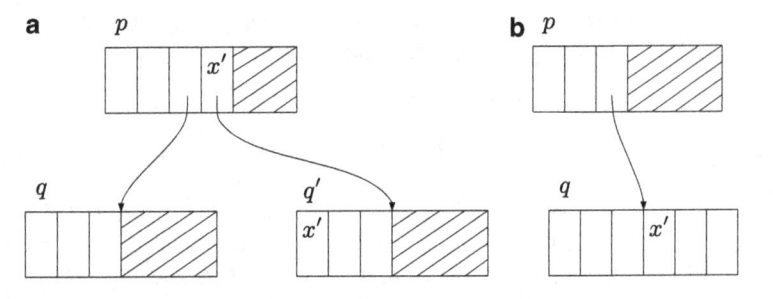

**Fig. 8.6** Merging a page. The records in page $q'$ are moved to its sibling page $q$ and page $q'$ is detached from the B-tree and deallocated. (**a**) Before merge (**b**) After merge

The page-merge modification is performed by the procedure call *page-merge*$(p, q, x', q')$, where the page-ids $p$, $q$, and $q'$ are given as input parameters and the least key in page $q'$ is returned in $x'$ (Algorithm 8.4). The precondition for this modification is that the parent page $p$ and its child pages $q$ and $q'$ are write-latched, $q'$ is the next younger sibling of $q$, the records in $q$ and $q'$ all fit into a single page, and the parent page $p$ is not about to underflow. The modification is logged with the redo-only log record

$$n : \langle page\text{-}merge, p, q, x', q', s', V' \rangle, \tag{8.4}$$

where $s'$ is the page-id of the space-map page that records the allocation of page $q'$ and $V'$ is the set of records that were moved from page $q'$ to page $q$. The LSN $n$ of the log record is stamped in the PAGE-LSN fields of pages $p$, $q$, $q'$, and $s'$. At the end, pages $s'$ and $q'$ are unlatched, but pages $p$ and $q$ remain write-latched.

---

**Algorithm 8.4** Procedure *page-merge*$(p, q, x', q')$

---

delete the index record $(x', q')$ from the parent page $p$
move the records in the younger child page $q'$ to the elder child page $q$
$V' \leftarrow$ the set of records moved
format $q'$ as a free page
$s' \leftarrow$ the page-id of the space-map page that records $q'$
*fix-and-write-latch*$(s')$
mark $q'$ as free in $s'$
$log(n, \langle page\text{-}merge, p, q, x', q', s, V' \rangle)$
PAGE-LSN$(p) \leftarrow$ PAGE-LSN$(q) \leftarrow$ PAGE-LSN$(q') \leftarrow$ PAGE-LSN$(s) \leftarrow n$
*unlatch-and-unfix*$(s')$
*unlatch-and-unfix*$(q')$

---

*Example 8.4* In the case of Example 8.2 (Fig. 8.2), the following log records are written:

$n_1 : \langle page\text{-}merge, p_2, p_3, z, p'_3, s'_3, V'_3 \rangle,$
$n_2 : \langle page\text{-}merge, p_3, p_4, y, p'_4, s'_4, V'_4 \rangle,$
$n_3 : \langle T, D, p_4, x, v, n' \rangle,$

where, respectively, $s'_3$ and $s'_4$ are the page-ids of the space-map pages for the deallocated pages $p'_3$ and $p'_4$, $V'_3$ and $V'_4$ are the sets of index records moved from page $p'_3$ to page $p_3$ and from page $p'_4$ to page $p_4$, and $z$ and $y$ are the least keys in pages $p'_3$ and $p'_4$. In the last log record, $T$ is the identifier of the transaction that did the deletion, and $n'$ is its previous UNDO-NEXT-LSN. Again observe that the B-tree is consistent and balanced after each of the three logged operations, provided that it is so initially.                                                                                     □

In the *records redistribute* of sibling pages $q$ and $q'$, which cannot be merged even though one of them is about to underflow, some records from the upper half of page $q$ are moved to page $q'$ (if $q$ contains more records than $q'$) or some records from the lower half of page $q'$ are moved to page $q$ (otherwise), and the key in the index record in page $p$ pointing to $q'$ is adjusted accordingly (Algorithm 8.5), as illustrated in Fig. 8.7.

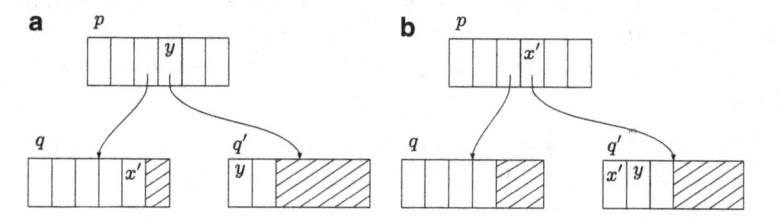

**Fig. 8.7** Redistributing the records in two sibling pages. Records from sibling page $q$ are moved to page $q'$ so as to make $q'$ not about to underflow. (**a**) Before redistribution (**b**) After redistribution

The records-redistribute modification is performed by the procedure call *records-redistribute*$(p, q, x', q')$, where the page-ids $p$, $q$, and $q'$ are given as input parameters (Algorithm 8.5). The precondition for this modification is that the parent page $p$ and its child pages $q$ and $q'$ are write-latched, $q'$ is the next younger sibling of $q$, and there are enough records in $q$ and $q'$ so that they can be distributed between $q$ and $q'$ such that neither page will be about to underflow. The modification is logged with the redo-only log record

$$n : \langle \textit{records-redistribute}, p, q, x', q', V, d \rangle, \tag{8.5}$$

where $V$ is the set of records that were moved, $x'$ is the least key in page $q'$ after moving the records, and $d$ is a bit that indicates the direction ($q \rightarrow q'$ or $q \leftarrow q'$) into which the records were moved. The LSN $n$ of the log record is stamped in the PAGE-LSN fields of pages $p$, $q$, and $q'$. The pages $p$, $q$, and $q'$ all remain write-latched.

---

**Algorithm 8.5** Procedure *records-redistribute*$(p, q, x', q')$

---
**if** page $q$ contains more records than its younger sibling page $q'$ **then**
    move some records from the upper half of page $q$ to page $q'$
**else**
    move some records from the lower half of page $q'$ to page $q$
**end if**
$V \leftarrow$ the set of records moved
$d \leftarrow$ the direction ($q \rightarrow q'$ or $q \leftarrow q'$) into which the records were moved
$x' \leftarrow$ the least key of the records in page $q'$
change the key to $x'$ in the index record for child page $q'$ in the parent page $p$
$log(n, \langle \textit{records-redistribute}, p, q, x', q', V, d \rangle)$
PAGE-LSN$(p) \leftarrow$ PAGE-LSN$(q) \leftarrow$ PAGE-LSN$(q') \leftarrow n$

---

**Lemma 8.5** *Assume that a set of concurrent tree-height increases, page splits, tree-height decreases, page merges and records redistributes (Algorithms 8.1–8.5), inserts into and deletes from leaf pages (Algorithms 3.1 and 3.2), and undoings of inserts and deletes (Algorithms 4.8 and 4.10) by different process threads are executed. Assume further that such threads are run on an initially consistent and balanced B-tree in such a way that for each operation the pages involved are write-latched and the precondition for the operation is satisfied at the start of the operation and the pages involved are unlatched at the end of the operation. Then the resulting B-tree is consistent and balanced.*

*Proof* From the algorithm for each operation mentioned in the lemma, it is seen that if the precondition for the operation is satisfied at the time, the operation is about to be executed, and then the algorithm retains the B-tree as consistent and balanced, provided that it is so initially. Thus, the same holds for any sequence of such operations that can be run on an initially consistent and balanced B-tree.                                                                                □

**Lemma 8.6** *The latching protocol applied in procedures tree-height-increase, page-split, tree-height-decrease, page-merge, and records-redistribute is deadlock-free.*

*Proof* In Algorithms 8.1–8.5, only write latches are acquired, and while a page is kept latched, no latch is requested on its parent or elder sibling or cousin. We also assume that in Algorithm 8.3, where we have to latch two space-map pages, those pages are latched in some prespecified order.                                      □

## 8.4  Sequences of Redo-Only Modifications

The preconditions stated for the structure modifications defined in the previous sections imply that if more than one page on an insertion path has to be split or more than one page on a deletion path has to be merged or its records redistributed, then those splits or merges or redistributes must be performed in top-down order along the path, and if a tree-height increase or decrease is needed, it must be the first modification to be performed.

As we have already noted, on the insertion path shown in Fig. 8.1a, where pages $p_3$ and $p_4$ are full but page $p_2$ has room for an index record, page $p_3$ must be split first, thus making space for the index record pointing to the new sibling page resulting from the split of child page $p_4$. Similarly, on the deletion path shown in Fig. 8.2a, where pages $p_3$ and $p_4$ are about to underflow but page $p_2$ is not, page $p_3$ and its sibling $p_3'$ are first merged, thus ensuring that page $p_3$ will not underflow when the index record pointing to child page $p_1'$ is deleted in merging $p_1'$ with its sibling $p_1$.

When in the procedure *find-page-for-insert* (Algorithm 7.8) it is found that the page covering the key $x$ of the tuple $(x, v)$ to be inserted has no room for the tuple, the procedure call *start-for-page-splits*$(p, x, v)$ (Algorithm 8.6) is issued so as to determine the starting page for page splits, that is, the highest-level page $p$ on the path down to the page covering $x$ that needs a structure modification (tree-height increase or page split) in order to accommodate the insertion. That page must satisfy the following conditions:

1. PAGE-LSN$(p) = path[i].page-lsn$ and $path[i].low-key \leq x < path[i].high-key$ or $p$ is a B-tree page with $x_1 \leq x \leq x_2$ for the keys $x_1$ and $x_2$ of some records in page $p$.

2. Page $p$, if not the root, has room for $(x, v)$ (if $p$ is a leaf page) or for an index record with a maximum-length key (if $p$ is a non-leaf page).

Again we first use just the information stored in the saved path to determine an approximation for such a page $p$, write-latch it, and check if it satisfies the above conditions; if not, we unlatch the page and probe the page at the next higher level on the saved path.

Once the starting page $p$ has been determined, a sequence of page splits, possibly preceded by a tree-height increase, is performed top-down along the path down to the page that currently covers key $x$. This is done by the procedure call *top-down-page-splits*$(p, x, v)$ (Algorithm 8.7), where $p$ is the page-id of the write-latched starting page. The call determines again the path down to the leaf page covering key $x$, performs the structure modifications needed level-by-level, and saves the path.

---

**Algorithm 8.6** Procedure *start-for-page-splits*$(p, x, v)$

---

$i \leftarrow$ the least index for which either *path*$[i]$.*page-id* is the page-id of the root or *path*$[i]$.*space-left* is sufficient for $(x, v)$ (if $i = 1$) or for an index record with a maximum-length key (if $i > 1$)
$p \leftarrow$ *path*$[i]$.*page-id*
*fix-and-write-latch*$(p)$
**while** page $p$ is not the root **do**
$\quad$ **if** PAGE-LSN$(p) =$ *path*$[i]$.*page-lsn* **and** *path*$[i]$.*low-key* $\leq x <$ *path*$[i]$.*high-key* **or** $p$ is a
$\quad$ B-tree page with $x_1 \leq x \leq x_2$ for the keys $x_1$ and $x_2$ of some records in page $p$ **then**
$\quad\quad$ **if** page $p$ has room for $(x, v)$ (if $p$ is a leaf page) or for an index record with a maximum-
$\quad\quad$ length key (if $p$ is a non-leaf page) **then**
$\quad\quad\quad$ exit the **while** loop
$\quad\quad$ **end if**
$\quad$ **end if**
$\quad$ *unlatch-and-unfix*$(p)$
$\quad$ $i \leftarrow i + 1$
$\quad$ $p \leftarrow$ *path*$[i]$.*page-id*
$\quad$ *fix-and-write-latch*$(p)$
**end while**
**if** page $p$ is the root **then**
$\quad$ *save-root*$(p)$
**end if**

---

---

**Algorithm 8.7** Procedure *top-down-page-splits*$(p, x, v)$

---

**if** page $p$ is the root and has no room for $(x, v)$ (if $p$ is a leaf page) or for an index record with a maximum-length key (if $p$ is a non-leaf page) **then**
    *tree-height-increase*$(p, q, x', q')$
    *save-root*$(p)$
    **if** $x < x'$ **then**
        *unlatch-and-unfix*$(q')$
    **else**
        *unlatch-and-unfix*$(q)$
        $q \leftarrow q'$
    **end if**
    *save-child*$(p, q)$
    *unlatch-and-unfix*$(p)$
    $p \leftarrow q$
**end if**
**while** page $p$ is a non-leaf page **do**
    $q \leftarrow$ the page-id of the child of $p$ that covers key $x$
    *fix-and-write-latch*$(q)$
    **if** page $q$ has no room for $(x, v)$ (if $q$ is a leaf page) or for an index record with a maximum-length key (if $p$ is a non-leaf page) **then**
        *page-split*$(p, q, x', q')$
        **if** $x < x'$ **then**
            *unlatch-and-unfix*$(q')$
        **else**
            *unlatch-and-unfix*$(q)$
            $q \leftarrow q'$
        **end if**
    **end if**
    *save-child*$(p, q)$
    *unlatch-and-unfix*$(p)$
    $p \leftarrow q$
**end while**

---

*Example 8.7* Consider executing the insert action $I[16, v']$ on the database of Fig. 7.1. Assume that the saved path has the initial value. In the procedure call *find-page-for-insert*$(p, p', 16, v', y)$, the optimistic traversal performed by the call *find-page-for-update*$(p, p', 16, \text{false})$ starts at the root page $p_1$ and goes down to the leaf page, $p_6$, that covers key 16, saving the path $p_1 \rightarrow p_3 \rightarrow p_6$ so traversed:

| | page-id | page-lsn | low-key | high-key | max-key | #records | space-left |
|---|---|---|---|---|---|---|---|
| 1: | $p_6$ | ... | 8 | 17 | 15 | 6 | 0 |
| 2: | $p_3$ | ... | 8 | 50 | 40 | 6 | 0 |
| 3: | $p_1$ | ... | $-\infty$ | $\infty$ | 99 | 6 | 0 |

Because page $p_6$ has no room for the tuple $(16, v')$, the write latch on the page is released, and the call *start-for-page-splits*$(p, 16, v')$ is performed. As all the pages up to and including the root page $p_1$ are full, the call returns with $p = p_1$ write-latched.

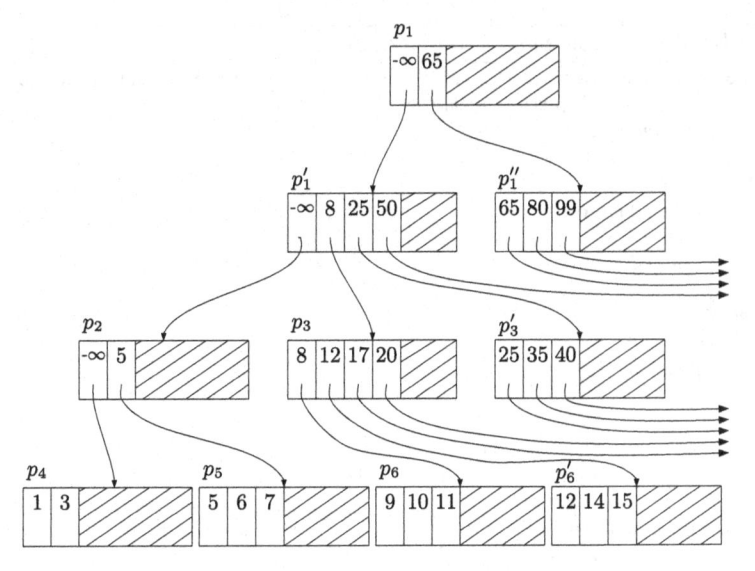

**Fig. 8.8**  The B-tree of Fig. 7.1 after the tree height has been increased and pages $p_3$ and $p_6$ have been split

Next the call *top-down-page-splits*$(p, 16, v')$ with $p = p_1$ is performed, resulting in the B-tree of Fig. 8.8. Here the height of the tree is increased by the call *tree-height-increase*$(p_1, q, x', q')$ that returns with $q = p_1'$, $x' = 65$ and $q' = p_1''$. Page $p_3$ is split by the call *page-split*$(p_1', p_3, x', q')$ that returns with $x' = 25$ and $q' = p_3'$. Page $p_6$ is split by the call *page-split*$(p_3, p_6, x', q')$ that returns with $x' = 12$ and $q' = p_6'$. The following log records are written:

$n_1 :$ ⟨*tree-height-increase*, $p_1, p_1', 65, p_1'', s_1, V_1, V_2, 3$⟩
$n_2 :$ ⟨*page-split*, $p_1', p_3, 25, p_3', s_2, V_3, 2$⟩
$n_3 :$ ⟨*page-split*, $p_3, p_6, 12, p_6', s_3, V_4, 1$⟩

where $p_1'$ and $p_1''$ are the new children of the root page $p_1$; $p_3'$ is the new sibling of page $p_3$; $p_6'$ is the new sibling of page $p_6$; $s_1$, $s_2$, and $s_3$ are the space-map pages that record the allocation of $p_1'$ and $p_1''$, and $p_3'$ and $p_6'$, respectively; and the sets of records moved are:

$V_1 = \{(-\infty, p_2), (8, p_3), (50, \ldots)\}$, from page $p_1$ to page $p_1'$.
$V_2 = \{(65, \ldots), (80, \ldots), (99, \ldots)\}$, from page $p_1$ to page $p_1''$.
$V_3 = \{(25, p_9), (35, \ldots), (40, \ldots)\}$, from page $p_3$ to page $p_3'$.
$V_4 = \{(12, \ldots), (14, \ldots), (15, \ldots)\}$, from page $p_6$ to page $p_6'$.

The call *top-down-page-splits*$(p, 16, v')$ saves the path $p_1 \rightarrow p_1' \rightarrow p_3 \rightarrow p_6'$ and returns with $p = p_6'$.

Page $p = p_6'$ has now room for the tuple $(16, v')$, but it does not contain the key next to 16. Hence, the page is unlatched and the call *find-page-for-insert*$(p, p', 16, \text{true})$ is performed. This call returns with page $p = p_6'$ write-latched and page $p' = p_7$ read-latched, so that the next key, 17, can be X-locked for short duration and the tuple $(16, v')$ can be inserted into $p_6'$. □

Similarly, when in the procedure *find-page-for-delete* (Algorithm 7.9) it is found that the page containing the tuple $(x, v)$ to be deleted is about to underflow, that is, would underflow by the deletion, the procedure call *start-for-page-merges*$(p, x)$ (Algorithm 8.8) is issued so as to determine the starting page for page merges, that is, the highest-level page $p$ on the path down to the page covering $x$ that needs a structure modification (tree-height decrease or page merge or records redistribution) in order to accommodate the deletion. That page must satisfy condition (1) above and the following condition:

3. The number of records in page $p$ is greater than the required minimum number of records in a leaf page (if $p$ is a leaf page) or that in a non-leaf page (if $p$ is a non-leaf page), unless $p$ is the root page.

Once the starting page $p$ has been determined, a sequence of page merges or records redistributes, possibly preceded by a tree-height decrease, is performed top-down along the path down to the page that currently holds the tuple $(x, v)$. This is done by the procedure call *top-down-page-merges*$(p, x)$ (Algorithm 8.9), where $p$ is the page-id of the write-latched starting page. The call determines again the path down to the leaf page containing $(x, v)$, performs the structure modifications needed level-by-level, and saves the path.

---

**Algorithm 8.8** Procedure *start-for-page-merges*$(p, x)$

---

$i \leftarrow$ the least index for which either *path*$[i]$.*page-id* is the page-id of the root or *path*$[i]$.*#records* is greater than the required minimum number of records in a non-root leaf page (if $i = 1$) or in a non-leaf page (if $i > 1$)
$p \leftarrow path[i].page\text{-}id$
*fix-and-write-latch*$(p)$
**while** page $p$ is not the root **do**
   **if** PAGE-LSN$(p) = path[i].page\text{-}lsn$ **and** $path[i].low\text{-}key \leq x < path[i].high\text{-}key$ **or** $p$ is a
   B-tree page with $x_1 \leq x \leq x_2$ for the keys $x_1$ and $x_2$ of some records in page $p$ **then**
      **if** the number of records in page $p$ is greater than the required minimum number of records
      in a non-root leaf page (if $p$ is a leaf page) or that in a non-leaf page (if $p$ is a non-leaf
      page) **then**
         exit the **while** loop
      **end if**
   **end if**
   *unlatch-and-unfix*$(p)$
   $i \leftarrow i + 1$
   $p \leftarrow path[i].page\text{-}id$
   *fix-and-write-latch*$(p)$
**end while**
**if** page $p$ is the root **then**
   *save-root*$(p)$
**end if**

---

---

**Algorithm 8.9** Procedure *top-down-page-merges*$(p, x)$

---

**while** page $p$ is a non-leaf page **do**
   $q \leftarrow$ the page-id of the child of $p$ that covers key $x$
   *fix-and-write-latch*$(q)$
   **if** page $q$ has only the required minimum number of records **then**
      **if** $q$ is the youngest child of $p$ **then**
         $q'' \leftarrow$ the page-id of the next elder sibling of $q$
         *unlatch-and-unfix*$(q)$
         *fix-and-write-latch*$(q'')$
         *fix-and-write-latch*$(q)$
         $q' \leftarrow q$
         $q \leftarrow q''$
      **else**
         $q' \leftarrow$ the page-id of the next younger sibling of $q$
         *fix-and-write-latch*$(q')$
      **end if**
      **if** the records in pages $q$ and $q'$ all fit into a single page **then**
         **if** $p$ is the root page and $q$ and $q'$ are its only children **then**
            *tree-height-decrease*$(p, q, x', q')$
            *save-root*$(p)$
         **else**
            *page-merge*$(p, q, x', q')$
            *save-child*$(p, q)$
            *unlatch-and-unfix*$(p)$
            $p \leftarrow q$
         **end if**
      **else**
         *records-redistribute*$(p, q, x', q')$
         **if** $x < x'$ **then**
            *unlatch-and-unfix*$(q')$
         **else**
            *unlatch-and-unfix*$(q)$
            $q \leftarrow q'$
         **end if**
         *save-child*$(p, q)$
         *unlatch-and-unfix*$(p)$
         $p \leftarrow q$
      **end if**
   **end if**
**end while**

---

*Example 8.8* Consider executing the delete action $D[3, v']$ on the database of Fig. 7.1. Assume that the saved path has the initial value. In the procedure call *find-page-for-delete*$(p, p', 3, y)$, the optimistic traversal performed by the call *find-page-for-update*$(p, p', 3, \text{false})$ starts at the root page $p_1$ and goes down to the leaf page, $p_4$, that covers key 3, saving the path $p_1 \rightarrow p_2 \rightarrow p_4$ so traversed:

| | page-id | page-lsn | low-key | high-key | max-key | #records | space-left |
|---|---|---|---|---|---|---|---|
| 1: | $p_4$ | ... | $-\infty$ | 5 | 3 | 2 | 4 records |
| 2: | $p_2$ | ... | $-\infty$ | 8 | 5 | 2 | 4 records |
| 3: | $p_1$ | ... | $-\infty$ | $\infty$ | 99 | 6 | 0 |

(Fig. 8.9a). Because page $p_4$ would underflow by the deletion of the tuple, the write latch on the page is released, and the call *start-for-page-merges*$(p, 3)$ is performed. As page $p_4$ and its parent page $p_2$ are about to underflow, but the grandparent $p_1$ is not, the call returns with $p = p_1$.

Next the call *top-down-page-merges*$(p, 3)$ with $p = p_1$ is performed. Here the records in page $p_2$ and its sibling page $p_3$ are redistributed by the call *records-redistribute*$(p_1, p_2, x', p_3)$ that returns $x' = 20$, and the page $p_5$ is merged into its sibling page $p_4$ by the call *page-merge*$(p_2, p_4, x', p_5)$ that returns $x' = 5$. The following log records are written:

$n_1 : \langle records\text{-}redistribute, p_1, p_2, 20, p_3, V_1, \leftarrow \rangle$,
$n_2 : \langle page\text{-}merge, p_2, p_4, 5, p_5, V_2 \rangle$,

where the sets of records moved are:

$V_1 = \{(8, p_6), (17, p_7)\}$, from page $p_3$ to page $p_2$ (direction $\leftarrow$),
$V_2 = \{(5, \ldots), (6, \ldots), (7, \ldots)\}$, from page $p_5$ to page $p_4$

(Fig. 8.9b). The call *top-down-page-merges*$(p, 3)$ saves the path $p_1 \rightarrow p_2 \rightarrow p_4$ and returns with $p = p_4$. The key, 5, next to 3 now resides in the same page; so it can be X-locked for commit duration and the tuple with key 3 be deleted from page $p_4$. □

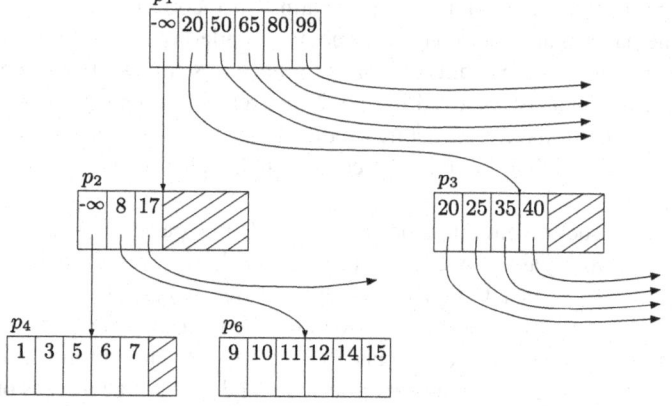

**Fig. 8.9** Delete action $D[3, v']$ on the B-tree of Fig. 7.1 triggers a redistribution of records between pages $p_2$ and $p_3$ and a merge of pages $p_4$ and $p_5$

**Lemma 8.9** *The latching protocol applied in the procedures start-for-page-splits, top-down-page-splits, start-for-page-merges, and top-down-page-merges is deadlock-free.*

*Proof* In Algorithms 8.6 and 8.8, only one latch (a write latch) is ever kept at a time, and in Algorithms 8.7 and 8.9, only write latches are acquired, and while a page is kept latched, no latch is requested on its parent or elder sibling or cousin, by Lemma 8.6.                                                                                     □

**Lemma 8.10** *The procedures start-for-page-splits and start-for-page-merges keep at most one page latched at a time and perform at most h write-latchings in all, where h is the height of the B-tree. The procedures top-down-page-splits and top-down-page-merges retain the logical database indexed by, and the consistency and balance of, an initially consistent and balanced B-tree. At most three B-tree pages and at most two space-map pages are kept latched simultaneously at any time. In the worst case, 3h write-latchings in all are performed in the procedure top-down-page-splits, and 4h write-latchings in all are performed in the procedure top-down-page-merges.*

*Proof* The procedures *start-for-page-splits* and *start-for-page-merges* in the worst case climb along the saved path from the leaf level up to the root, write-latching $h$ pages.

From Algorithm 8.7, we see that in the procedure *top-down-page-splits*, the preconditions for the calls of *tree-height-increase* and *page-split* are satisfied, and from Algorithm 8.9, we see that in the procedure *top-down-page-merges* the preconditions for the calls of *tree-height-decrease*, *page-merge*, and *records-redistribute* are satisfied, provided that the B-tree is consistent and balanced initially. In particular, we note that in the procedure *top-down-page-merges*, the constraints imposed in Sect. 7.1 on the prespecified constants $e$, $d$, $E$, and $D$ ensure that if two sibling pages cannot be merged, then distributing the records of those pages equally between the pages makes both pages not about to underflow. Thus, the consistency and balance of the B-tree at the end of the procedures follows from Lemma 8.5. That the logical database indexed by the B-tree remains unchanged follows from the fact that none of the five structure modifications mentioned above changes the set of tuples in the leaf pages of the B-tree (although tuples are moved from one leaf page to another).

All the five structure modifications keep three B-tree pages write-latched simultaneously. The worst case of five simultaneous write latches occurs in the procedure *tree-height-decrease*: in addition to the three B-tree pages, also two space-map pages in the worst case must be kept write-latched. In the procedure *top-down-page-splits*, one tree-height-increase and $h$ page splits in all, and in the procedure *top-down-page-merges*, one tree-height-decrease and $h - 1$ page merges or records redistributes in all, are performed in the worst case. When the merge path goes along the rightmost side of the tree, performing a merge on the youngest sibling requires that first the youngest sibling is unlatched (so as to avoid deadlocks), then the next elder sibling is latched, and then the youngest sibling is latched again, thus adding the count of latches by $h - 1$.                                                □

**Theorem 8.11** *The time complexity of performing a read action $R[x, \theta z, v]$, an insert action $I[x, v]$, a delete action $D[x, v]$, an undo-insert action $I^{-1}[x, v]$, or an undo-delete action by the procedures read($T, x, \theta z, v$), insert($T, x, v$), delete($T, x, v$), undo-insert($n, T, p, x, n'$), and undo-delete($n, T, p, x, v, n'$), using the B-tree algorithms above, is $O(\log N)$, where $N$ is the total number of tuples in a database indexed by a consistent and balanced B-tree, provided that (1) no update action or structure modification triggered by any transaction other than $T$ is in progress at the same time and that (2) the locks requested for the action are granted immediately without wait.*

*Proof* The theorem follows from Lemmas 7.3, 7.6, 7.8, 7.10, 7.12, and 8.10. Provisions (1) and (2) together exclude the possibility of repeated traversals for page splits or merges or for locating the target page for the action and the keys to be locked.                                                                        □

**Theorem 8.12** *Let $b$ be a consistent and balanced B-tree and let $H$ be any history of forward-rolling, backward-rolling, committed, and rolled-back transactions that can be executed on $b$ using the above algorithms. Then $H$ is possible under the key-range locking protocol, and the execution results in a consistent and balanced B-tree $b'$ with*

$$db(b') = H(db(b)),$$

*that is, the logical database indexed by $b'$ is the database produced by running $H$ on the database indexed by $b$.*

*Proof* That $H$ is possible under the key-range locking protocol follows from the facts that the read, insert, and delete algorithms in Sect. 6.7 follow the rules of the key-range locking protocol and that the traversal algorithms in Chap. 7 correctly locate the data page covering the key to be read, inserted, or deleted and the data page holding the key next to the key to be inserted or deleted. The consistency and balance of the B-tree $b'$ follows from Lemma 8.5. Finally, the fact that the logical database indexed by $b'$ is the database produced by $H$ on the database indexed by the initial B-tree $b$ follows from the correctness of insert and delete algorithms in Sect. 6.7, the undo-insert and undo-delete algorithms in Sect. 4.3, the traversal algorithms in Chap. 7, and the structure-modification algorithms in this chapter.                                                                        □

**Theorem 8.13** *Let b be a consistent and balanced B-tree and let H be any history of forward-rolling, backward-rolling, committed, and rolled-back transactions that can be run on the database db(b) under the key-range locking protocol. Then the history H can be executed in time*

$$O(|H| \log N)$$

*by a set of concurrent process threads, with one thread for each transaction, resulting in a consistent and balanced B-tree b' with*

$$db(b') = H(db(b)),$$

*where |H| denotes the length of H and N is the greatest number of tuples in any database H'(db(b)), where H' is a prefix of H.*

*Proof* In one such execution for the sequence of actions in history $H$, only one action is in execution at a time, so that at any time, only the process thread is active whose action is currently being executed, while the other process threads are not executing any action, and hence are not holding any latches, meaning that all the latches requested are granted immediately. Because history $H$ is assumed to be possible under the key-range locking protocol, then, by definition, all the locks requested conditionally or unconditionally are granted immediately. The complexity bound for the execution follows from Theorem 8.11.                                    □

## 8.5 Redoing B-Tree Structure Modifications

In Sect. 4.11 we defined the concept of "selective redoing" and gave a generic algorithm (Algorithm 4.18) for redoing the effect of a structure modification selectively on any of the pages involved in the modification. Now we can specialize the generic redo algorithm for our five redo-only structure modifications defined above.

The redo algorithm for page splits is given as Algorithm 8.10. Observe here how selective redoing works: to redo the effect of the split onto one of the pages involved, only that page is latched, and the information therein plus the information contained in the log record are used in the redoing. The other four redo algorithms are very similar and are left to the exercises.

---

**Algorithm 8.10** Procedure *redo-page-split*($r$)

---

**if** $r$ is of the form "$n : \langle page\text{-}split, p, q, x', q', s', V', h \rangle$" **then**
  **if** REC-LSN($p$) $\leq n$ **then**
    *fix-and-write-latch*($p$)
    **if** page $p$ is unmodified **then**
      REC-LSN($p$) $\leftarrow$ max{REC-LSN($p$), PAGE-LSN($p$) $+ 1$}
    **end if**
    **if** PAGE-LSN($p$) $< n$ **then**
      insert the index record $(x', q')$ into page $p$
      PAGE-LSN($p$) $\leftarrow n$
    **end if**
    *unlatch-and-unfix*($p$)
  **end if**
  **if** REC-LSN($q$) $\leq n$ **then**
    *fix-and-write-latch*($q$)
    **if** page $q$ is unmodified **then**
      REC-LSN($q$) $\leftarrow$ max{REC-LSN($q$), PAGE-LSN($q$) $+ 1$}
    **end if**
    **if** PAGE-LSN($q$) $< n$ **then**
      delete from page $q$ all records with keys $\geq x'$
      PAGE-LSN($q$) $\leftarrow n$
    **end if**
    *unlatch-and-unfix*($q$)
  **end if**
  **if** REC-LSN($q'$) $\leq n$ **then**
    *fix-and-write-latch*($q'$)
    **if** page $q'$ is unmodified **then**
      REC-LSN($q'$) $\leftarrow$ max{REC-LSN($q'$), PAGE-LSN($q'$) $+ 1$}
    **end if**
    **if** PAGE-LSN($q'$) $< n$ **then**
      format page $q'$ as an empty B-tree page of height $h$
      insert the records in $V'$ into page $q'$
      PAGE-LSN($q'$) $\leftarrow n$
    **end if**
    *unlatch-and-unfix*($q'$)
  **end if**
  **if** REC-LSN($s'$) $\leq n$ **then**
    *fix-and-write-latch*($s'$)
    **if** page $s'$ is unmodified **then**
      REC-LSN($s'$) $\leftarrow$ max{REC-LSN($s'$), PAGE-LSN($s'$) $+ 1$}
    **end if**
    **if** PAGE-LSN($s'$) $< n$ **then**
      mark page $q'$ as allocated in page $s'$
      PAGE-LSN($s'$) $\leftarrow n$
    **end if**
    *unlatch-and-unfix*($s'$)
  **end if**
**end if**

---

With redo-only structure modifications, each log record must carry all the information needed to redo the effects on all the pages involved in the modification, which means that such a log record tends to be large in size. The log record for a page split occupies about half a page, that for a records redistribute at most half a page, that for a tree-height increase a full page, and those for a page merge and tree-height decrease at most a full page. Thus, in each case it is possible to fit each log record in a single log-file page, even if no compression is used.

We may also allow a long log record to span two log-file pages, provided that the header of the record gives the total record length or other indication that makes it possible to find out if the record resides in its entirety in a single page or continues in the next page. Now if in recovery from a failure it is found that only a prefix of the last log record has survived on the log disk, then that log record is simply omitted in the recovery process. Naturally, with write-ahead logging, no updated page can appear on the database disk before the entire log record for the update appears on the log disk.

**Theorem 8.14** *Let b be a consistent and balanced B-tree and let H be any history of forward-rolling, backward-rolling, committed, and rolled-back transactions that can be executed on b using the above algorithms. Then in the event of a process failure or a system crash the analysis pass followed by the redo pass of the* ARIES *recovery algorithm transforms b into a consistent and balanced B-tree b' with*

$$db(b') = H'(db(b)),$$

*where H' is the prefix of H up to the last action whose entire log record is found on the log disk after the failure. Assuming that no transaction in H' does any dirty write, the undo pass of* ARIES *recovery transforms b' into a consistent and balanced B-tree b'' with*

$$db(b'') = H''(db(b')) = H'H''(db(b)),$$

*where H'' is any shuffle of the action sequences that are used to complete the active transactions in H' to rolled-back transactions.*

*Proof* Because each structure modification is logged with a single log record that makes it possible to selectively redo physically the effect of the modification on each page involved in the modification and because each insert, delete, undo-insert, or undo-delete action of any transaction on a data page is logged with a log record that makes it possible to redo the update physically on the page, and because write-ahead logging is followed, the redo pass of ARIES recovery transforms the B-tree $b$ into a B-tree $b'$ that existed at the time of writing the last log record that survived on the log disk. By Theorem 8.12, $b'$ is consistent and balanced and indexes the database current at that time.

Because of the absence of dirty writes, the history can be completed with the sequences of undo actions needed to roll back the active transactions, where the sequences can be shuffled in any order, by Theorem 5.25. Because $b'$ is consistent and balanced, the undo pass of ARIES can be safely executed on it. The correctness of the undo-insert and undo-delete algorithms in Sect. 4.3, the B-tree traversal algorithms in Chap. 7, and the structure-modification algorithms in this chapter then imply that the resulting B-tree $b''$ is consistent and balanced and indexes the database produced by the completed history.                                    □

## 8.6  Bottom-Up Structure Modifications

The traditional approach to B-tree structure modifications is to perform the modifications bottom-up, so that the leaf page is modified first and then the modification is propagated up the insertion or deletion path, level by level, as high as needed. The advantage of this approach over our top-down approach is that no unnecessary modifications occur. Note that in the top-down approach, even when the saved path is utilized, we may unnecessarily split a page on the insertion path. This may happen when the page does not have room for a record with a maximum-length key or when other transactions delete on the same path between the optimistic and pessimistic traversals (see Problem 8.6).

However, bottom-up modifications have a major disadvantage. When pages at more than one level have to be split or merged, the B-tree cannot be kept in a consistent state between the modifications. Therefore, the entire sequence of structure modifications needed on an insertion or a deletion path must be made a single atomic unit of work. The usual approach is to log the structure modifications with several redo-undo log records written for a special system-generated *structure-modification transaction* that commits as soon as the entire sequence of modifications is completed.

In recovery from a process failure or a system crash, it may happen that for a structure-modification transaction that encompasses several modifications and is logged with several log records, only some but not all of those log records are found on the log disk. Such a structure modification must be rolled back As the modifications are all physical operations, only physical undo is possible.

*Example 8.15* Consider executing the insert action $I[16, v']$ on the database of Fig. 7.1 when pages are split in bottom-up order, in contrast to the top-down order used in Example 8.7. As the index record pointing to the new sibling page created in a split cannot be inserted into a full parent page, the insertion of the index record must be made a separate modification, to be executed when the parent page has been split. Thus, we split $p_6$ creating $p'_6$, then split $p_3$ creating $p'_3$, then insert the index record $(12, p'_6)$ into page $p_3$, then increase the height of the tree, and finally insert the index record $(65, p'_3)$ into page $p'_1$. The following log records are written:

$n_1 : \langle S, begin\text{-}smo \rangle,$
$n_2 : \langle S, page\text{-}split, p_6, 12, p'_6, s_3, V_4, 1, n_1 \rangle,$
$n_3 : \langle S, page\text{-}split, p_3, 25, p'_3, s_2, V_3, 2, n_2 \rangle,$
$n_4 : \langle S, insert\text{-}index\text{-}record, p_3, 12, p'_6, n_3 \rangle,$

$n_5 : \langle S, \textit{tree-height-increase}, p_1, p'_1, 65, p''_1, s_1, V_1, V_2, 3, n_4 \rangle$,
$n_6 : \langle S, \textit{insert-index-record}, p'_1, 25, p'_3, n_5 \rangle$,
$n_7 : \langle S, \textit{commit-smo} \rangle$,

where $S$ is the identifier of the system-generated structure-modification transaction and the sets $V_1$, $V_2$, $V_3$, and $V_4$ are as in Example 8.7. When the commit log record for the structure modification (*commit-smo*) has been written, the modification is regarded as committed and the latches are released.

On the contrary, assume that a process failure or a system crash occurs and that in recovery it is found that log records up to and including the one with LSN $n_4$ have survived on the log disk. Obviously, after performing the redo pass, the modifications logged with LSNs $n_4$, $n_3$, and $n_2$ must be undone in the undo pass, so that the following log records are written:

$n_5 : \langle S, \textit{abort-smo} \rangle$,
$n_6 : \langle S, \textit{undo-insert-index-record}, p_3, 12, n_3 \rangle$,
$n_7 : \langle S, \textit{undo-page-split}, p_3, 25, p'_3, s_2, V_3, n_2 \rangle$,
$n_8 : \langle S, \textit{undo-page-split}, p_6, 12, p'_6, s_3, V_4, n_1 \rangle$,
$n_9 : \langle S, \textit{commit-smo} \rangle$.

To accomplish this, the pages mentioned in the log records are write-latched, and the inverse modifications of the logged modifications are performed. Naturally, this is possible only if the page contents are exactly in the states in which they were left by the modifications. □

To make physical undo possible, while a structure-modification transaction is rolling forward, other processes must be prevented from updating the modified pages until the structure-modification transaction commits. A safe way to accomplish this is to retain read latches on the modified pages until the structure-modification transaction commits.

*Example 8.16*  In Example 8.15, B-tree pages are latched as follows:

1. Pages $p_1$, $p_3$, and $p_6$ are write-latched (in this order).
2. The split of $p_6$: page $p'_6$ is write-latched.
3. The write latches of $p_6$ and $p'_6$ are downgraded to read latches.
4. The split of $p_3$: page $p'_3$ is write-latched.
5. The insertion into $p_3$ of the index record $(12, p'_6)$.
6. The write latches of $p_3$ and $p'_3$ are downgraded to read latches.
7. The tree-height increase: pages $p'_1$ and $p''_1$ are write-latched.
8. The insertion into $p'_1$ of the index record $(25, p'_3)$.
9. The commit of the structure modification: all the latches are released.

Observe that we have to first write-latch all the pages on the insertion path that need modification, because we have to keep all the modified pages at least read-latched until the commit of the entire sequence of splits (to make possible the rollback of the modification), and in the same time we have to obey the rule that pages must be latched in the first-parent-then-child order and that we cannot upgrade read latches to write latches (to prevent undetectable deadlocks). □

**Fig. 8.10** Transaction $T_1$ is performing the action $W[x, u, v]$. Another transaction $T_2$ wants to perform the action $I[y, v']$, where $y$ is covered by page $p_4$

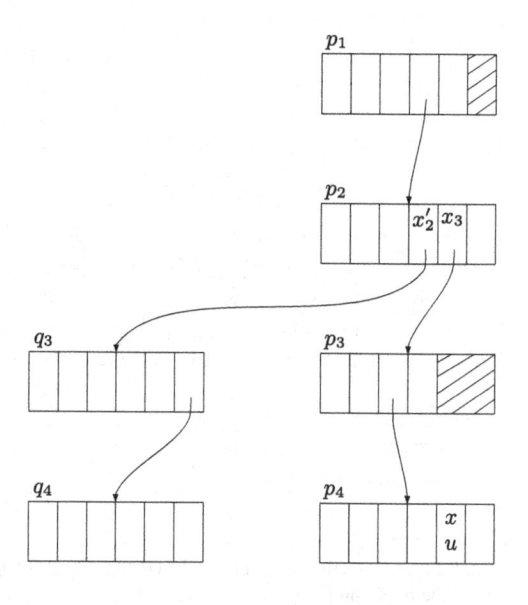

A further complication arises when in the backward scan of the log in the undo pass we encounter the log record of an update action that has to be undone logically on the B-tree that, as the result of the redo pass, is still in an inconsistent state because of a yet-undone incomplete structure modification.

*Example 8.17* A transaction $T_1$ wants to perform the action $W[x, u, v]$. For that purpose, the process thread executing $T_1$ latch-couples the path $p_1 \rightarrow p_2 \rightarrow p_3 \rightarrow p_4$ from the root $p_1$ of a B-tree down to the leaf page $p_4$ that covers key $x$ (see Fig. 8.10). Once it has reached page $p_4$, the thread executing $T_1$ has this page write-latched.

Another transaction $T_2$ now wants to perform the action $I[y, v']$, where page $p_4$ covers key $y \neq x$. For this action, the thread executing $T_2$ latch-couples the path $p_1 \rightarrow p_2 \rightarrow p_3$ and is put to sleep waiting for a write latch on page $p_4$. Page $p_4$ is still write-latched by the thread executing $T_2$.

A third transaction $T_3$ is performing an insertion into page $q_4$, which is full, causing splits along the path $p_1, p_2, q_3, q_4$ all the way up to page $p_2$. Assume that the sequence of splits is almost complete so that only the insertion of the index record $(x_2', p_2')$ into page $p_1$ is still missing (see Fig. 8.11). The following log records have been written:

$n_1 : \langle T_1, B \rangle$,
$n_2 : \langle T_2, B \rangle$,
$n_3 : \langle T_3, B \rangle$,
$n_4 : \langle S, begin\text{-}smo \rangle$,
$n_5 : \langle S, page\text{-}split, q_4, x_4', q_4', s_4, V_4, 1, n_4 \rangle$,
$n_6 : \langle S, page\text{-}split, q_3, x_3', q_3', s_3, V_3, 2, n_5 \rangle$,
$n_7 : \langle S, insert\text{-}index\text{-}record, q_3', x_4', q_4', n_6 \rangle$,
$n_8 : \langle S, page\text{-}split, p_2, x_2', p_2', s_2, V_2, 3, n_7 \rangle$.

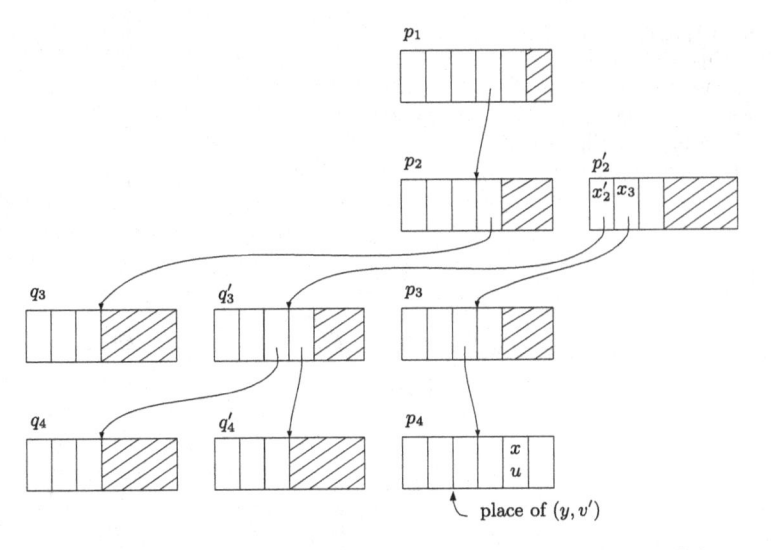

**Fig. 8.11** Transaction $T_3$ wants to insert into the full page $q_4$ and is therefore splitting pages up along the insertion path

Meanwhile, the thread executing $T_1$ performs the action $W[x, u, v]$ on page $p_4$, writes the log record

$n_9 : \langle T_1, W, p_4, x, u, v, n_1 \rangle$

and unlatches page $p_4$. Now the thread executing $T_2$ gets a latch on page $p_4$. The tuple $(y, v')$ to be inserted does not fit into page $p_4$, so the page must be split. Therefore $T_2$ allocates a new page $p_4'$, moves half of the tuples on page $p_4$ to this page, inserts the tuple, and commits (thus flushing the log). The tuple $(x, v)$ is among those that were moved (see Fig. 8.12). The following log records are written:

$n_{10} : \langle S', begin\text{-}smo \rangle,$
$n_{11} : \langle S', page\text{-}split, p_4, y_4', p_4', s', V', 1, n_{10} \rangle,$
$n_{12} : \langle S', insert\text{-}index\text{-}record, p_3, y_4', p_4', n_{11} \rangle,$
$n_{13} : \langle S', commit\text{-}smo \rangle,$
$n_{14} : \langle T_2, I, p_4, y, v', n_2 \rangle,$
$n_{15} : \langle T_2, C \rangle.$

Transactions $T_1$ and $T_3$ are still active. At this point, a failure occurs. As a result of the redo pass of recovery, the B-tree will be in the same state as it was at the time of failure—but this state is inconsistent: there is no way to reach page $p_2'$ from the root $p_1$, and thus also pages $p_3$, $p_4$, and $p_4'$ are unreachable.

The undo pass of recovery should roll back transactions $T_1$ and $T_3$ and the incomplete structure modification $S$. The action $W[x, u, v]$ by $T_1$ has been logged later than the splits by $S$. Thus, the undo action $W^{-1}[x, u, v]$ is performed before the splits are undone.

The action $W^{-1}[x, u, v]$ is first attempted physically on page $p_4$, which is the page mentioned in the log record for $W[x, u, v]$. This is not possible, because after

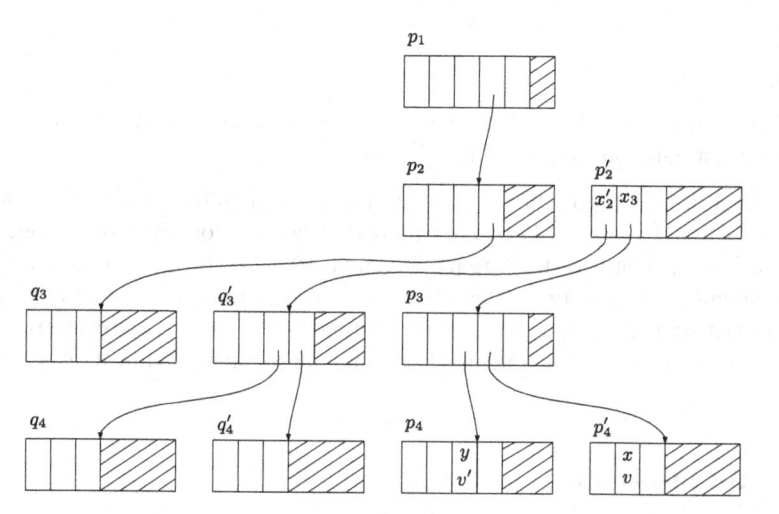

**Fig. 8.12** Transaction $T_1$ has performed the action $W[x, u, v]$. Transaction $T_2$ has split the full page $p_4$ and performed the action $I[y, v']$

latching page $p_4$, it is found out that the tuple $(x, v)$ is no longer there. Thus, we have to resort to logical undo. For logical undo, the key $x$ is searched for by a traversal that starts from the root of the tree. However, this is not possible either, because the tree does not contain the router to page $p_4'$ that covers $x$. In conclusion, the undo pass fails.                                                                                    □

The problem of the previous example can be solved using a *two-pass undo pass*:

1. By a backward scan of the log, roll back all uncommitted structure modifications.
2. By another backward scan of the log, roll back all active user transactions.

*Example 8.18*  Continuing with the previous example, the first pass of the undo pass of recovery rolls back the uncommitted structure modification $S$, writing the following log records:

$n_{16} : \langle S, abort\text{-}smo \rangle$.
$n_{17} : \langle S, undo\text{-}page\text{-}split, p_2, x_2', p_2', s_2, V_2, n_7 \rangle$.
$n_{18} : \langle S, undo\text{-}insert\text{-}index\text{-}record, q_3', x_4', n_6 \rangle$,
$n_{19} : \langle S, undo\text{-}page\text{-}split, q_3, x_3', q_3', s_3, V_3, n_5 \rangle$,
$n_{20} : \langle S, undo\text{-}page\text{-}split, q_4, x_4', q_4', s_4, V_4, n_4 \rangle$,
$n_{21} : \langle S, commit\text{-}smo \rangle$.

The second pass of the undo pass aborts and rolls back the active transactions $T_1$ and $T_3$, writing the log records:

$n_{22} : \langle T_1, A \rangle$,
$n_{23} : \langle T_3, A \rangle$,
$n_{24} : \langle T_1, W^{-1}, p_4', x, u, n_1 \rangle$,

$n_{25} : \langle T_3, C \rangle,$
$n_{26} : \langle T_1, C \rangle.$

Undoing the action $W[x, u, v]$ logically (on page $p'_4$) succeeds, because the B-tree is in a consistent state after the rollback of $S$.                                              □

Instead of using two scans of the log during the undo pass, we can also solve the problem of Example 8.17 by preventing the scenario from occurring altogether. The problem is that two structure modifications triggered by different transactions are ongoing at the same time. This can be prevented by requiring that every process thread that starts a structure modification on the B-tree must acquire an exclusive *tree latch* on the B-tree and hold that latch until the structure modification commits.


## 8.7  B-Link Trees

A *B-link tree* is a B-tree with the addition that the pages at each level of the tree are *sideways linked* with unidirectional links from left to right. Accompanied with this addition, it is convenient to also store the high key of each page explicitly in the page, as suggested in Problem 7.7. The sideways link in page $p$ can then be represented as a record $(y, p')$, where $p'$ is the page-id of the page next to $p$ at the same level and $y = high\text{-}key(p) = low\text{-}key(p')$. For the last page at each level, this record is $(\infty, \perp)$, where $\perp$ denotes the null link. An example of such a B-link tree is shown in Fig. 8.13, which indexes the set of tuples with keys 1, 3, 5, 6, 7, 9, 10, 11, 15, 16, 17, 19, 26, 27, 28, and 29.

Because of the sideways links, the consistency condition of B-link trees can be relaxed, so that we do not require explicit parent-to-child links to exist at all times, except for the parent-to-child link for the eldest child, which must always be present. The B-link tree of Fig. 8.13 exemplifies this relaxation in that the index record $(5, p_5)$ is missing from page $p_2$ and that the index record $(15, p_7)$ is missing from page $p_3$. Note that in search for, say, key 6, we arrive at leaf page $p_4$, where we find that the searched key, if it exists, must reside in the sibling page $p_5$, because $6 \geq 5 = high\text{-}key(p_4)$ and $6 < 8 = high\text{-}key(p_2)$ (cf. Fig. 8.14).

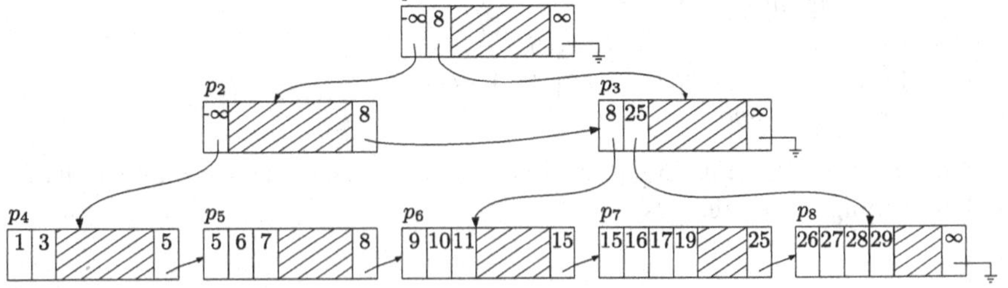

**Fig. 8.13**  A B-link tree

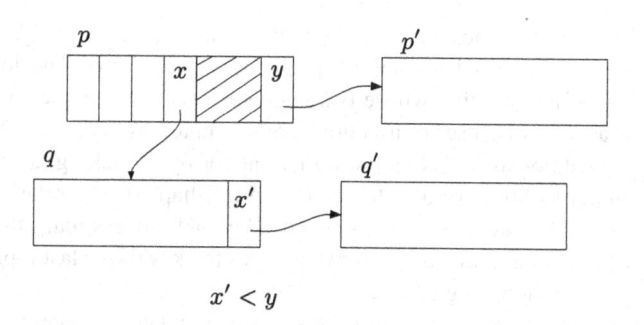

$$x' < y$$

**Fig. 8.14** Finding out when a parent-to-child link is missing in a B-link tree. Here $y = $ *high-key*$(p) = $ *low-key*$(p')$ and $x' = $ *high-key*$(q) = $ *low-key*$(q')$

The advantage of the relaxed consistency of a B-link tree is that the structure modifications can be made smaller atomic units, namely, ones that modify only a single level of the tree, as opposed to the structure modifications in our top-down approach which modify pages at two adjacent levels. For example, the split of a page $q$ can be made to consist only of the allocation of a new sibling page $q'$, moving the upper half of the records in page $q$ to page $q'$ and inserting into page $q$ the sideways-link record $(y, q')$, where $y$ is least key moved to page $q'$. The insertion of the parent-to-child record $(y, q')$ into the parent of page $q$ is a separate structure modification (an atomic unit), which can be performed later, for instance, when a traversal for some insertion or deletion next time visits page $q'$ via the sideways link from the sibling page $q$, then the parent-to-child link $(y, q')$ is inserted into the parent page.

Different issues pertaining to B-link trees are considered in the Problems section.

## 8.8 Loading of a B-Tree

A database is often initialized by loading data from an external source file to an initially empty relation. Although it is always possible to implement the load process using the standard tuple-wise insert actions, we wish to make the load more efficient by inserting more tuples at a time and by exploiting the natural assumption that during the load process user transactions may not access the relation, which is made visible only after the load has been completed.

We view the load process as a task initialized and controlled by the database administrator rather than executed as a part of user transactions. We also note that the data coming from the source file may have to be cleansed and filtered before it can be loaded into the database—an issue we will not discuss here.

When the physical structure of the relation is a B-tree, we may also wish that the pages are populated using a preset *fill-factor*, stating, for example, that each leaf page should filled 75% full of records. Note that when a set of tuples is inserted using the algorithms thus far presented, the result is a B-tree with all leaf pages except the last filled only to 50%.

For simplicity, we assume that the tuples to be loaded into relation $r(\underline{X}, V)$ reside in a source file $f$ that has been sorted in ascending key order. The load begins with an initially empty relation whose B-tree thus consists of a single, empty page. Simultaneous accesses by user transactions are prevented by retaining the relation $r$ in the system catalog as invisible to transactions (or by X-locking the relation by the multi-granular locking protocol to be defined in Chap. 9). The relation is made visible only after the load has been completed. The load process scans the file $f$ in chunks of records, maintaining a cursor $z$ that stores the key of the last tuple inserted into the relation. The load begins with $z = -\infty$.

Since no processes other than the load process can update the relation, and since the tuples are inserted in ascending key order, the insertions always go to the last leaf page, which is split after being filled. To guarantee the balance of the B-tree during the load, we must assume that the fill-factor used is somewhat less than 100%, so that when a full leaf page is split at least two tuples go to the new sibling page. This is an easy modification to the *page-split* algorithm (Algorithm 8.2).

We wish to save in logging, so that instead of logging all the tuples inserted into a leaf page, we log only the key range (i.e., the least and greatest keys) of the inserted tuples. More specifically, when the records within key range $[x, y]$ in file $f$ are inserted into leaf page $p$ of the B-tree, the insertion is logged with the redo-only log record

$$\langle \textit{populate-leaf-page}, p, f, x, a_x, y \rangle, \tag{8.6}$$

where $p$ and $f$ are the identifiers of the page and the file, respectively, and $a_x$ is the address in file $f$ of the record that contains the tuple with key $x$. The LSN of the log record is stamped in the PAGE-LSN field of page $p$. If in the redo pass of recovery we have to redo the insertion of those tuples, we retrieve from the file $f$ the set of tuples in the logged key range $[x, y]$ and insert the tuples into the page $p$.

Naturally, the use of a light-weight log record such as the above implies that we have to keep available and unchanged the additional external information (the file $f$) necessary for its interpretation, as long as the effect of the operation is not yet reflected in the disk version of the page. As such dependence of database durability on external information may be considered undesirable or harmful, it is wise to arrange that the leaf pages created by the load process are flushed onto disk as soon as possible. If taking a complete checkpoint is out of the question, we may at least push a leaf page to the front of the LRU chain as soon as it has been created, so that it becomes the first page to be evicted from the buffer.

To cope with failures, the load process is made *restartable* as follows. When the load process starts, it first traverses to the last leaf page so as to find out the greatest key $z$ already inserted. Then it writes the log record

$$\langle \textit{start-load}, r, z, f \rangle, \tag{8.7}$$

where $r$ and $f$ are the identifiers of the relation and the file, respectively. Then the load process begins inserting tuples $(x, v)$ with keys $x > z$, repeatedly filling the last leaf page with the next not-yet-inserted tuples, advancing the cursor $z$ to the key of the last tuple inserted into the page, and causing a split at the next attempt to insert more of the remaining tuples into a full leaf page.

If the load process succeeds to reach the end of the file $f$, it writes the log record

$$\langle end\text{-}load, r, f \rangle. \tag{8.8}$$

To help tracing errors in the load process, the *start-load* and *end-load* log records might also carry the timestamp and other descriptive information about the source file $f$ such as the number of tuples with keys greater than $z$, and the *end-load* log record might carry aggregate information about the load such as the number of inserted tuples.

To avoid the dependence of database durability on the file $f$, we can always log the inserted tuples themselves, so that instead of the log record (8.6) we use the log record

$$\langle populate\text{-}leaf\text{-}page, p, V \rangle, \tag{8.9}$$

where $V$ is the set of tuples inserted into page $p$.

As the tuple insertions performed by the load process are all redo-only operations, the loading can be undone only by running a transaction that effectively undoes the insertions. This can be done efficiently using a bulk-delete action that deletes from the relation all tuples in the entire key space $(-\infty, \infty]$ (see Chap. 10).

## Problems

**8.1** Consider executing the transaction $T_2$:

$BD[6]D[7]AD^{-1}[7]D^{-1}[6]C$

on the database indexed by the B-tree of Fig. 7.1. Assuming that the saved path has its initial value set by the call *initialize-saved-path*() and that no other transactions are in progress simultaneously, what page latchings and unlatchings of B-tree pages occur during the execution of the transaction? What structure modifications are performed? We assume redo-only modifications are used. Show the B-tree after performing the forward-rolling phase of $T_2$ and after performing the whole of $T_2$. What log records are generated?

**8.2** By outlining algorithms similar to *redo-page-split* (Algorithm 8.10), verify that the log records for the other four redo-only structure modifications (tree-height increase, tree-height decrease, page merge, and records redistribute) allow for selective redoing.

**8.3** A simple and practical compression method called *prefix truncation* can be used to save space in B-tree pages. For each leaf or non-leaf page $p$, we store the longest common prefix $x$ of *low-key*$(p)$ and *high-key*$(p)$ once in the page, and for every record $(xy, v)$ with key $xy$ in the page, we store the truncated record $(y, v)$. When the key range of a page changes in a B-tree structure modification, the compression within the page must be recomputed. In a structure modification that shrinks the key range, such as a page split, tree-height increase, or a records redistribute, the longest common prefix may get longer, and in a structure modification that enlarges the key range, such as a page merge, tree-height decrease, or records redistribute, the longest common prefix may get shorter. Work out how this compression can be accomplished, if possible, with the structure modifications presented in Sects. 8.2 and 8.3. Note that there is a problem with determining exactly the longest common prefix of *low-key*$(p)$ and *high-key*$(p)$ when $p$ is the youngest child of its parent and its key range changes in a situation when the information in the saved path cannot be trusted. Show that this problem does not occur if we store *high-key*$(p)$ explicitly in page $p$, as suggested in Problem 7.7.

**8.4** Some practical implementations of the B-tree relax the balance conditions such that pages are allowed to become empty until they are detached from the B-tree and deallocated. Adapt the algorithms in this chapter to this *free-at-empty* policy. Note however that we cannot leave completely empty pages hanging around in the B-tree; if a page becomes empty, it must be detached from the B-tree within the same structure modification that empties it. Explain why. What might be the actual gains achieved by this policy?

**8.5** A page split includes the allocation of the new sibling page and thus the updating of the corresponding space-map page in the same redo-only atomic structure modification. Similarly, the deallocation of the emptied sibling is included in a page merge, the allocations of the two child pages are included in a tree-height decrease, and the deallocations of the two child pages are included in a tree-height decrease. The contention by different processes on the space-map pages, which must be kept write-latched during the entire structure modification, may degrade concurrency. Consider means of alleviating this problem. What if the allocations and deallocations are made redo-only modifications of their own? What kind of log records are then used?

**8.6** With top-down structure modifications, some unnecessary page splits and merges may occur when the keys vary in length keys or when other concurrent transactions perform insertions or deletions on the same path between the optimistic and pessimistic traversals. Give examples of this. Also show that with fixed-length keys and in the absence of concurrent updates on the same path no unnecessary splits or merges are performed.

**8.7** Outline a procedure *bottom-up-page-merges* that performs the sequence of page merges or records redistributes (possibly preceded by a tree-height decrease) triggered by a delete action, when bottom-up structure modifications are used. What redo-undo log records are generated for such a sequence? Also outline an algorithm for undoing such a sequence when recovering from a system crash. What redo-only log records are generated?

**8.8** Consider executing the transaction

$$BR[x_1, > 9, v_1] R[x_2, > x_1, v_2] R[x_3, > x_2, v_3] R[x_4, > x_3, v_4] C$$

on the database indexed by the B-link tree of Fig. 8.13. Assuming that the saved path has its initial value set by the call *initialize-saved-path*() and that no other transactions are in progress simultaneously, what page latchings and unlatchings of B-link tree pages occur during the execution of the transaction?

**8.9** Give algorithms for redo-only structure modifications for B-link trees. Each modification should only modify a single level of the tree and retain the (relaxed) consistency defined for B-link trees, so that each child page of each page of the tree is accessible, if not directly via a parent-to-child link, via a parent-to-child link to the eldest child and the sideways link chain. How are these modifications used in the execution of the following transaction on the B-link tree of Fig 8.13?

$$BI[18] I[20] I[21] D[1] D[3] C.$$

**8.10** Thanks to the relaxed balance conditions of the B-link tree, there exists a very simple method to load the contents of a file $f$ into an initially empty relation $r$ implemented as a B-link tree. In the spirit of the load process suggested for B-trees in Sect. 8.8, outline an algorithm for loading a B-link tree and for repairing its imbalance from which it may initially suffer.

**8.11** Along the lines discussed in Sect. 8.8, outline a restartable algorithm for loading a B-tree. For triggering the page splits and tree-height increases needed, use the procedure *find-page-for-insert*$(p, p', x, v, y)$ simplified so that the page $p'$ and the next key $y'$ are not determined. We also assume that the procedures *page-split*$(p, q, x', q')$ and *tree-height-increase*$(p, q, x', q')$ are changed so that they move to page $q'$ only two records.

# Bibliographical Notes

Traditional B-tree algorithms perform structure modifications bottom-up along the insertion or deletion path, as described in many textbooks on database management. In many early studies on B-tree concurrency control, transactional access and recovery issues are either omitted or discussed very briefly. Biliris [1987] makes the important observation that the commit of a structure modification is independent of the commit of the transaction that triggered the modification, so that a structure modification (that retains the consistency of the B-tree) can be allowed to commit even though the triggering transaction eventually aborts and rolls back. Fu and Kameda [1989] consider concurrent access to B-trees in the context of nested multi-action transactions.

Mohan [1990a, 1996a], Mohan and Levine [1992], and Lomet and Salzberg [1992, 1997] were the first to publish detailed algorithms for managing transactions on database relations indexed by B-trees, compatible with the ARIES recovery algorithm. In the ARIES/KVL and ARIES/IM algorithms [Mohan, 1990a, 1996a, Mohan and Levine, 1992], structure modifications are performed bottom-up, with the help of a saved path and other clever techniques that avoid both deadlocks and simultaneous write-latching of all the pages on the modification path. The "free-at-empty" policy [Gray and Reuter, 1993, Johnson and Shasha, 1993] is applied, allowing pages to become empty until they are detached from the B-tree and deallocated.

In the ARIES/KVL and ARIES/IM algorithms, a structure modification interrupted by a process failure or a system crash may have to be rolled back; the problem associated with this and discussed in Sect. 8.6 was first observed by Mohan [1990a] and Mohan and Levine [1992], who present a preventive approach (tree latch) as a solution. Lomet [1998] also discusses the problem and presents an alternative solution (two-phase undo pass) [Lomet, 1992]. The interaction of search-tree structure modifications and transactional access is also discussed by Sippu and Soisalon-Soininen [2001].

The idea of performing structure modifications top-down appears in the B-tree algorithms by Mond and Raz [1985]: using lock-coupling along the search path, a traversal for an insert action splits in advance all full pages encountered, and a traversal for a delete action merges or redistributes in advance all about-to-underflow pages encountered. However, with these algorithms pages may be split or merged unnecessarily even with fixed-length keys, unlike with our traversal algorithms that use the saved path so as to shorten the search path and to avoid unnecessary structure modifications in most cases. Moreover, the algorithms of Mond and Raz [1985] do not include key-range locking or recovery.

The B-link tree (Sect. 8.7) was originally presented by Lehman and Yao [1981]. A generalization called the $\Pi$-tree was defined by Lomet and Salzberg [1992, 1997]. The original motivation for the B-link tree was the claim that no latch-coupling (actually lock-coupling) would be needed, so the tree could be traversed while holding only a single page latched at a time. However, this imposes restrictions on how pages emptied by delete actions can be detached and deallocated. Symmetric treatment of insert and delete actions needs latch-coupling (or lock-coupling) [Lanin and Shasha, 1986]. A better motivation for the B-link tree is that structure modifications can be defined as small redo-only operations that modify pages at a single level only, thus both increasing concurrency and making recovery easier. This property is shared by the Foster B-tree of Graefe et al. [2012].

The balance conditions of the tree are relaxed in the B-link tree algorithms of Lehman and Yao [1981], Lanin and Shasha [1986], Lomet and Salzberg [1992, 1997], and Lomet [2004], as well as in the Foster-B-tree algorithms of Graefe et al. [2012], so that logarithmic search-path length is not guaranteed to be maintained under all circumstances. Jaluta [2002] and Jaluta et al. [2005] present B-link-tree algorithms in which deletions are handled uniformly with insertions, structure modifications are redo-only, and the tree is kept in balance under process failures and system crashes. These algorithms use no saved path; instead, latches of three modes (read, write, update) are utilized. For ordinary B-trees, similar algorithms using redo-only structure modifications are presented by Jaluta et al. [2003] (without saved paths) and by Jaluta et al. [2006] (for page-server systems, with saved paths).

Jaluta and Majumda [2006] present a technique for implementing B-tree page allocations and deallocations as redo-undo modifications separate from redo-only page-splits and page-merges in such a way that every page that remains allocated after recovery from a failure is guaranteed to be part of the B-tree, even though the failure occurred between page allocation and page split or between page merge and page deallocation. The technique thus enhances concurrency while eliminating the need for garbage collection for pages that are allocated but not part of the B-tree (cf. Problem 8.5).

Compression techniques for B-trees, such as prefix truncation (see Problem 8.3), and their effect on B-tree algorithms are discussed by Bayer and Unterauer [1977], Lomet [2001], and Graefe [2011], among others. Recovery and concurrency-control algorithms for multidimensional index structures have been presented by Evangelidis et al. [1997] (for hB$^{\Pi}$-trees), Kornacker et al. [1997] (for generalized search trees), and Haapasalo et al. [2013] (for ordinary R-trees), among others. Recovery techniques for B-trees are surveyed by Graefe [2012].

# Chapter 9
# Advanced Locking Protocols

The locking protocols presented thus far assume that the lockable units of the database are single tuples. Such a choice is appropriate to transactions that access a few tuples only. If a transaction accesses many tuples, it must also acquire many locks. Each such access incurs the computational overhead of requesting and perhaps waiting for the granting of the lock, and, in the case of a huge number of commit-duration locks, the storage overhead on storing the locks in the lock table until the commit of the transaction.

Using coarser units of locking such as whole relations is probably convenient for a transaction that accesses many tuples. On the other hand, such coarse locks discriminate against transactions that only want to lock one or two tuples of the relation. Obviously, we need a locking protocol that allows a choice between multiple granularities of lockable units. The problem with locks on units of different granules is how to detect efficiently lock incompatibilities. The solution is to introduce additional lock modes, called *intention locks*.

In this chapter we present the classical *multi-granular locking protocol* adapted to our key-range locking protocol. We also argue why it is better to use units of the logical database (i.e., relations and tuples) rather than units of the physical database (i.e., pages and records) as lockable units.

For avoiding deadlocks caused by lock upgrades, we present a locking protocol based on *update-mode locks*. This protocol is most suitable in implementing SQL cursors in a deadlock-free way. We also discuss different ways to reduce the number of commit-duration shared locks. The locking protocol for SQL cursors gives rise to a new isolation level, located between the read-committed and repeatable-read levels and called cursor stability.

A technique called COMMIT-LSN can be used to reduce the number of short-duration locks for transactions run at the read-committed isolation level. In connection with ARIES recovery, this technique can also be used to reduce the time a database system recovering from a failure is unable to accept new transactions.

© Springer International Publishing Switzerland 2014
S. Sippu, E. Soisalon-Soininen, *Transaction Processing*, Data-Centric Systems
and Applications, DOI 10.1007/978-3-319-12292-2_9

## 9.1   Key-Range Locking for Multiple Granules

In this section we discuss how the key-range locking protocol defined in Sect. 6.4
generalizes to the situation in which the logical database consists of a hierarchy
of data items of different granules, as discussed in Sect. 1.10. For instance, the
entire logical database may be composed of databases of multiple users, each user
database of multiple relations, and each relation of multiple tuples. In this setting it
is useful if larger-granule data items can be locked (with one or two locks) without
the need to acquire explicit locks on all the smaller-granule data items contained
in it, when we want to operate on all the contained smaller-granule data items. For
instance, if we want to read all tuples from a relation, then it certainly seems more
practical to acquire just a single S lock on the relation instead of locks on every
tuple in the relation.

We assume that the units at all levels or at some specified levels of the granule
hierarchy are termed *lockable units* that can be locked. In the case of a three-level
granule hierarchy, when the units at all the levels are lockable, this means that
locks can be acquired (1) on the whole database $b$, (2) on individual relations $r$
in database $b$, and (3) on keys $x$ of tuples in relation $r$. The names of these locks
are, respectively, $b$, $(b, r)$, and $(b, r, x)$ or actually 4-byte hash values calculated
from them.

For the action $R[b, r', \theta r, R']$ of browsing the schema of a relation $r'$ in
database $b$, transaction $T$ must acquire a commit-duration S lock on the relation
identifier $(b, r')$. For the action $I[b, r, R]$ of creating a new relation $r$, $T$ must
acquire a commit-duration X lock on $(b, r)$ and a short-duration X lock on relation
identifier $(b, r')$, where $r'$ is the relation next to $r$ in database $b$. For the action
$D[b, r, R]$ of dropping a relation $r$, $T$ must acquire a short-duration X lock on $(b, r)$
and a commit-duration X lock on the next relation $(b, r')$. This protocol prevents
dirty creations and droppings of relations and dirty and unrepeatable schema reads
(including phantom relations).

Isolation anomalies must now also be defined between actions on a whole and its
parts, such as a relation and its tuples. For example, consider a transaction $T_1$ that
creates a new relation $r$ or drops a relation $r$ in database $b$. While $T_1$ is still active,
another transaction $T_2$ reads tuples of relation $r$ or inserts or deletes tuples in $r$. The
read action by $T_2$ must be defined as a dirty read and the insert and delete actions as
dirty writes.

In the example, $T_1$ will protect its actions by acquiring X locks on the relation
$(b, r)$ and on the next relation $(b, r')$, following the key-range locking protocol
applied at the relation level. The X lock on relation $(b, r)$ must mean that $T_1$ has
also locked (implicitly) all tuples in $r$. Likewise, an S lock on relation $(b, r)$ must
mean that all tuples in $r$ are S-locked.

The problem is how to detect the incompatibility of locks on a relation $(b, r)$
and on a key $(b, r, x)$. One solution could be to organize the lock table into a
data structure that stores paths of the granule hierarchy. For example, when a
transaction $T$ requests an $m$ lock on relation $(b, r)$, the lock manager checks that

no other transaction holds an incompatible lock on $b$, $(b, r)$ or $(b, r, x)$ where $x$ is any locked key in relation $r$.

However, this solution cannot be made as efficient as is necessary, especially since we stated earlier that, for reasons of efficiency, the lock names must be short hash values rather than the original variable-length keys. Thus, for efficiency, all locks must be stored in a simple hash structure indexed by the lock names of data items to be locked.

## 9.2 Intention Locks and Multi-Granular Locking

A practical solution to the problem of detecting lock incompatibilities between data items of different granularities is to use new kinds of locks, called *intention locks*. Such locks are of three different modes: IS (intention-shared), IX (intention-exclusive), and SIX (shared + intention-exclusive).

An *intention-shared lock* or an *IS lock* for transaction $T$ on a larger-granule data item $x$ means that $T$ is allowed to acquire S locks on the smaller-granule data items contained in $x$ (and additional IS locks, if the data items in $x$ are further subdivided into smaller data items). When $T$ wants to read only some of the data items contained in $x$, it IS-locks $x$ and then S-locks those contained data items it actually reads.

An *intention-exclusive lock* or an *IX lock* for transaction $T$ on a larger-granule data item $x$ means that $T$ may acquire X and S locks on the smaller-granule data items contained in $x$ (and additional IX, IS, and SIX locks, if the data items in $x$ are further subdivided). When $T$ wants to update only some of the data items contained in $x$, it IX-locks $x$ and then X-locks those contained data items it actually updates.

A *shared and intention-exclusive lock* or a *SIX lock* is a combination of an S lock and an IX lock. A SIX lock for transaction $T$ on data item $x$ means that $T$ is allowed to read all of $x$ (under the protection of the S lock included in the SIX lock) and to acquire X and IX locks on (and hence to update) smaller-granule data items contained in $x$ (by the IX lock included in SIX).

An S lock on a larger-granule data item $x$ *implicitly locks* in S mode all the smaller-granule data items contained in $x$, recursively down into the smallest-granule data items, without the need to lock them explicitly. Likewise, an X lock on $x$ implicitly X-locks all smaller-granule data items contained in it. If a transaction holds an implicit S lock (resp. X lock) on $x$, it must not request an explicit S lock (resp. X lock) on $x$.

An X lock is incompatible with an IS lock and with an IX lock. Thus, if transaction $T_1$ holds an X lock on a larger-granule data item $x$, then no other transaction $T_2$ can simultaneously hold an S lock (resp. an X lock) on any data item contained in $x$, because then $T_2$ would have to hold an IS lock (resp. IX lock) on $x$. Accordingly, $T_2$ is prevented from reading or updating data items contained in a data item being updated by $T_1$.

An S lock is incompatible with an IX lock. Thus, if transaction $T_1$ holds an S lock on $x$, then no other transaction $T_2$ can simultaneously hold an X lock on any data item contained in $x$, because then $T_2$ would have to hold an IX lock on $x$. Accordingly, $T_2$ is prevented from updating data items contained in a data item being read by $T_1$.

An IS lock and an IX lock are compatible, and so are two IX locks and, naturally, two IS locks. This is because it is natural to allow both $T_1$ and $T_2$ to IS-lock or IX-lock $x$ with the intention of reading or updating different data items contained in $x$. On the other hand, if $T_1$ and $T_2$ both hold IX locks on $x$ and want to update the same data item contained in $x$, then they would have to hold X locks on that data item, which is impossible because of the incompatibility of two X locks.

The lock mode SIX is needed for the case that a transaction wants to read all the data items contained in a larger-granule data item $x$ but to update only some of them. We want to have a single lock to cover this situation; otherwise the transaction would have to hold both an S lock and an IX lock on $x$. Because of the S lock (resp. IX lock) contained in a SIX lock, a SIX lock is incompatible with an IX lock (resp. an S lock).

The *multi-granular locking protocol* combined with key-range locking works as follows. At each level of granularity, key-range locking is used. When a transaction $T$ explicitly wants to S- or X-lock a data item $x$ with a $d$-duration lock, it must first acquire the appropriate $d$-duration intention locks on every larger-granule data item on the path from the root of the granule hierarchy down to $x$. Thus, locks are acquired first on the larger-granule data items and only then on the contained smaller-granule ones, according to the granularity hierarchy. Releasing a $d$-duration lock on $x$ is possible only if the transaction does not hold any $d$-duration locks on data items contained in $x$. Thus, locks are released in reverse order.

*Example 9.1* Assuming a three-level granule hierarchy consisting of a database, relations, and tuples, the following locks are acquired for the insert action $I[b, r, x, v]$ by transaction $T$:

$(T, b, \text{IX}, \text{commit duration})$
$(T, (b, r), \text{IX}, \text{commit duration})$
$(T, (b, r, x), \text{X}, \text{commit duration})$
$(T, (b, r, y), \text{X}, \text{short duration})$

First, the database $b$ is locked with a commit-duration IX lock, then the relation $r$ contained in $b$ is locked with a commit-duration IX lock, and finally the key $x$ of the tuple to be inserted into $r$ is locked with a commit-duration X lock and its next key $y$ with a short-duration X lock.                                                □

The lock-compatibility matrix for the lock modes specified for multi-granular locking is shown in Fig. 9.1. The lock-upgrade matrix is shown in Fig. 9.2.

The mutual exclusivity order for the lock modes is a partial order, specifically a lattice with X (the most exclusive lock mode) at the top and IS (the least exclusive lock mode) at the bottom (Fig. 9.3):

IS < IX < SIX < X.
IS < S < SIX < X.

**Fig. 9.1** Lock-compatibility matrix for multi-granular locking with the lock modes S, X, IS, IX, SIX

| lock requested by $T_2$ | lock held by $T_1$ | | | | |
|---|---|---|---|---|---|
| | IS | IX | S | SIX | X |
| IS | true | true | true | true | false |
| IX | true | true | false | false | false |
| S | true | false | true | false | false |
| SIX | true | false | false | false | false |
| X | false | false | false | false | false |

**Fig. 9.2** Lock-upgrade matrix for multi-granular locking with the lock modes S, X, IS, IX, SIX

| lock requested by $T$ | lock held by $T$ | | | | |
|---|---|---|---|---|---|
| | IS | IX | S | SIX | X |
| IS | IS | IX | S | SIX | X |
| IX | IX | IX | SIX | SIX | X |
| S | S | SIX | S | SIX | X |
| SIX | SIX | SIX | SIX | SIX | X |
| X | X | X | X | X | X |

**Fig. 9.3** The lattice of the multi-granular lock modes S, X, IS, IX, SIX

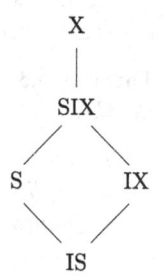

There is no ordering relation between the lock modes IX and S. Their supremum (smallest common upper bound) in the lattice is SIX.

*Example 9.2*  The following two scenarios are possible in the multi-granular locking protocol (Fig. 9.4a, b):

(a) Transaction $T_1$ holds an IS lock on relation $r$ and an S lock on the key $x$ of a tuple in $r$. At the same time, another transaction $T_2$ holds an IX lock on $r$, an S lock on $x$, and an X lock on another key $y$ that is also present in $r$. In addition, a third transaction $T_3$ holds an IX lock on $r$ and an X lock on a third key $z$ that also exists in $r$.

(b) Transactions $T_1$ and $T_2$ hold IS locks on relation $r$ and S locks on a key $x$ in $r$. At the same time, transaction $T_3$ holds a SIX lock on $r$ and an X lock on another key $y$ that is also present in $r$.

The following scenario is not possible (Fig. 9.4c):

(c) Transaction $T_1$ holds an IX lock on relation $r$ and an X lock on key $x$ in $r$. At the same time $T_2$ holds a SIX lock on $r$.

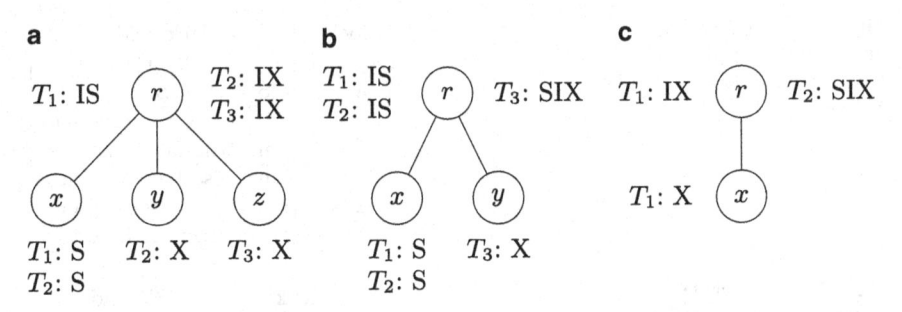

**Fig. 9.4** Possible scenarios (**a, b**) and an impossible scenario (**c**) in multi-granular locking

Scenario (c) is prevented by the locking protocol, because the lock modes IX and SIX are incompatible, or, actually, because IX and S are incompatible (recall that SIX = S + IX).                                                                                             □

The following theorem states the fundamental property of the multi-granular locking protocol, namely, that the incompatibility of implicit locks is detected:

**Theorem 9.3** *Assume the multi-granular locking protocol with lock modes S, X, IS, IX, SIX.*

(a) *If transaction $T_1$ holds an explicit lock on data item $x_1$ and transaction $T_2$ holds an explicit or implicit X lock on data item $x_2$, then $x_1 \neq x_2$ and $x_1$ are not contained in $x_2$.*

(b) *If transaction $T_1$ holds an explicit X lock on data item $x_1$ and transaction $T_2$ holds an explicit S lock or SIX lock or an implicit S lock on data item $x_2$, then $x_1 \neq x_2$ and $x_1$ are not contained in $x_2$.*

*Proof* Because $T_2$ holds an explicit or implicit lock on $x_2$, there must exist a data item $x_2'$ explicitly locked by $T_2$ such that $x_2' = x_2$ or $x_2'$ contains $x_2$. For the sake of contradiction, assume that $x_1 = x_2$ or $x_1$ is contained in $x_2$. Then $x_2'$ is on the path from the root of the granule hierarchy down to $x_1$. In case (a), the explicit lock on $x_2'$ held by $T_2$ is an X lock. By the locking protocol, because $T_1$ holds an explicit lock on $x_1$, it should also hold some lock on $x_2'$, which however is not possible, because the lock mode X is incompatible with all lock modes. In case (b), the explicit lock on $x_2'$ held by $T_2$ is an S lock or a SIX lock. By the locking protocol, $T_1$ should also hold an IX lock (or a SIX lock) on $x_2'$, which however is not possible, because the lock modes IX and SIX are incompatible with the lock modes S and SIX.                        □

The granule hierarchy of lockable units can also consist of parts of the physical database, or logical and physical parts can be mixed up. For instance, the hierarchy could be system, database, file, page, and tuple.

Sometimes, intention locks are not required all the way from the top of the hierarchy (i.e., the whole system): it is possible to start from a lower level (such as a database) and proceed downwards. If the creation and destroying of databases are specified to be special operations that happen outside normal user transactions, even the database level does not need to be part of the granule hierarchy of a normal transaction. Usually, an administrator's manual for the database management system specifies what kinds of locks are acquired on data items at which level of the granule hierarchy and for which actions.

*Example 9.4* A past database management system used a multi-granular locking protocol with a page as the finest granule to be locked. According to that protocol, locks and intention locks were acquired using the following rules:

1. For an SQL **select** query that touches relations $r_1, \ldots, r_n$, an IS lock is acquired on each $r_i$ and an S lock on data pages to be read from $r_1, \ldots, r_n$. However, if the query touches over 10 (or a specified maximum number of) pages of relation $r_i$, the IS lock on $r_i$ is upgraded to an S lock and the S locks on the pages are released.
2. For the SQL operations **update**, **insert**, or **delete**, an IX lock is acquired on the relation $r$ to be updated and an X lock on the data pages to be updated of $r$. If the update touches more than 10 (or a specified maximum number of) pages of $r$, the IX lock on $r$ is upgraded to an X lock and the X locks on the pages are released. Possible other pages or relations that the operation touches (but does not update) are IS- or S-locked as in a **select** query.
3. For an operation that changes the physical structure of relation $r$ (e.g., from a hash table into a B-tree), an X lock is acquired on $r$.
4. For the operation **create table** $r(\ldots)$, an X lock is acquired on the relation $r$ to be created.
5. For the operation **drop table** $r$, an X lock is acquired on the relation $r$ to be dropped.

The lock upgrades in (1) and (2) are called *lock escalation*. □

## 9.3 Logical Locking vs. Physical Locking

Logical locking is natural, because the purpose of locking is to protect the read and update actions done by transactions on the data items of the logical database. Logical locks are completely independent of the physical location of the data item to be locked. Thus, a logically locked data item can be moved without having to modify the lock at all: when a tuple $(x, v)$ on a B-tree leaf page is logically locked by a transaction $T$, and while $T$ is still active, the tuple is moved to another page in a structure modification such as a page split (either as a result of $T$'s own or some other transaction's actions), the logical lock held by $T$ on $x$ remains unchanged.

Another advantage of logical locks is that they can often be acquired before the target data item is actually located or stored in the physical database. The commit-duration X lock needed for a write action $W[x, u, v]$ can be acquired at the start of the execution of the action, before the location of the tuple $(x, u)$ is known. For the insert action $I[x, v]$, the commit-duration X lock can also be acquired at the start of the operation, but the short-duration X lock on the next key can be acquired only after the page that covers key $x$ has been fixed and write-latched and the key next to it (which may or may not reside on the same page) has been found.

Instead of an actual attribute value of the tuple, logical locking can also be based on a logical *surrogate key*, such as a number $n$ that says that this was the $n$th tuple that ever was inserted into the relation. However, when using surrogates, the lock can be acquired only after the tuple and the surrogate attribute contained in it has been found.

Still, some existing database management systems use physical locking instead of logical locking. For instance, data pages or individual record positions in them are locked. In *page locking*, the lock names are page identifiers or hash values calculated from them. A lock on data page $p$ locks all those data items of the logical database that are currently present in $p$ or that logically belong to $p$. In *record locking* or *record-identifier locking*, the lock names are record identifiers, that is, pairs $(p, i)$ of page-ids and record positions (or hash values calculated from them). A lock on record-id $(p, i)$ locks the data item at position $i$ in page $p$ or a data item that belongs there.

The granularity of record locking is close to that of locking unique keys. Page locking is unnecessarily coarse, if the target of the operation is only a single data item in the page. A page or record lock can be acquired only after the physical location of the data item to be operated on is known. Page locks and record locks held by a transaction $T$ also have to be changed if the data items are moved to another place while $T$ is active. For instance, when $T$ inserts a set of tuples into a relation structured as a sparse B-tree and a leaf has to be split because of this insertion, part of the tuples in the page is moved to another page. The page or record locks have to be changed so as to reflect the new locations of the tuples.

Fine-granular physical locking (record locking) is feasible when the tuples of a relation are stored in a data structure such as an unordered sequential file (heap) where their locations do not change in normal update actions. A dense index is then created on the base structure, for example, a B-tree whose leaf pages contain records that point to the actual tuples. If the base structure needs to be reorganized, a coarse-grained lock that is independent of the locations of the tuples (a relation lock) is acquired.

## 9.4  Update-Mode Locks

Deadlocks caused by lock upgrades can be prevented by using *update-mode locks* or *U locks*, for short. A transaction will acquire a U lock on data item $x$ instead of an S lock when it is possible that the lock needs to be upgraded to an X lock later. In this protocol, an S lock cannot be upgraded to an X lock nor to a U lock.

A U lock on data item $x$ can be acquired when no other transaction holds an X lock or a U lock on $x$ (i.e., the same as for an S lock). When a transaction has a

U lock on $x$, other transactions cannot get any new locks on $x$, but they can keep the S locks that they may already have on $x$.

When a transaction that holds a U lock needs to update the locked data item $x$, it must upgrade its U lock to an X lock. To do this, the transaction must wait for all of the readers of $x$ to release their S locks. Thus, when a transaction $T$ holds a U lock on $x$, the operation $lock(T, x, X, d)$ possibly causes $T$ to wait for the granting of the lock.

The automatic acquiring of U locks requires that the query language has a way of hinting the system about this. In embedded SQL, when a cursor has been created with the attribute **for update**, U locks instead of S locks are acquired on the tuples touched by the cursor. If the current tuple fetched by the cursor is updated, the U lock is upgraded to an X lock that is held until the transaction commits. If the current tuple is not updated and the cursor is advanced to fetch the next tuple, the U lock is either downgraded to an S lock (when the repeatable-read isolation level is maintained) or released altogether (at the read-committed isolation level).

The mutual exclusivity order for the lock modes S, U, and X is

$$S < U < X.$$

The lock-compatibility matrix for the set of lock modes $\{S, U, X\}$ is asymmetric (see Fig. 9.5). In the lock-upgrade matrix (Fig. 9.6), the value "−" means that the lock upgrade is not permitted.

**Theorem 9.5** *The locking protocol with S, U, and X locks prevents deadlocks that would be caused by lock upgrades.*

*Proof* According to the protocol, for any data item $x$, only one transaction at a time can hold a U lock on $x$. Thus, only one transaction can be waiting to upgrade a U lock on $x$ into an X lock. On the other hand, upgrades of S locks are not permitted at all. □

| lock requested by $T_2$ | lock held by $T_1$ | | |
|---|---|---|---|
| | S | U | X |
| S | true | false | false |
| U | true | false | false |
| X | false | false | false |

**Fig. 9.5** Lock-compatibility matrix for update-mode locking with the lock modes S, U, X

| lock requested by $T$ | lock held by $T$ | | |
|---|---|---|---|
| | S | U | X |
| S | S | U | X |
| U | − | U | X |
| X | − | X | X |

**Fig. 9.6** Lock-upgrade matrix for update-mode locking with the lock modes S, U, X

## 9.5   Cursor Stability

Requiring serializability (isolation level 4°), or preventing all isolation anomalies, does not always allow for enough concurrency. A transaction $T$ that produces a report of a major part of the database will prevent concurrent writes to all tuples of the database that $T$ has already processed: $T$ will acquire S locks on the keys $x$ of the tuples $(x, v)$ that it reads, and the locking protocol requires that all the locks are kept until $T$ commits. Another transaction $T'$ can read or delete the tuples that $T$ has read, or insert new tuples in between them, only after $T$ has committed or completed rollback.

However, the transaction $T$ reads each tuple only once and does not update the database, and thus, allowing unrepeatable reads might perhaps not cause great harm. The isolation level of transaction $T$ could be dropped down to 2° (or read-committed level) by changing the read locks in $T$ to short-duration locks. In this case, however, it is possible that the report produced by $T$ does not represent a transaction-consistent state of the database.

Consider the following special case of an unrepeatable read. A *lost update* happens when a transaction $T_1$ does not observe an update performed by another transaction $T_2$. More specifically, $T_1$ reads a key range, after which $T_2$ updates a tuple with key $x$ in this range and commits. Then $T_1$ also updates $x$:

$$H = \ldots R_1[y, \theta z] \ldots o_2[x] \ldots C_2 \ldots o_1[x] \ldots$$

Here $o_1[x]$ and $o_2[x]$ are updates (inserts, deletes, or writes) on key $x$, and $y \geq x \ \theta \ z$. Note that transaction $T_1$ may base its update on the result of the read action $R_1[y, \theta z]$. Such a lost update is possible if the S lock acquired on key $y$ for the action $R_1[y, \theta z]$ is of short duration. After the S lock is released at the end of the action, $T_2$ may get X locks on key $x$ and its next key for its action $o_2[x]$.

*Example 9.6*   Consider the following transactions on a banking database:

$$T_1 = BR[x, v_1]W[x, u_1, v_1 - 100]C.$$
$$T_2 = BR[x, v_2]W[x, u_2, v_2 - 200]C.$$

Transaction $T_1$ can be interpreted as "withdraw 100 euro from account $x$" and transaction $T_2$ as "withdraw 200 euro from account $x$."

The following history can be run on every database that contains the tuple $(x, v)$:

| $T_1$: | $BR[x, v]$ | | $W[x, v - 200, v - 100]C$ |
|---|---|---|---|
| $T_2$: | | $BR[x, v]W[x, v, v - 200]C$ | |

The only isolation anomaly present in this history is the unrepeatable read $R[x, v]$ by $T_1$. This history becomes possible if the S lock on key $x$ acquired by $T_1$ for the read action $R[x, v]$ is released immediately after performing the action. After this history, the balance for account $x$ will be $v - 100$, where $v$ is the original balance. Transaction $T_1$ based its update on an old balance, and the update by $T_2$ was overwritten.   □

How can we be sure that the update will always be based on the fresh value of the tuple $(x, v)$? The solution is to read $(x, v)$ and keep the S lock on key $x$ until it is found out whether or not the tuple needs to be updated. If the tuple needs to be updated, the S lock on key $x$ is upgraded to an X lock, and this lock is kept until the transaction commits. If there is no need for the update, the S lock on $x$ can be released.

The above solution prevents lost updates. However, other types of unrepeatable reads can of course still be present. The isolation level that prevents lost updates is called *cursor stability* in the literature, and it lies between levels 2° and 3°.

A locking protocol that ensures cursor stability but otherwise allows unrepeatable reads can be automatically implemented only if there exists a suitable programming language structure, such as the cursor mechanism of SQL. There, the key of the current tuple (current of cursor) is S-locked, and the lock is upgraded to a commit-duration X lock if the tuple is updated. Otherwise, the S lock is released when the next tuple is fetched. Reads that are not done using a cursor have to be protected with commit-duration S locks. To prevent deadlocks caused by upgrading S locks, update-mode locks (U locks) can be used instead of S locks: the U lock is upgraded to an X lock if the tuple is updated; otherwise, it is released.

## 9.6   Reads Without Locks

Assume that transaction $T$ is to be run at isolation level 2° (read committed). With the locking protocols presented above, this level is maintained when a commit-duration X lock is acquired for each update action and a short-duration S lock for each read action by $T$. The only purpose of the S lock is to prevent a dirty read, that is, make sure that the tuple to be read is in a committed state during the read action.

However, the transactions acquire just as many locks in this scheme as in the locking protocol that ensures level-3° isolation. Besides, releasing short-duration S locks individually may take more time than releasing the same number of commit-duration locks all at once at transaction commit.

The following simple idea can be used to reduce the number of short-duration S locks needed. Let COMMIT-LSN be the minimum of the LSNs of the begin-transaction log records of all currently active transactions. That is, COMMIT-LSN is the LSN of the begin-transaction log record $\langle T', B \rangle$ for the oldest transaction $T'$ that is still active.

For maintaining the COMMIT-LSN, the active-transaction table needs to include for each active transaction—in addition to its identifier, state (forward- or backward-rolling), and UNDO-NEXT-LSN and LAST-LOCK fields—a BEGIN-LSN field that contains the LSN of the begin-transaction log record of the transaction.

Assume that transaction $T$, running at isolation level 2°, wants to read tuples from data page $p$, which has already been fixed and read-latched by the server-process thread that executes $T$. Now assume that

$$\text{PAGE-LSN}(p) < \text{COMMIT-LSN}.$$

In this case page $p$ cannot contain any updates by active transactions, not even by $T$ itself. If some transaction $T'$ had inserted, deleted, or updated some tuple in page $p$, then PAGE-LSN$(p)$ would be greater than the LSN of the begin-transaction log record of $T'$ and hence, if $T'$ is still active, greater than COMMIT-LSN.

It is possible that some transaction $T'$ already holds an X lock on a tuple in page $p$ that will soon be updated, but this update has not yet been done, because PAGE-LSN$(p)$ is smaller than the COMMIT-LSN and hence smaller than the LSN of the begin-transaction log record of $T'$. Moreover, the update cannot be done before the thread executing $T$ has released its read latch on page $p$.

Thus, regardless of whether some $T'$ holds such an X lock, no dirty read can happen, even if $T$ is allowed to read all tuples in $p$ without acquiring any locks on them. After $T$ has read all of $p$'s tuples, the page is unlatched and unfixed.

This scheme can give significant savings in locking operations if $T$ reads all tuples from multiple data pages. An S lock must be acquired only on the tuples in such pages $p'$ for which PAGE-LSN$(p')$ is greater than the COMMIT-LSN. With multigranular locking, intention locks (IS, IX, SIX) on lockable units larger than tuples are acquired as usual.

The COMMIT-LSN scheme is made even more effective by writing the begin-transaction log record for a transaction only just before its first update and not writing begin-transaction or commit log records at all for read-only transactions.

## 9.7 Recovery and Concurrent Transactions

Usually, it is necessary to apply a locking protocol (or some other concurrency-control scheme) only during normal transaction processing, when transactions are generated by multiple concurrent server-process threads. During the redo pass of restart recovery, no locks need to be acquired, because actions in the redo pass are always performed in the original order, to repeat a history from the log that had earlier been accepted. Similarly, it is not necessary to acquire locks for the undo pass of recovery. Also recall the important principle shared by all locking protocols that no locks are acquired for undo actions when performing transaction rollbacks.

In Sect. 5.6 we observed that if dirty writes are prevented during normal transaction processing, active transactions can always be completed to rolled-back transactions with any interleaving of the undo actions in the undo phase. There will be no dirty writes in such a completed history nor any other isolation anomalies that were not already present in the original history. In the undo pass of the ARIES recovery algorithm, the undo actions are done in reverse chronological order of the forward-rolling actions, by a single backward scan of the log. This is only one of the possible orderings (though the most natural).

Normally, no new transactions are allowed to enter the system during recovery. However, recovery can take a long time, and thus, the data in the database can be unavailable to the users for a long time. To maintain *data availability*, new transactions should be accepted into the system as soon as possible after a system crash.

It is possible to implement the undo pass of recovery in such a way that individual transactions are rolled back in the same manner as in normal transaction processing, and the undo actions can be executed concurrently with new transactions. In this method, one or more processes or threads are created to execute the undo passes of the transactions to be rolled back.

The method will work only if the necessary X locks are first acquired for all of the backward-rolling transactions. In read-write locking (for the read-write transaction model), this is easily done by looking at the log records written for the $W[x, u, v]$ actions. In key-range locking, however, the delete action $D[x, v]$ must also log the next key $y$ of key $x$, so that a commit-duration X lock on it could be acquired from the information in the log.

An alternative to acquiring the locks is to use the COMMIT-LSN mechanism during the undo pass of recovery. Here, the undo pass is executed in the original manner, in a single backward scan of the log and without acquiring any locks. In the analysis pass of recovery, the COMMIT-LSN is set to the LSN of the begin-transaction log record of the oldest active transaction.

New transactions, which have to acquire locks in the normal way, are allowed to enter the system as soon as the undo pass begins, but a new transaction is allowed to access page $p$ only if PAGE-LSN($p$) is less than the COMMIT-LSN. If PAGE-LSN($p$) is greater than the COMMIT-LSN, the new transaction must wait for the undo pass of recovery to finish.

*Example 9.7*  Assume that transaction $T$ performs an action $D[x, v]$ on leaf page $p$ of a B-tree, where the least key $y$ in the database resides in the page, $p'$, next to page $p$. Then some other transaction commits, and after flushing the log the system crashes.

As $T$ was active at the time of the crash and had updated page $p$, we have PAGE-LSN($p$) > COMMIT-LSN. Hence any new transaction $T'$ allowed to the system during the undo pass is denied access to page $p$ until the undo pass has been completed.

But assume that page $p'$ had no uncommitted updates at the time of the crash, so that PAGE-LSN($p'$) < COMMIT-LSN. Recall that when transaction $T$ accessed the page $p'$ so as to determine the key $y$ next to key $x$, the page $p'$ was only read-latched and not updated.

In the undo pass, the transactions that were active at the time of the crash are allowed to roll back without holding any locks. Thus, $T$ will roll back and perform the undo action $D^{-1}[x, v]$ without holding the commit-duration X lock on the next key $y$.

A new transaction $T'$ can now access page $p'$ and perform actions such as $I[x']$, $D[y]$, and $R[x', \theta z]$, where $y > x' \geq z \geq$ *low-key*($p'$). If done before $T$ completes its rollback, these actions were not permitted by the key-range locking protocol during normal processing. However, none of these actions exhibits an isolation anomaly.                                                                                      □

## Problems

**9.1** Consider a multi-granular key-range locking protocol on a four-level granule hierarchy: the system, databases, relations (and indexes), and tuples. The units at all the four levels are lockable. What locks are acquired by the transactions generated by the following program fragments? All the transactions operate on a relation $r(\underline{X}, V)$ of database $b$ in system $s$. The transactions are run at the serializable isolation level, but at the same time as much concurrency as possible should be allowed, while saving on the number of locks.

(a) **create table** $r(X, V)$; **commit.**
(b) **alter table** $r$ **add primary key** $(X)$; **commit.**
(c) **create index** $I$ **on** $r(X)$; **commit.**
(d) **drop table** $r$; **commit.**
(e) **select** $*$ **from** $r$ **where** $x_1 < X$ **and** $X < x_2$;
     **insert into** $r$ **values** $(y, v)$; **commit.**
(f) **select count** $(*)$ **from** $r$;
     **delete from** $r$ **where** $X = x$; **commit.**
(g) **update** $r$ **set** $V = V + 1$ **where** $X = x$; **commit.**
(h) **update** $r$ **set** $V = V + 1$ **where** $x_1 < X$ **and** $X < x_2$; **commit.**
(i) **delete from** $r$ **where** $X < x$; **commit.**

In case (c), a dense index is created. In cases (e) to (i), there exists a sparse (primary) B-tree index to relation $r$ on key $X$.

**9.2** Consider the four-level granule hierarchy of the previous problem. Assume that only the units at the two lowest levels, namely, relations and tuples, are termed lockable, so that no locks are acquired at the two highest levels (system and databases) of the granule hierarchy. What problems may arise? What if some intermediate level, such as the database level, is omitted from the set of lockable units?

**9.3** Utilizing the property that an IX lock is more permissive than an X lock, we can relax the basic (single-granular) key-range locking protocol, so that for an insert action $I[x, v]$, a short-duration IX lock instead of a short-duration X lock is acquired on the key $y$ next to key $x$. Show that this relaxation works, that is, the relaxed protocol still prevents all isolation anomalies. Symmetrically, we might think that for a delete action $D[x, v]$, it would suffice to acquire only a short-duration IX lock (instead of a short-duration X lock) on key $x$. However, this relaxation does not work. Show why not.

**9.4** Assume that the different-granule units of the database form an acyclic directed graph instead of a tree-like hierarchy. Think of an object-oriented database in which an object may be a member of more than one collection object. Also in a relational database, the lockable units may form an acyclic graph instead of a tree when index-specific locking is used. With *index-specific locking*, indexes to a relation are lockable units.

How is the multi-granular locking protocol generalized from trees to acyclic graphs? Consider what nodes on the paths from the root unit down to the target unit need to be locked (a) for reading the target unit and (b) for updating the target unit.

Assuming that there is a sparse B-tree index to relation $r(\underline{X}, Y, V)$ on the primary key $X$ and a dense B-tree index to $r$ on the attribute $Y$, what locks are acquired for the following transactions under index-specific multi-granular key-range locking?

$T_1$ : **select** $X$, $V$ **from** $r$ **where** $Y = y'$; **commit**.
$T_2$ : **insert into** $r$ **values** $(x, y, v)$; **commit**.

**9.5** Assuming a multi-granular locking protocol with a three-level granule hierarchy (database, relations, tuples), what locks are acquired for the actions generated from the following application program fragment when the transaction is run (a) at the serializable isolation level and (b) at cursor stability?

```
exec sql declare cursor C for
select * from r where :y < X and X < :z for update
exec sql open C
while sqlstate = OK do
   exec sql fetch C into :x, :v
   if f(:v) then
      update r set V = g(:v) where current of C
   end if
end while
exec sql close C
exec sql commit
```

Here $f$ and $g$ are some functions on the domain of attribute $V$ of relation $r(\underline{X}, V)$.

**9.6** Define a locking protocol that is a combination of multi-granular key-range locking and update-mode locking. Give both the lock-compatibility and lock-upgrade matrices. Apply the protocol to the transaction of the previous problem.

**9.7** Reconsider Problem 6.7 regarding the starvation problem present in Algorithms 6.6–6.8. Assume that the algorithms are used to read, insert, and delete tuples in relation $r$. Show how the multi-granular locking protocol can be used to prevent starvation. To that end, work out simple modifications to the algorithms.

**9.8** In Sect. 9.7 we explained how the COMMIT-LSN mechanism makes it possible to accept new transactions to the system while the undo pass of restart recovery is still in progress. Actually, using the same mechanism, we can in some cases run new transactions already during the redo pass. Explain how.

**9.9** A *real-time database system* processes transactions with time constraints such as *deadlines*. The system tries to minimize the number of transactions that miss their deadlines. To that end, the locking protocols should be modified so as to observe the deadlines of transactions.

Consider the following simple modification of a locking protocol with lock modes S and X. The queue of transactions that are waiting for a lock on a data item is kept in ascending deadline order. A transaction that holds a lock but has already missed its deadline is aborted and rolled back. If a transaction requests an S lock on a data item currently S-locked by another transaction, the lock is not granted immediately if there is some other transaction with an earlier deadline waiting for an X lock on the data item; in that case the requesting transaction is put to wait for the S lock.

What changes are needed in the design of the lock table and the locking algorithms to implement the above-modified locking protocol? Find cases in which the protocol works poorly.

## Bibliographical Notes

Multi-granular locks already appear in the logical locking protocol applied in System R and described by Astrahan et al. [1976]: locks can be acquired on granules such as segment, relation, and tuple. System R's concurrency-control algorithm is reviewed by Mohan [1996a]. The multi-granular locking protocol with lock modes S, X, IS, IX, and SIX for tree hierarchies (Sect. 9.2) and its generalization for directed acyclic graphs (Problem 9.4) come from Gray et al. [1975, 1976].

The refined key-range locking protocol that uses IX locks (Problem 9.3) is a simplified version of the protocols presented by Mohan [1990a, 1996a]. The multi-granular locking protocol for acyclic graphs coupled with update-mode locks and key-range locking is treated in detail by Gray and Reuter [1993].

The COMMIT-LSN mechanism comes from Mohan [1990b, 1996b], and its use for improving data availability during the redo and undo passes of restart recovery is explained by Mohan [1993b] (cf. Problem 9.8). The cursor-stability isolation level of SQL is discussed by several authors, including Gray and Reuter [1993], Berenson et al. [1995], and Kifer et al. [2006].

Locking protocols for transactions with deadlines (see Problem 9.9) are considered by Abbott and Garcia-Molina [1988a,b, 1989] and Agrawal et al. [1995], among others.

# Chapter 10
# Bulk Operations on B-Trees

Processing data in *bulks* of many data tuples is usually more efficient than processing each tuple individually. A bulk of tuples to be inserted into a relation or a set of keys of tuples to be read or deleted can be sorted in key order before accessing the B-tree-indexed relation. In this chapter we show that processing tuples in key order on a B-tree is far more efficient than in random order, even when using the standard algorithms presented in the previous chapters.

In transaction processing, bulks may appear in different ways. A *system-specific bulk* appears as a set of single-item insertions or deletions by different transactions deferred to be finally processed in one large operation. An *application-specific bulk* is created by a specific *bulk action* performed by a single transaction so as to insert a set of tuples or to read or update or delete a set of tuples with given keys or within given key ranges. The final processing of a system-specific bulk is usually a series of reorganizations (structure modifications) on the B-tree due to installing buffered input or physically deleting tuples marked to be deleted, and this occurs outside the originating transactions, while the processing of an application-specific bulk is part of the transaction's application logic (except that structure modifications, of course, commit independently of the transaction).

The discussion in this chapter is mainly about application-specific bulks that appear in applications such as data-warehouse maintenance, where large bulks of data are regularly inserted and deleted. For protecting bulk actions on a relation whose key space can naturally be partitioned into fixed subranges, such as with the multi-attribute key of a warehouse fact table, we present a kind of multi-granular key-range locking protocol in which (logical) partitions of the relation can be locked. Some of the techniques to be presented also apply to environments such as embedded systems in which throughput can be improved by collecting very frequently incoming updates as (system-specific) bulks that are periodically applied to the B-tree. Moreover, in Chap. 11 we present methods for online index construction applying bulk reads and bulk insertion.

© Springer International Publishing Switzerland 2014
S. Sippu, E. Soisalon-Soininen, *Transaction Processing*, Data-Centric Systems and Applications, DOI 10.1007/978-3-319-12292-2_10

## 10.1   Transactions with Bulk Actions

We begin by defining a transaction model for *bulk-action transactions*, that is, transactions with bulk actions. Such a transaction can contain the following forward-rolling *bulk actions*:

1. *Bulk-read actions* of the form $R[s_X, s_{XV}]$. Such an action takes as input a set $s_X$ of non-overlapping open, closed, or half-open key ranges $(z_1, z_2)$, $[z_1, z_2]$, $(z_1, z_2]$, or $[z_1, z_2)$ and produces as output a tuple set $s_{XV}$ consisting of all tuples $(x, v) \in r$ with key $x$ in one of the ranges in $s_X$.
2. *Bulk-insert actions* of the form $I[s_{XV}]$ that take as input a set $s_{XV}$ of tuples with keys not appearing in $r$ and insert those tuples into $r$.
3. *Bulk-delete actions* of the form $D[s_X, s_{XV}]$. Such an action takes as input a set $s_X$ of non-overlapping open, closed, or half-open key ranges and deletes from $r$ all tuples $(x, v)$ with key $x$ in one of the ranges in $s_X$. The deleted tuples are returned in the set $s_{XV}$.
4. *Bulk-update actions* or *bulk-write actions* of the form $W[s_X, f, s_{XVV'}]$. Such an action takes as input a set $s_X$ of non-overlapping open, closed, or half-open key ranges and a function $f$ on tuple values and replaces, for each key $x$ in one of the ranges, the tuple $(x, v)$ with key $x$ by the tuple $(x, f(v))$. The set $s_{XVV'}$ returned consists of the tuples $(x, v, f(v))$.

The undo actions for the bulk-insert and bulk-delete actions are defined in the obvious way:

5. The undo action $I^{-1}[s_{XV}]$ for bulk-insert action $I[s_{XV}]$ deletes from $r$ the tuples in $s_{XV}$.
6. The undo action $D^{-1}[s_X, s_{XV}]$ for bulk-delete action $D[s_X, s_{XV}]$ inserts into $r$ the tuples in $s_{XV}$.
7. The undo action $W^{-1}[s_X, f, s_{XVV'}]$ for bulk-update action $W[s_X, f, s_{XVV'}]$ replaces in $r$ each tuple $(x, v')$ with $(x, v, v') \in s_{XVV'}$ by the tuple $(x, v)$.

The bulk-read, bulk-delete, and bulk-update actions model SQL statements such as

> **select** ∗ **from** $r$ **where** $C(X)$,
> **delete from** $r$ **where** $C(X)$,
> **update** $r$ **set** $V = f(V)$ **where** $C(X)$,

where $C(X)$ is a disjunction of *range predicates* of the form $z_1 \, \theta_1 \, X$ **and** $X \, \theta_2 \, z_2$ with $\theta_1, \theta_2 \in \{<, \leq\}$ on the primary key $X$ of relation $r$ and $f$ is a function on $V$. The bulk-read and bulk-insert actions together can be used to model SQL statements such as

> **insert into** $r$ **select** ∗ **from** $r'$ **where** $C(X)$ **order by** $X$,

where $r'$ is a relation with a schema compatible with that of $r$.

In SQL, bulk reads, deletes, and updates on relation $r(XYV)$ can take more complicated forms such as

**select** $*$ **from** $r$ **where** $C_1(X)$ **and** $C_2(XY)$,
**delete from** $r$ **where** $C_1(X)$ **and** $C_2(XY)$,
**update** $r$ **set** $V = f(V)$ **where** $C_1(X)$ **and** $C_2(XY)$,

where $C_1(X)$ is a disjunction of range predicates on the primary key $X$ and $C_2(XY)$ is a *data predicate* that depends on the values of the attributes in $XY$ of the tuple to be deleted. Moreover, bulk deletes and updates are often *cursor-based* such as in the following:

**exec sql declare** $t$ **cursor for**
**select** $*$ **from** $r$ **where** $C_1(X)$ **and** $C_2(XY)$
**exec sql open** $t$
**while** unfetched tuples exist **do**
　　**exec sql fetch** $t$ **into** $:x, :y, :v$
　　**if** $C_3(XYV)$ holds for $x$, $y$ and $v$ **then**
　　　　**exec sql delete from** $r$ **where current of** $t$
　　**end if**
**end while**

Here $C_3(XYV)$ is a *residual predicate* that depends both on the attribute values of the fetched tuple and on the values of some host-program variables (or, in the case of an internal *system cursor*, on some query-processor variables).

In all these cases, however, if a B-tree index exists on the primary-key attribute $X$, the algorithms to be presented for the bulk actions in our transaction model can easily be augmented to also evaluate any data predicates $C_2(XY)$ or residual predicates $C_3(XYV)$ during the index scan.

The model of bulk-action transactions differs significantly from our key-range transaction model. A single bulk action can span any number of data pages and hence can no longer be implemented as an atomic action accomplished by latching one data page or two successive data pages of a sparse B-tree index. A bulk-insert, bulk-delete, or bulk-update action must be logged with several log records. The state of a transaction cannot be represented with sufficient precision as a sequence of bulk actions.

As a bulk action can be left incomplete due to a failure, we need to fix an order in which the tuples in the bulk are processed when executing the action. Obviously, the only meaningful order is the key order. Thus, the execution of a bulk-read action $R[s_X, s_{XV}]$, bulk-delete action $D[s_X, s_{XV}]$, or bulk-update action $W[s_X, f, s_{XVV'}]$ involves sorting the key set $s_X$, and the execution of a bulk-insert action $I[s_{XV}]$ involves sorting the tuple set $s_{XV}$ in key order before starting the B-tree traversal.

In defining the state of a bulk-action transaction $T$ as a sequence of actions executed, we view each bulk action as *expanded* into a sequence of tuple-wise actions, that is, single-tuple reads, inserts, and deletes. In such a state, the last bulk-action may be incomplete, with only a prefix of the key-ordered tuple-wise action sequence executed.

*Example 10.1* For the bulk-action transaction

$$T = BR[s_X, s_{XV}]I[s'_{XV}]C$$

with $s_{XV} = \{(x_1, v_1), \ldots, (x_n, v_n)\}$ and $s'_{XV} = \{(y_1, w_1), \ldots, (y_m, w_m)\}$, possible states include

$BR[x_1, v_1] \ldots R[x_i, v_i]$,
$BR[x_1, v_1] \ldots R[x_n, v_n]I[y_1, w_1] \ldots I[y_j, w_j]$,
$BR[x_1, v_1] \ldots R[x_n, v_n]I[y_1, w_1] \ldots I[y_m, w_m]C$,
$BR[x_1, v_1] \ldots R[x_n, v_n]I[y_1, w_1] \ldots I[y_j, w_j]AI^{-1}[y_j, w_j] \ldots I^{-1}[y_{j'}, w_{j'}]$,
$BR[x_1, v_1] \ldots R[x_n, v_n]I[y_1, w_1] \ldots I[y_j, w_j]AI^{-1}[y_j, w_j] \ldots I^{-1}[y_1, w_1]C$,

where $i \leq n$, $j' \leq j \leq m$, $x_k < x_{k+1}$ for $k = 1, \ldots, n - 1$, and $y_k < y_{k+1}$ for $k = 1, \ldots, m - 1$. □

Concurrency of bulk-action transactions is similarly modeled with *expanded histories*, that is, shuffles of expanded transaction states, thus allowing concurrency within bulk actions. However, for representing isolation anomalies between bulk-action transactions via expanded transaction histories, the above mapping of bulk-read actions is inadequate.

*Example 10.2* Assume that the database $r$ is empty initially. In the history

$T_1 : \quad BR[\{[1, 3]\}, \emptyset] \qquad \cdots$
$T_2 : \qquad\qquad\qquad\quad BI[\{(2, v)\}] \quad \cdots$

the bulk-read action by $T_1$ should be considered an unrepeatable read, although no read action appears in the corresponding expanded history. Similarly, in the history

$T'_1 : \qquad\qquad\qquad\qquad\qquad BR[\{[1, 3]\}, \emptyset] \quad \cdots$
$T'_2 : \quad BD[\{[2, 2]\}, \{(2, v)\}] \qquad\qquad\qquad\qquad \cdots$

with $r = \{(2, v)\}$ initially, the bulk-read action by $T'_1$ should be considered a dirty read. □

For reducing the definitions of isolation anomalies between bulk-action transactions to those between transactions in the key-range model, we map bulk-read actions $R[s_X, s_{XV}]$ as follows. Each key range $(z_1, z_2)$, $[z_1, z_2)$, $(z_1, z_2]$, or $[z_1, z_2]$ in $s_X$ is mapped to the sequence of key-range-read actions

$$R[x_1, \theta z_1, v_1]R[x_2, > x_1, v_2] \ldots R[x_k, > x_{k-1}, v_k]R[x_{k+1}, > x_k, v_{k+1}],$$

where $x_1, x_2, \ldots, x_k$ are the keys in the range to be read, and the last action, $R[x_{k+1}, > x_k, v_{k+1}]$, is only present if $x_k < z_2$. In the case $R[x_{k+1}, > x_k, v_{k+1}]$ is present, its output is not considered to be part of the output of $R[s_X, s_{XV}]$. The comparison operator $\theta$ is either $>$ or $\geq$ depending on whether the range to be read is open or closed on the left end. In the special case that there are no keys in the range read, the last (and only) action in the mapped sequence is defined to be $R[x_1, \theta z_1, v_1]$.

*Example 10.3* With the above mapping of bulk-read actions, the histories of Example 10.2 are mapped to the following expanded histories:

$$T_1 : \quad BR[\infty, > 1, 0] \qquad \ldots$$
$$T_2 : \qquad\qquad\qquad BI[2, v] \quad \ldots$$

This exhibits an unrepeatable read.

$$T_1' : \qquad\qquad\qquad BR[\infty, > 1, 0] \quad \ldots$$
$$T_2' : \quad BD[2, v] \qquad\qquad \ldots$$

This exhibits a dirty read. □

Uncommitted updates are defined as in Sect. 5.2 for the expanded transaction histories.

*Example 10.4* In the expanded history

$$BR[x_1, v_1] \ldots R[x_n, v_n] I[y_1, w_1] \ldots I[y_j, w_j] A I^{-1}[y_j, w_j] \ldots I^{-1}[y_{j'}, w_{j'}],$$

the keys $y_1$ to $y_{j'-1}$ have an uncommitted update, while the keys $y_{j'}$ to $y_j$ do not. □

Dirty writes, dirty reads, and unrepeatable reads can now be defined in the usual way for pairs of tuple-wise read, insert, and delete actions in an expanded history.

## 10.2 Locking Range Partitions

Having reduced the definitions of isolation anomalies between bulk-action transactions to those between transactions in the key-range model, we have the option of using the standard key-range locking protocol of Sect. 6.4 to control the concurrency of bulk-action transactions. However, in the case of a bulk action involving thousands or even millions of tuples, making that many calls to the lock manager is not feasible. The tuple-level locks must be escalated to locks of a coarser granularity, usually meaning that the entire relation is locked or, if the relation is physically range-partitioned into subranges of the key space of the primary key, the relation is intention-locked and the partition fragments whose key ranges intersect a range to be accessed are locked explicitly.

A typical target for bulk actions is a data-warehouse *fact table* $r(XV)$, where the primary key $X$ consists of foreign keys referencing dimension tables.

*Example 10.5* As a running example we will use the fact table

*sales(sales-date, store-id, item-id, sales-amount),*

which contains, for the years 2010–2015, the total sales amount in euros of each item sold at any store of a chain in any day the store was open. Assuming that there are 200 stores and that an average store is kept open for 300 days a year and that

1,000 different items are sold per day on the average, the table will contain about $(6 \times 300) \times 200 \times 1000 = 360$ million tuples at the end of the year 2015.

A typical bulk-insert action on the *sales* table inserts new sales tuples representing the sales amounts of items sold at a single store during the previous day, such as

**insert into** *sales*
**select** ∗ **from** *sales-2015-09-25-S123*
**order by** *sales-date, store-id, item-id,*

where the relation (or view) *sales-2015-09-25-S123* contains the sales data of store S123 from September 25, 2015 (about 1,000 tuples).

A typical bulk-delete action undoes a committed bulk insert of incorrect data, or periodically deletes some of the oldest data from the table, such as the data for the first quarter of 2010:

**delete from** *sales* **where** *sales-date* < '2010-04-01'

(about $(300/4) \times 200 \times 1000 = 15$ million tuples).

The above actions insert or delete a set of tuples belonging to a subrange of the key space of the primary key of the table, where the component attributes of the key appear in the order *sales-date, store-id, item-id.*                                              □

Given that typical actions on the fact table are key-range actions on the primary-index key of the table, it is tempting to try to define a locking protocol that uses a single X lock to lock the entire key range to be inserted or deleted and a single S lock to lock the entire key range to be read. A simple solution would be to resort to *predicate locking* (see Sect. 6.8), which in this case means that the lock table would store predicates that specify a key range of the primary key, and checking two locks of incompatible modes for compatibility would mean testing the intersection of two ranges for emptiness.

However, as explained earlier, in efficient lock management, the lock names are short (4-byte) hash values computed from the key or identifier of the data item to be locked, the lock table is organized as a hash table, and the compatibility of two locks of incompatible modes is tested by comparing two lock names for inequality. We wish to define a locking protocol that satisfies all these properties. Moreover, we wish to maintain the natural requirement that locks once granted to a transaction never need be changed due to a structure modification (such as a page split or merge) that affects the locked part of the database while the transaction is still active. This rules out using locks whose names are derived from dynamic physical structures, such as page-ids of B-tree pages (possibly used to name a lock on the key range currently covered by the page).

Our solution is to allow for the database administrator to define a hierarchy of lockable *partition fragments* of different granularities for a relation by specifying *logical range partitions* on the relation. The specification would need a simple extension of the SQL **create table** and **alter table** statements.

*Example 10.6* For our *sales* table we might specify:

> **alter table** *sales*
> > **logical partition** *sales-p1* **by equality on**
> > > *year(sales-date), quarter(sales-date)*
> >
> > **logical partition** *sales-p2* **by equality on**
> > > *year(sales-date), month(sales-date), day(sales-date)*
> >
> > **logical partition** *sales-p3* **by equality on**
> > > *year(sales-date), month(sales-date), day(sales-date),*
> > > *store-id*

Here *year*, *quarter*, *month*, and *day* are functions on date values that extract the year, the quarter (1–4), the month (1–12), and the day (1–31) from a date value. In this case three partitions are specified: a partition by quarters (*sales-p1*), a partition by days (*sales-p2*), and a partition by days and stores (*sales-p3*). A fragment in partition *sales-p1* consists of the *sales* tuples for a single quarter of a single year, a fragment in partition *sales-p2* consists of the *sales* tuples for a single day, and a fragment in partition *sales-p3* consists of the *sales* tuples for a single day and store.                                                                                         □

We emphasize that the partitions so specified are purely *logical*; they do not in any way affect the storage of the tuples of the relation in the leaf pages of the B-tree nor the structure or contents of the non-leaf pages of the B-tree. Accordingly, the **alter table** statement in Example 10.6 in no way affects the current contents of the B-tree; it just inserts the partition specifications into the system catalog of the database. The logical partitions are also completely independent of any physical partition that the B-tree may have.

In general, logical range partitions for a relation $r(XV)$ are specified by defining a sequence of functions $g_1, \ldots, g_m$ on the key space of the primary key $X$ such that the following conditions are satisfied:

1. Each $g_i$ is monotone, so that $g_i(x) \leq g_i(y)$ for all keys $x$ and $y$ with $x \leq y$ in the ordering defined by the primary key.
2. Each $g_{i+1}$ defines a subpartition of the partition defined by $g_i$, that is, $g_i(x) = g_i(y)$ for all keys $x$ and $y$ with $g_{i+1}(x) = g_{i+1}(y)$, $i = 1, \ldots, m-1$.

The root of the partition hierarchy is the partition with the entire relation $r$ as its only fragment, defined by $g_0(x) = ()$. The finest-grained partition, with single tuples as fragments, is defined by $g_{m+1}(x) = x$. The functions $g_i$ are called *partition granularities*.

The partition granularities $g_i$, $i = 1, \ldots, m$, are defined using methods of standard SQL or user-defined data types. Fragments in the partition defined by $g_i$ are uniquely identified by their *keys*, that is, the values $g_i(x)$.

We say that fragment $g_i(x)$ is *contained* in fragment $g_{i-1}(x)$ and *contains* fragment $g_{i+1}(x)$ and that fragment $g_i(x)$ *covers* all keys $y$ with $g_i(y) = g_i(x)$.

*Example 10.7* The partitions *sales-p1*, *sales-p2*, and *sales-p3* in Example 10.6 are defined by the partition granularities $g_1$, $g_2$, and $g_3$ given below, where $d$ is a date, $s$ is a store-id, and $i$ is an item-id:

$$g_0(d,s,i) = ()$$
$$g_1(d,s,i) = (year(d), quarter(d))$$
$$g_2(d,s,i) = (year(d), month(d), day(d))$$
$$g_3(d,s,i) = (year(d), month(d), day(d), s)$$
$$g_4(d,s,i) = (year(d), month(d), day(d), s, i)$$

The key of a fragment in the *sales-p1* partition is of the form $(y,q)$, where $y$ is a year and $q$ is a quarter number (1–4). The key of a fragment in the *sales-p3* partition is of the form $(y,m,d,s)$, where $y$ is a year, $m$ is a month number (1–12), $d$ is a day number (1–31), and $s$ is a store-id.                                             $\square$

The partitions of the above example were chosen so as to conform to the logical structure of the primary key, but this is not a necessity. For example, for an integer key $X$, we might define a partition hierarchy by $g_1(x) = x \div 100000$, $g_2(x) = x \div 10000$, $g_3(x) = x \div 1000$.

The *partition-based key-range locking protocol* is based on locking keys of partition fragments. To distinguish between locks on fragments in different relations, each key is prefixed by the identifier of the relation. All lock names are 4-byte hash values computed from the prefixed keys. For example, the name of a lock on the fragment for the first quarter of year 2010 in the *sales-p1* partition is a hash value computed from the tuple $(sales, 2010, 1)$.

The lock modes available are the conventional ones for multi-granular locking: S, X, IS, IX, and SIX. Other lock modes, such as U (update-mode), could also be employed. The locking granularity to be used for an operation is decided by the query processor.

A bulk-read action $R[s_X, s_{XV}]$ with $[z_1, z_2] \in s_X$ has to S-lock the smallest partition fragment that covers the keys $z_1$ and $z_2$ and to IS-lock the containing fragments. In other words, when

$$k = \max\{i \mid 0 \le i \le m, g_i(z_1) = g_i(z_2)\},$$

the fragments $g_0(z_1), g_1(z_1), \ldots, g_{k-1}(z_1)$ are IS-locked and the fragment $g_k(z_1)$ is S-locked. From the mapping of bulk-read actions to single-tuple read actions discussed in the previous section, it follows that to achieve repeatable-read-level isolation, if the tuple with the least key $y$ greater than the keys of the tuples in the range $[z_1, z_2]$ does not belong to the fragment $g_k(z_1)$, that is, $g_k(y) \ne g_k(z_1)$, then the key $y$ must also be S-locked for commit duration and the containing fragments $g_i(y)$ IS-locked.

*Example 10.8* In the case of the query

**select sum**(*sales-amount*) **from** *sales*
**where** *sales-date* ≥ "2015-07-15" **and** *sales-date* ≤ "2015-08-15"

we have, with the partition granularities of Example 10.7,

$$g_1(2015\text{-}07\text{-}15, s, i) = (2015, 3) = g_1(2015\text{-}08\text{-}15, s', i'), \text{ but}$$
$$g_2(2015\text{-}07\text{-}15, s, i) = (2015, 7, 15) \neq (2015, 8, 15) = g_2(2015\text{-}08\text{-}15, s', i').$$

Thus, the *sales* relation is IS-locked and the fragment (*sales*, 2015, 3) is S-locked. In this case the key of the tuple next to the last tuple read belongs to the S-locked fragment, so that no additional locks need be acquired.                               □

When a key range spans two coarse-granule fragments, an unnecessarily wide range may be locked. Maintaining knowledge about the structure of multi-attribute keys in the system catalog may help the query processor to split such a range predicate into a disjunction of range predicates that only involve finer-granule fragments.

*Example 10.9* In the case of the query

**select sum**(*sales-amount*) **from** *sales*
**where** *sales-date* ≥ "2015-12-15" **and** *sales-date* ≤ "2016-01-15"

we have

$$g_0(2015\text{-}12\text{-}15, s, i) = () = g_0(2016\text{-}01\text{-}15, s', i'), \text{ but}$$
$$g_1(2015\text{-}12\text{-}15, s, i) = (2015, 4) \neq (2016, 1) = g_2(2016\text{-}01\text{-}15, s', i').$$

Thus, the entire *sales* relation would be S-locked. However, with suitable information stored in the system catalog, it can easily be inferred from the partition specifications and the known properties of the SQL data type *date* that it is wise to rewrite the selection predicate as the following disjunction:

*sales-date* ≥ "2015-12-15" **and** *sales-date* ≤ "2015-12-31"
**or** *sales-date* ≥ "2016-01-01" **and** *sales-date* ≤ "2016-01-15".

The former disjunct leads to S-locking the fragment (*sales*, 2015, 4) and the latter the fragment (*sales*, 2016, 1).                               □

In the case of a bulk-insert action $I[s_{XV}]$, the source of the insertion, the relation $s_{XV}$, is queried in order to find out the fragments that need be locked. To that end, for each partition granularity $g_i$, $i = 0, \ldots, m - 1$, we determine a temporary relation *temp-s_i* that tells the keys of $g_i$-fragments and the number of contained $g_{i+1}$-fragments: each tuple in *temp-s_i* consists of the key of one fragment and the count of tuples in the input belonging to that fragment.

Using the above temporary relations together with a specified upper limit on the number of X locks to be acquired, we can prune the sets of fragments until the number of X locks to be acquired falls below the limit. The pruning proceeds from finer-granule fragments to coarser-granule fragments. First, if a fragment $g_i(x)$ contains more $g_{i+1}$-fragments than the specified limit, then those $g_{i+1}$-fragments will not be X-locked. Second, of two different fragments $g_i(x)$ and $g_i(y)$ with $i <$ $m$, the one containing more $g_{i+1}$-fragments is more suitable for X-locking than the other.

We apply the relaxed key-range locking protocol discussed in Problem 9.3. According to this protocol, the short-duration lock to be acquired on the next key is only an IX lock instead of an X lock.

*Example 10.10*  In the case of the bulk-insert action

> **insert into** *sales*
> **select** * **from** *sales-2015-09-25-S123*
> **order by** *sales-date, store-id, item-id,*

the query for the $g_3$ partition is

> **insert into** *temp-sales-2015-09-25-S123₃*
> **select** *year(sales-date), month(sales-date), day(sales-date), store-id,*
>     **count**(*)
> **from** *sales-2015-09-25-S123*
> **group by** *year(sales-date), month(sales-date), day(sales-date), store-id,*

which will reveal that the tuples to be inserted all belong to a single $g_3$-fragment (*sales*, 2015, 9, 25, S123) and that there are 1,000 such tuples. Thus, commit-duration IX locks are acquired on the *sales* table and the fragments (*sales*, 2015, 3) and (*sales*, 2015, 9, 25), and a commit-duration X lock is acquired on the fragment (*sales*, 2015, 9, 25, S123). The only locks that remain to be acquired during the actual insertion are the short-duration IX lock on the key of the *sales* tuple next to the last tuple inserted and the IX locks on the fragments containing this next tuple. □

In the general case, a bulk-insert action X-locks for commit duration a sequence of fragments and IX-locks for commit duration the containing fragments, where the X-locked fragments together cover the keys of the tuples to be inserted. In addition to these locks, in order to follow the principle of key-range locking, the key of every existing tuple next to an inserted tuple that is not covered by an X-locked fragment must be IX-locked for short duration, and the containing fragments must be IX-locked for commit duration. All these locks are acquired in ascending key order during a left-to-right scan of the insertion points of the tuples to be inserted.

A bulk-delete action $D[s_X, s_{XV}]$ with $[z_1, z_2] \in s_X$ must X-lock for short duration the smallest partition fragment that covers the keys $z_1$ and $z_2$ and to IX-lock for short duration the containing fragments. In other words, when

$$k = \max\{i \mid 0 \le i \le m, g_i(z_1) = g_i(z_2)\},$$

the fragments $g_0(z_1), g_1(z_1), \ldots, g_{k-1}(z_1)$ are IX-locked and the fragment $g_k(z_1)$ is X-locked. Moreover, following the principle of key-range locking, a commit-duration X-lock must be acquired on the key of the tuple next to the last deleted tuple in the range $[z_1, z_2]$, and commit-duration IX locks must be acquired on the containing fragments.

## 10.3 Bulk Reading

The call $bulk\text{-}read(T, s_X, s_{XV})$ (Algorithm 10.1) implements the bulk-read action $R[s_X, s_{XV}]$ for transaction $T$. For simplicity, we consider only closed ranges $[z_1, z_2]$. As with the single-tuple read action (Algorithm 6.6), we assume that a saved path is maintained. With bulk actions this is all the more important because such actions usually touch many tuples in the same leaf page.

The idea is to process the key ranges of $s_X$ in ascending key order and for each range to traverse to the leaf page that covers the lower bound of the range, to read from the page all the tuples that belong to the range, and then to move to the next leaf page until the upper bound of the range is reached. In processing a range $[z_1, z_2]$, the variable $z$ takes $z_1$ as its initial value and, after reading from the leaf page that covers $z$ all tuples $(x, v)$ with $z \leq x \leq z_2$, advances to the key $y$ next to the greatest key read (Fig. 10.1).

The call $find\text{-}page\text{-}for\text{-}bulk\text{-}read(p, p', z, z_2)$ is used to traverse the B-tree to the leaf page $p$ that covers key $z$ and to the leaf page $p'$ that covers $high\text{-}key(p)$. The call returns with the pages $p$ and $p'$ read-latched. Page $p'$ is determined only if $z_2 \geq high\text{-}key(p)$; otherwise, the call returns with $p' = p$. The algorithm for $find\text{-}page\text{-}for\text{-}bulk\text{-}read(p, p', z, z_2)$ is similar to that for $find\text{-}page\text{-}for\text{-}fetch(p, p', \theta z, both)$ (Algorithm 7.4) with $both = $ true.

Before the B-tree traversal for a range $[z_1, z_2]$ is started, the smallest partition fragment that covers the keys $z_1$ and $z_2$ is S-locked and the containing fragments are IS-locked, for commit duration. The locks on the next keys are acquired, using conditional locking, as the keys are encountered during the left-to-right scan of the tuples in the range. In the algorithm, $keys(p)$ denotes the set of keys of tuples in leaf page $p$ and $keys(S)$ denotes the set of keys of tuples in tuple set $S$.

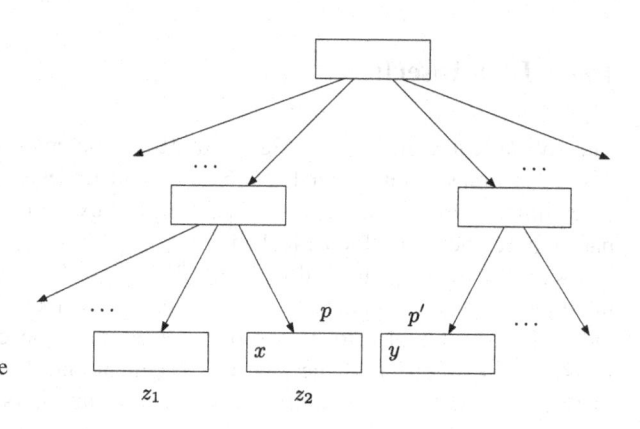

**Fig. 10.1** Bulk-reading tuples $(x, v)$ with keys $x$ in the range $[z_1, z_2]$. The least key $y$ greater than $z_2$ is in the next page $p'$

---

**Algorithm 10.1** Procedure *bulk-read*$(T, s_X, s_{XV})$

---

sort the key ranges in $s_X$ in ascending key order,
$s_{XV} \leftarrow \emptyset$
**for** all ranges $[z_1, z_2] \in s_X$ **do**
  $k \leftarrow \max\{i \mid 0 \leq i \leq m, g_i(z_1) = g_i(z_2)\}$
  **for** $i = 0$ to $k - 1$ **do**
    $lock(T, g_i(z_1), \text{IS}, \text{commit-duration})$
  **end for**
  $lock(T, g_k(z_1), \text{S}, \text{commit-duration})$
  $z \leftarrow z_1$
  **while** $z \leq z_2$ **do**
    *find-page-for-bulk-read*$(p, p', z, z_2)$
    $S \leftarrow \{(x, v) \in p \mid z \leq x \leq z_2\}$
    $y \leftarrow$ the least key in $keys(p) \cup keys(p') \cup \{\infty\}$ greater than those in $keys(S)$
    **if** $y > z_2$ **and** for all $i = 0, \ldots, m$: $g_i(y) \neq g_k(z_1)$ **then**
      **for** $i = 0$ to $m - 1$ **do**
        *conditionally-lock*$(T, g_i(y), \text{IS}, \text{commit-duration})$
      **end for**
      *conditionally-lock*$(T, y, \text{S}, \text{commit-duration})$
      **if** some of the requested locks were not granted **then**
        *unlatch-and-unfix*$(p, p')$
        **for** $i = 0$ to $m - 1$ **do**
          $lock(T, g_i(y), \text{IS}, \text{commit-duration})$
        **end for**
        $lock(T, y, \text{S}, \text{commit-duration})$
        continue with the **while** loop
      **end if**
    **end if**
    $s_{XV} \leftarrow s_{XV} \cup S$
    *unlatch-and-unfix*$(p, p')$
    $z \leftarrow y$
  **end while**
**end for**

---

## 10.4   Bulk Insertion

The call *bulk-insert*$(T, s_{XV})$ (Algorithm 10.2) implements the bulk-insert action $I[s_{XV}]$ for transaction $T$. First, the bulk $s_{XV}$ of tuples to be inserted is sorted in ascending key order and the sorted sequence is examined so as to find out what partition fragments need to be locked.

Using the method outlined in Sect. 10.2, the call *choose-locks*$(s_{XV}, L)$ determines a reasonably small set of fragments to be locked that together contain the bulk $s_{XV}$ of tuples to be inserted. Given $s_{XV}$, the call returns a list $L$ of pairs $(f, M)$, where $f$ is the key of a fragment and $M$ is the mode (X or IX) of the lock on $f$ to be acquired. The order of the pairs in $L$ is the top-down,

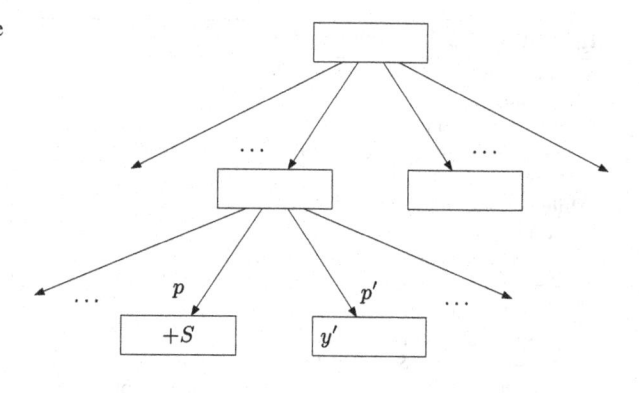

**Fig. 10.2** Bulk-inserting the tuples of $S$ into leaf page $p$. The least key $y'$ in the next page $p'$ may be among the next keys that must be IX-locked for short duration

left-to-right order determined by the partition hierarchy. The locks are then acquired in this order.

Next the sorted sequence of tuples to be inserted is scanned, with the tuple variable $(x, v)$ denoting the first not-yet-inserted tuple in the sequence. The leaf page $p$ that covers key $x$ is located and write-latched and as many tuples that fit into page $p$ are inserted into it, and the next keys of the inserted keys that are not covered by the already locked fragments are IX-locked for short duration, again using conditional locking. Then the tuple variable $(x, v)$ is advanced to the tuple next to the last inserted tuple in the sequence of tuples to be inserted.

Given a tuple $(x, v) \in s_{XV}$, the call *find-page-for-bulk-insert*$(p, h, p', x, v)$ traverses the B-tree and determines the leaf page $p$ that covers key $x$, the high-key $h$ of $p$, the page $p'$ next to $p$ (if any), write-latching $p$, and read-latching $p'$. The call also arranges, with appropriate structure modifications, that page $p$ has room at least for the tuple $(x, v)$. If no next page $p'$ exists (i.e., $h = \infty$), then $p' = p$. The algorithm for the call is similar to Algorithm 7.8.

In the algorithm, $S$ denotes the subsequence of the sorted sequence of tuples to be inserted into page $p$. The keys in *keys*$(p)$ are next keys to be locked that immediately follow some key in *keys*$(S)$ in the merged sequence of tuples from page $p$ and set $S$. Also, the least key $y'$ in the next page $p'$ is a next key to be locked if the last tuple in the merged sequence is a tuple from $S$ (Fig. 10.2). Moreover, $\infty$ is a next key if no next page $p'$ exists. The call *insert-bulk-into-page*$(T, p, S)$ (Algorithm 10.4) is used to insert the tuples in $S$ into page $p$ and to log the insertions with the redo-undo log records (3.9).

---

**Algorithm 10.2** Procedure *bulk-insert*$(T, s_{XV})$

---

sort the tuples in $s_{XV}$ in ascending key order
*choose-locks*$(s_{XV}, L)$
**for** $(f, M) \in L$ **do**
   *lock*$(T, f, M,$ commit-duration$)$
**end for**
**while** $s_{XV}$ not empty **do**
   $(x, v) \leftarrow$ the first tuple in $s_{XV}$
   *find-page-for-bulk-insert*$(p, h, p', x, v)$
   $S \leftarrow$ the longest prefix of $s_{XV}$ of tuples with keys less than $h$ that fit in page $p$
   $y' \leftarrow$ the least key in page $p'$
   $Y' \leftarrow keys(p) \cup \{y'\} \cup \{\infty\}$
   $W \leftarrow Y' \cup keys(S)$ merged in ascending order
   $Y \leftarrow$ the keys in $Y'$ that are next keys of some tuples in $S$ in the sequence $W$
   **for** all $y \in Y$ **do**
     **if** for all $i = 0, \ldots, m$: $L$ contains neither $(g_i(y), \text{IX})$ nor $(g_i(y), \text{X})$ **then**
       **for** $i = 0$ to $m$ **do**
         *conditionally-lock*$(T, g_i(y), \text{IX},$ short-duration$)$
       **end for**
       **if** some of the requested locks were not granted **then**
         *unlatch-and-unfix*$(p, p')$
         **for** $i = 0$ to $m$ **do**
           *lock*$(T, g_i(y), \text{IX},$ short-duration$)$
         **end for**
         continue with the **while** loop
       **end if**
     **end if**
   **end for**
   *insert-bulk-into-page*$(T, p, S)$
   $s_{XV} \leftarrow s_{XV} \setminus S$
   *unlatch-and-unfix*$(p, p')$
   release the short-duration locks if any
**end while**

---

## 10.5   Bulk Deletion

The call *bulk-delete*$(T, s_X, s_{XV})$ (Algorithm 10.3) implements the bulk-delete action $D[s_X, s_{XV}]$ for transaction $T$. Again, for simplicity, we consider only closed ranges $[z_1, z_2]$. As with the bulk-read action, the key ranges of $s_X$ are first sorted in ascending key order, and the processing of a range $[z_1, z_2]$ in $s_X$ starts with locking the smallest partition fragment that covers the keys $z_1$ and $z_2$. The smallest covering fragment is X-locked and the containing fragments are IX-locked, for short duration.

In processing a range $[z_1, z_2]$, the variable $z$, with initial value $z_1$, is advanced, step by step, up to $z_2$. At each step, while holding write-latched the leaf page $p$ that covers $z$, as many tuples are deleted from $p$ that is possible without the underflow of $p$, after which the variable $z$ is advanced to the key $y$ next to the last tuple deleted in the step. When $y$ exceeds $z_2$, it is X-locked, and the containing fragments are IX-locked, for commit duration.

Given a key range $[z, z_2]$, the call *find-page-for-bulk-delete*$(p, p', z, z_2)$ traverses the B-tree and determines the leaf page $p$ that covers the key $z$ and, if $z_2$ is greater than the greatest key in $p$, the page $p'$ next to $p$, write-latching $p$ and read-latching $p'$ (if any). The call also arranges, with appropriate structure modifications, that the page $p$ will not underflow when a tuple is deleted. The set $S$ of tuples to be deleted from $p$ at this time contains a maximum number of tuples in the range $[z, z_2]$ that can be deleted without causing the page to underflow. The call *delete-bulk-from-page*$(T, p, S)$ (Algorithm 10.5) is used to delete the tuples in $S$ from page $p$ and to log the deletions with the redo-undo log records (3.10).

If tuples remain in page $p$ that qualify for deletion, then in the next iteration of the **while** loop in Algorithm 10.3, *find-page-for-bulk-delete*$(p, p', z, z_2)$ is called with $z$ being the key of the first not-yet-deleted tuple in page $p$. As the page now is about to underflow, the call performs page merges or records redistributes, possibly preceded by a tree-height decrease, so that the page holding the tuple with key $z$ will tolerate the deletion of more tuples.

---

**Algorithm 10.3** Procedure *bulk-delete*$(T, s_X, s_{XV})$

sort the key ranges in $s_X$ in ascending key order
$s_{XV} \leftarrow \emptyset$
**for** all ranges $[z_1, z_2] \in s_X$ **do**
  $k \leftarrow \max\{i \mid 0 \le i \le m, g_i(z_1) = g_i(z_2)\}$
  **for** $i = 0$ to $k - 1$ **do**
    $lock(T, g_i(z_1), \text{IX, short-duration})$
  **end for**
  $lock(T, g_k(z_1), \text{X, short-duration})$
  $z \leftarrow z_1$
  **while** $z \le z_2$ **do**
    *find-page-for-bulk-delete*$(p, p', z, z_2)$
    $S' \leftarrow$ the tuples in page $p$ with keys in $[z, z_2]$, sorted in ascending key order
    $S \leftarrow$ the longest prefix of $S'$ that can be deleted without page $p$ underflowing
    $y \leftarrow$ the least key in $keys(p) \cup keys(p') \cup \{\infty\}$ greater than the keys in $S$
    **if** $y > z_2$ **then**
      **for** $i = 0$ to $m - 1$ **do**
        *conditionally-lock*$(T, g_i(y), \text{IX, commit-duration})$
      **end for**
      *conditionally-lock*$(T, y, \text{X, commit-duration})$
      **if** some of the requested locks were not granted **then**
        *unlatch-and-unfix*$(p, p')$
        **for** $i = 0$ to $m - 1$ **do**
          $lock(T, g_i(y), \text{IX, commit-duration})$
        **end for**
        $lock(T, y, \text{X, commit-duration})$
        continue with the **while** loop
      **end if**
    **end if**
    *delete-bulk-from-page*$(T, p, S)$
    *unlatch-and-unfix*$(p, p')$
    $s_{XV} \leftarrow s_{XV} \cup S$
    $z \leftarrow y$
  **end while**
  release the short-duration locks
**end for**

## 10.6   Logging and Undoing Bulk Actions

In Algorithm 10.2, tuples from a bulk are inserted into leaf page $p$ using the call *insert-bulk-into-page*$(T, p, S)$, which could be implemented by calling *insert-into-page*$(T, p, x, v)$ (Algorithm 3.1) on all tuples $(x, v)$ in $S$. In Algorithm 10.3, tuples are deleted from leaf page $p$ using the call *delete-bulk-from-page*$(T, p, S)$, which could similarly be implemented by calling *delete-from-page*$(T, p, x)$ (Algorithm 3.2) on the keys $x$ of tuples in $S$. The log then shows just an expanded history of tuple-wise actions, which can be undone, if so needed, by the procedures *undo-insert*$(n, T, p, x, n')$ (Algorithm 4.7) and *undo-delete*$(n, T, p, x, v, n')$ (Algorithm 4.9).

To save in the number of assignments of the PAGE-LSN of the page to be updated and of the UNDO-NEXT-LSN of the transaction, we can use Algorithm 10.4 to insert the tuples in $S$ into page $p$ and Algorithm 10.5 to delete the tuples in $S$ from page $p$. These algorithms log the tuple-wise actions as in *insert-into-page*$(T, p, x, v)$ and *delete-from-page*$(T, p, x, v)$, but only the LSN of the last tuple-wise update needs to be stamped into the PAGE-LSN of page $p$.

---

**Algorithm 10.4** Procedure *insert-bulk-into-page*$(T, p, S)$

---
$n' \leftarrow$ UNDO-NEXT-LSN$(T)$
**for** all tuples $(x, v) \in S$ **do**
    insert $(x, v)$ into page $p$
    $log(n, \langle T, I, p, x, v, n' \rangle)$
    $n' \leftarrow n$
**end for**
PAGE-LSN$(p) \leftarrow n$
UNDO-NEXT-LSN$(T) \leftarrow n$

---

**Algorithm 10.5** Procedure *delete-bulk-from-page*$(T, p, S)$

---
$n' \leftarrow$ UNDO-NEXT-LSN$(T)$
**for** all tuples $(x, v) \in S$ **do**
    delete $(x, v)$ from page $p$
    $log(n, \langle T, D, p, x, v, n' \rangle)$
    $n' \leftarrow n$
**end for**
PAGE-LSN$(p) \leftarrow n$
UNDO-NEXT-LSN$(T) \leftarrow n$

---

## 10.7  Complexity of Scanning a Key Range

In this section we analyze the complexity of the bulk-read, bulk-insert, and bulk-delete algorithms presented above. In all these actions the main target for optimization is the phase of searching for the leaf pages on which the actions need to be applied. We first notice that, due to the lack of sideways linking of the leaf pages in the B-tree, the traversal algorithms use the saved-path strategy in order to avoid the repeated search from the root page.

We first consider the bulk-read action. Let $s_X$ be a set of non-overlapping key ranges to be read. We wish to prove that in the absence of concurrent updating transactions, the total number of page latchings performed in bulk-reading the tuples in the range set $s_X$ is

$$O(\log n + \sum_{i=2}^{k} \log(1 + dist(l_{i-1}, l_i))), \qquad (10.1)$$

where $n$ is the number of all leaf pages, $l_1, l_2, \dots, l_k$ is the ordered sequence of leaf pages containing tuples in some range of $s_X$ (so that the keys in $l_{i-1}$ are less than those in $l_i$), and $dist(l_{i-1}, l_i)$ denotes the number of leaf pages between $l_{i-1}$ and $l_i$.

In order to prove the result, we begin with some definitions and lemmas. For any two leaf pages $p$ and $q$, let $M(p, q)$ denote the number of pages that are on the path from the root to $q$ but not on the path from the root to $p$; see Fig. 10.3.

Assume that the B-tree has $n$ leaf pages, numbered $0, 1, \dots, n - 1$ from left to right, and consider $k$ of them with numbers $p_1 < p_2 < \cdots < p_k$. Let $M$ denote the number of pages (including the pages $p_1, \dots, p_k$) which lie on the paths from

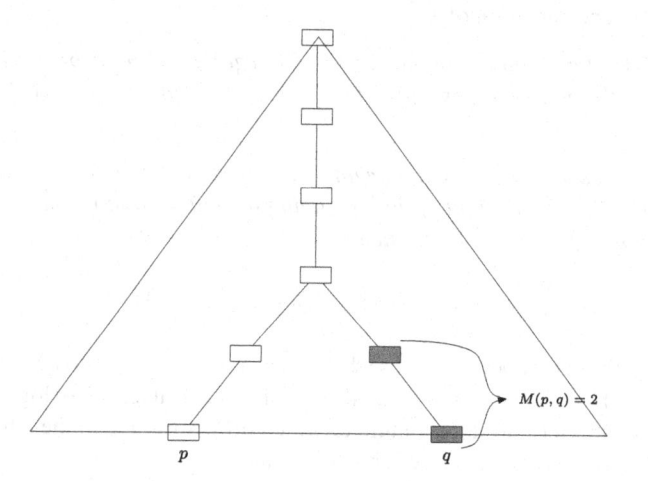

**Fig. 10.3**  The number in $M(p, q)$ is the number of pages not on the path from $p$ to the root but on the path from $q$ to the root

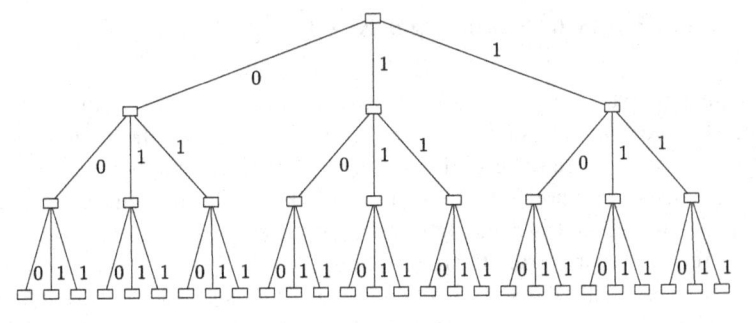

**Fig. 10.4** The labels of consecutive leaves grow at most by 1

the root to some $p_i$. Since the path from the root to $p_1$ contains at most $\lceil \log n \rceil + 1$ pages, we have

$$M \leq \lceil \log n \rceil + 1 + \sum_{i=2}^{k} M(p_{i-i}, p_i). \qquad (10.2)$$

We define a *label b* for each leaf as follows. For any non-leaf page, $q$ we label the leftmost edge from $q$ with 0 and all other edges with 1. Then the label $b(p)$ of a leaf page $p$ is the integer whose binary representation is the sequence of 0's and 1's on the path from the root to $p$.

Clearly, $b(p_i) - b(p_{i-1}) \leq p_i - p_{i-1}$ for $i = 2, \ldots, k$. This is because the labels of consecutive leaves grow at most by 1; cf. Fig. 10.4.

In order to operate with the label sequences, we need some results from the binary arithmetic. For any nonnegative integer $x$, we denote by $v(x)$ the number of one bits in the binary representation of $x$.

**Lemma 10.11** *Let $x$ and $y$ be nonnegative integers, and let $c$ be the number of carries when the binary representations of $x$ and $y$ are added. Then $v(x) + v(y) = v(x + y) + c$.* ☐

**Lemma 10.12** *Let $x$ and $y$ be nonnegative integers, such that $x < y$, and let $i$ be the number of bits to the right of and including the leftmost bit in which the binary representations of $x$ and $y$ differ. Then*

$$i \leq v(x) - v(y) + 2\lceil \log(y - x + 1) \rceil.$$

*Proof* Let $c$ be the number of carries when $x$ and $y - x$ are added. By Lemma 10.11, $v(x) + v(y - x) = v(y) + c$. When $x$ and $y - x$ are added, at least $i - \lceil \log(y - x + 1) \rceil$ carries are required to produce a number which differs from $x$ in the $i$th bit. Thus, $i - \lceil \log(y - x + 1) \rceil \leq c$, and we conclude that

$$i \leq c + \lceil \log(y - x + 1) \rceil = v(x) + v(y - x) - v(y) + \lceil \log(y - x + 1) \rceil$$
$$\leq v(x) - v(y) + 2\lceil \log(y - x + 1) \rceil,$$

(10.3)

as desired.

$\square$

Now we are ready to prove:

**Lemma 10.13** *The number $M$ of pages that lie on the paths from the root to leaves with numbers $p_1 < p_2 < \cdots < p_k$ fulfills the condition*

$$M \leq 2\left(\lceil \log n \rceil + \sum_{i=2}^{k} \lceil \log(p_i - p_{i-1} + 1) \rceil\right).$$

*Proof* Consider any two leaves $p_{i-1}$ and $p_i$. Denote by $lca(p_{i-1}, p_i)$ the lowest common ancestor of $p_{i-1}$ and $p_i$. There are two cases to consider.

*Case 1.* The edge from $lca(p_{i-1}, p_i)$ toward $p_{i-1}$ is labeled by 0 and the edge toward $p_i$ is labeled by 1. Then $b(p_i) > b(p_{i-1})$ (cf. Fig. 10.4). Furthermore, $M(p_{i-1}, p_i)$ being the number of pages on the path from $lca(p_{i-1}, p_i)$ to $p_i$ not including $lca(p_{i-1}, p_i)$ is equal to the number of bits to the right most and the leftmost bit in which the binary representations of $b(p_{i-1})$ and $b(p_i)$ differ. By Lemma 10.12,

$$M(p_{i-1}, p_i) \leq v(b(p_{i-1})) - v(b(p_i)) + 2\lceil \log(b(p_i) - b(p_{i-1}) + 1) \rceil$$
$$\leq v(b(p_{i-1})) - v(b(p_i)) + 2\lceil \log(p_i - p_{i-1} + 1) \rceil,$$

(10.4)

because $b(p_i) - b(p_{i-1}) \leq p_i - p_{i-1}$.

*Case 2.* The edges from $lca(p_{i-1}, p_i)$ toward $p_{i-1}$ and $p_i$ are both labeled by 1. Denote by $b'$ the binary number obtained from $b(p_{i-1})$ by changing the 1-bit corresponding the edge from $lca(p_{i-1}, p_i)$ toward $p_{i-1}$ to 0. Then $b(p_i) - b' \leq p_i - p_{i-1}$ and $b(p_i) > b'$. Furthermore $M(p_{i-1}, p_i)$ is equal to the number of bits to the right most and the leftmost bit in which $b'$ and $b(p_i)$ differ. By Lemma 10.12,

$$M(p_{i-1}, p_i) \leq v(b') - v(b(p_i)) + 2\lceil \log(b(p_i) - b' + 1) \rceil$$
$$\leq v(b') - v(b(p_i)) + 2\lceil \log(p_i - p_{i-1} + 1) \rceil$$
$$\leq v(b(p_{i-1})) - v(b(p_i)) + 2\lceil \log(p_i - p_{i-1} + 1) \rceil,$$

(10.5)

because $v(b(p_{i-1})) = v(b') + 1$.

Substituting into (10.2) what is given above yields

$$M \leq \lceil \log n \rceil + 1 + \sum_{i=2}^{k} (v(b(p_{i-1})) - v(b(p_i)) + 2\lceil \log(p_i - p_{i-1} + 1)\rceil)$$

$$= \lceil \log n \rceil + 1 + v(b(p_1)) - v(b(p_k)) + 2\sum_{i=2}^{k}\lceil \log(p_i - p_{i-1} + 1)\rceil)$$

$$\leq 2\lceil \log n \rceil + \sum_{i=2}^{k}\lceil \log(p_i - p_{i-1} + 1)\rceil).$$

$$(10.6)$$

The final inequality follows from the fact that $v(b(p_k)) \geq 1$ and $v(b(p_1)) \leq \lceil \log n \rceil$ unless $k = 1$. ☐

From Lemma 10.13 we conclude that the total number of pages on the paths from the root to the leaves containing keys in the set $s_X$ of key ranges is bounded by (10.1). The use of the saved-path strategy in Algorithm 10.1 then implies:

**Theorem 10.14** *Assume a B-tree with n leaf pages and a set $s_X$ of non-overlapping key ranges. Then, in the absence of concurrent updating transactions, the total number of page latching performed in a bulk-read($T, s_X, s_{XV}$) call is bounded by (10.1).* ☐

The bound (10.1) on the number of page latchings also holds for the bulk-insert and bulk-delete actions as given in Algorithms 10.2 and 10.3. This result directly follows from the algorithms: observe how the bulk insertion progresses; cf. Fig. 10.2. Notice, however, that for the bulk insertion, the pages to be latched are those lying on the paths from the root to the leaves after performing the bulk insertion.

## Problems

**10.1** With the partition-based key-range locking protocol, what locks must be acquired for a bulk-update action $W[s_X, f, s_{XVV'}]$?

**10.2** Give an algorithm for the call *find-page-for-bulk-read*($p, p', z, z_2$) used in implementing the bulk-read action (Algorithm 10.1).

**10.3** Give algorithms for the calls *find-page-for-bulk-insert*($p, h, p', x, v$) and *find-page-for-bulk-delete*($p, p', z, z_2$) used in implementing the bulk-insert and bulk-delete actions (Algorithms 10.2 and 10.3).

**10.4** In Algorithms 10.1 and 10.3, only closed key ranges $[z_1, z_2]$ are considered. What changes (if any) are needed if also open and half-open ranges are allowed?

**10.5** Give an algorithm for the bulk-update action $W[s_X, f, s_{XVV'}]$.

**10.6** To save in the number of accesses to the log manager, it might be tempting to log the insertion of a set $S$ of tuples into B-tree leaf page $p$ with a single redo-undo log record of the form $\langle T, I, p, S, n' \rangle$ and the deletion of a set $S$ of tuples from $p$ with a single redo-undo log record $\langle T, D, p, S, n' \rangle$. But this would cause major difficulties with our logging protocol. Explain.

**10.7** Assume that a transaction scans a key range $[y, z]$ by performing first the read action $R[x_1, \geq y, v_1]$ and then repeatedly the read action $R[x_{i+1}, > x_i, v_{i+1}]$ for $i = 1, 2, \ldots$ until $x_{i+1}$ exceeds $z$ or reaches $\infty$, using the *read* calls of Sect. 6.7. Show that in the absence of concurrent updating transactions, the total number of page latchings performed in all the *read* calls is

$$O(K + \log N),$$

where $K$ is the number of tuples with keys in the scanned key range $[y, z]$ and $N$ is the number of all tuples in the database.

**10.8** A *partitioned B-tree* is an ordinary B-tree with an artificial leading key attribute added. More specifically, if $X$ is the key of the relation to be indexed, the partitioned B-tree uses the key $(P, X)$, where the leading key attribute $P$ takes *partition numbers* as values. Partition 0 is the regular partition in which most of the tuples usually reside.

When a new bulk of tuples is to be inserted, a new partition number $i > 0$ is reserved and the tuples $(x, v)$ from the bulk are inserted into the B-tree as tuples $(i, x, v)$. This insertion is executed as part of the inserting user transaction. Later when the transaction has committed, tuples in partitions $i > 0$ may gradually be migrated to partition 0, using system transactions (redo-only structure modifications). Conversely, when a bulk of tuples is to be deleted, system transactions may first be run to migrate the tuples to a new partition, which is then deleted as part of a normal user transaction.

What are the obvious advantages and disadvantages of the partitioned B-tree for bulk actions and single-tuple actions? Give a multi-granular locking protocol for bulk-read, bulk-insert, and bulk-delete actions on the partitioned B-tree.

## Bibliographical Notes

Several authors have developed algorithms for bulk operations on B-trees and designed index structures specially tailored for bulk operations. Pollari-Malmi et al. [1996] present an optimal bulk-insertion algorithm on B-tree indexes with concurrency control and discuss its applications to full-text indexing. Mohan [2002] presents a method for performing bulk deletes and updates in key-range scans on an index, with additional predicates restricting the set of qualifying tuples; the method minimizes the number of lock-table accesses needed when tuple-level key-range locking is employed.

The partition-based key-range locking protocol of Sect. 10.2 is from Lilja et al. [2007], who use it to protect bulk-delete actions on B-trees. They also present a bulk-delete algorithm that avoids visiting leaf pages of subtrees whose tuples are all deleted and instead detaches such subtrees from the B-tree. The partitioned B-tree structure with an artificial leading key considered in Problem 10.8 is from Graefe [2003a,b]. Logging and recovery techniques for partitioned B-trees are discussed by Graefe [2012].

The complexity result of searching for a sorted set of keys as stated in Lemma 10.13 is from Brown and Tarjan [1980]; Lemma 10.11 used in proving the result is from Knuth [1973].

# Chapter 11
# Online Index Construction and Maintenance

In the preceding chapters, we have assumed that tuples in a relation are accessed via a sparse B-tree index on the primary key of the relation. In this chapter we extend our database and transaction model with read, delete, and update actions based on ranges of non-primary-key attributes. To accelerate these actions, *secondary indexes* must be constructed. A secondary index, as considered here, is a *dense B-tree* whose leaf pages contain index records that point to the tuples stored in leaf pages of the sparse primary B-tree index of the relation.

A relation can have many secondary indexes on different attributes or combinations of attributes, thus making it possible for the query optimizer to find efficient execution plans for many kinds of queries. However, this comes with the price of an overhead on updates: when a transaction inserts or deletes a tuple, all the indexes must also be updated within the same transaction.

In database performance tuning, it is common that new secondary indexes are constructed when it is found that some regularly performed SQL statements need acceleration, while secondary indexes that are found to be seldom used in query-execution plans are dropped. Index construction for a large relation is a heavy and time-consuming operation; in many cases we cannot afford the relation to be inaccessible for transactions during index construction.

In this chapter we present an *online index-construction* algorithm that allows transactions to update the relation while a secondary index is being constructed. When the index construction is completed, the new index can also be used in query-execution plans. We also briefly address the question of online maintenance of the physical clustering of B-tree indexes.

© Springer International Publishing Switzerland 2014
S. Sippu, E. Soisalon-Soininen, *Transaction Processing*, Data-Centric Systems
and Applications, DOI 10.1007/978-3-319-12292-2_11

## 11.1  Secondary-Key-Based Actions

We now assume that our logical database consists of a relation

$$r(\underline{X}, Y, V),$$

where $X$ is the *primary key* (with unique values as before) and $Y$ is sequence of
attributes (not necessarily with unique values) used as a *secondary key* on which
key-range-read and bulk-delete and bulk-update actions can also be based. Note
that insertions cannot, of course, be performed, but using the primary key can be
    The primary-key-based actions now take the following forms:

1. Single-tuple read actions $R[x, \theta z, y, v]$ for given primary key $x$
2. Single-tuple insert actions $I[x, y, v]$
3. Single-tuple delete actions $D[x, y, v]$ for given primary key $x$
4. Single-tuple write actions $W[x, y, v, y', v']$ for given $x$, $y'$ and $v'$
5. Bulk-read actions $R_X[s_X, s_{XYV}]$ for given set $s_X$ of primary-key ranges
6. Bulk-insert actions $I[s_{XYV}]$
7. Bulk-delete actions $D_X[s_X, s_{XYV}]$ for given set $s_X$ of primary-key ranges
8. Bulk-update actions $W[s_X, f, s_{XYVY'V'}]$ for given set $s_X$ of primary-key ranges
   and function $f(Y, V)$

Besides these actions we now allow transactions to contain the following
secondary-key-based actions:

(a) Single-tuple read actions $R[x, y, \theta z, v]$ for given secondary key $z$
(b) Bulk-read actions $R_Y[s_Y, s_{XYV}]$ for given set $s_Y$ of secondary-key ranges
(c) Bulk-delete actions $D_Y[s_Y, s_{XYV}]$ for given set $s_Y$ of secondary-key ranges
(d) Bulk-update (or bulk-write) actions $W_Y[s_Y, f, s_{XYVV'}]$ for given set $s_Y$ of
    secondary-key ranges and function $f(V)$

Here action (a) is only defined for unique secondary keys, and it retrieves the tuple
$(x, y, v)$ with the least key $y \theta z$. Actions (b)–(d) also apply to non-unique keys.
Action (b) returns the set $s_{XYV}$ of tuples $(x, y, v)$ with keys $y$ in one of the ranges
in $s_Y$, action (c) deletes the tuples $(x, y, v)$ with keys $y$ in one of the ranges in $s_Y$
and returns the deleted tuples in set $s_{XYV}$, and action (d) replaces each tuple $(x, y, v)$
with key $y$ in one of the ranges in $s_Y$ by the tuple $(x, y, f(v))$ and returns the set
$s_{XYVV'}$ of tuples $(x, y, v, f(v))$.
    Expanded histories are used to model the interleaved execution of transactions
containing the above secondary-key-based actions besides the primary-key-based
ones. This is done along the lines explained in Sect. 10.1. The single-tuple actions
generated from actions (c) and (d) are logged in the obvious way with redo-undo
log records and their undo actions with redo-only log records, as usual.
    The definitions of the isolation anomalies that can appear in an expanded history
must be extended so as to cover the secondary-key-based actions. For example, dirty
writes can also occur between two secondary-key-based delete or write actions or

between a primary-key-based insert, delete, or write action and a secondary-key-based delete or write action. Secondary-key-based read actions cause new kinds of dirty and unrepeatable reads. We leave the details of the definitions as exercises (see the Problems section).

The key-range locking protocol must be generalized accordingly. In *key-specific* (or *index-specific*) *key-range locking*, names of finest-granular locks on a relation $r$ are triples $(r, X, x)$ (or actually 4-byte hash values computed from them), where $r$ is the identifier of the relation, $X$ is the identifier of the attribute sequence used as the key, and $x$ is a key value. In the case of our example relation $r(\underline{X}, Y, V)$, lock names are thus of the forms $(r, X, x)$ and $(r, Y, y)$.

Key-specific locks make the granule hierarchy of lockable data items a directed acyclic graph: in the case of a relation $r$ with primary key $X$ and secondary keys $Y_1, \ldots, Y_n$, there are $n + 1$ (logical) access paths from $r$ to any tuple in $r$. Read actions can be protected with locks on granules along a single path to the target tuple, namely, the one defined by the key whose range is read, if only insert, delete, and write actions are protected by locks along every path to the target tuple.

With key-specific locking, a primary-key-based read action $R[x, \theta z, y, v]$ on relation $r$ is protected by an IS lock on $r$ and an S lock on $(r, X, x)$, and a secondary-key-based single-tuple read action $R[x, y, \theta z, v]$ (for unique key $Y$) is protected by an IS lock on $r$ and an S lock on $(r, Y, y)$.

For a primary-key-based bulk-read action $R_X[s_X, s_{XYV}]$, an IS lock is acquired on $r$ and, for each range in $s_X$, S locks are acquired on all $(r, X, x)$ with $(x, y, v) \in r$ and $x$ in the range. Moreover, if no tuple in $r$ has the upper bound of the range as its primary key, an S lock is also acquired on $(r, X, x')$ where $x'$ is the key next to the upper bound of the range.

Similarly, for a secondary-key-based bulk-read action $R_Y[s_Y, s_{XYV}]$, an IS lock is acquired on $r$ and, for each range in $s_Y$, S locks are acquired on all $(r, Y, y)$ with $(x, y, v) \in r$ and $y$ in the range, and, if no tuple in $r$ has the upper bound of the range as its secondary key, an S lock is also acquired on $(r, Y, y')$ where $y'$ is the key next to the upper bound of the range.

For an insert action $I[x, y, v]$, a commit-duration IX lock is acquired on $r$, commit-duration X locks are acquired on $(r, X, x)$ and $(r, Y, y)$, and short-duration X locks (or just IX locks, cf. Problem 9.3) are acquired on $(r, X, x')$ and $(r, Y, y')$, where $x'$ is the key next to $x$ and $y'$ is the key next to $y$.

For a delete action $D[x, y, v]$, a commit-duration IX lock is acquired on $r$, short-duration X locks are acquired on $(r, X, x)$ and $(r, Y, y)$, and commit-duration X locks are acquired on $(r, X, x')$ and $(r, Y, y')$, where $x'$ is the key next to $x$ and $y'$ is the key next to $y$.

*Example 11.1* Assume that the contents of the database are initially

$$\{(1, 40, v_1), (3, 20, v_3), (4, 50, v_4), (5, 60, v_5), (6, 60, v_6)\}.$$

The history

$T_1$:     $BR[3, > 1, 20, v_3]$                                              $\dots$                    $C$

$T_2$:     $BR_Y[\{(40, 70)\}, \{(4, 50, v_4), (5, 60, v_5), (6, 60, v_6)\}]$           $\dots$                    $C$

$T_3$:                                                                  $BI[2, 50, v_2]$   $C$

contains two isolation anomalies: both of the two read actions (whose mutual order is irrelevant) are unrepeatable reads. In fact, in terms of the corresponding expanded history of single-tuple actions, there are four unrepeatable reads.

The history is not possible under the key-specific key-range locking protocol, when the transactions are run at the serializable isolation level, because $T_1$ protects its read action with a commit-duration S lock on $(r, X, 3)$ and $T_2$ protects its read action with commit-duration S locks on $(r, Y, 50)$, $(r, Y, 60)$, and $(r, Y, \infty)$ and because $T_3$ protects its insert action by commit-duration X locks on $(r, X, 2)$ and $(r, Y, 50)$ and by short-duration X locks (or IX locks) on $(r, X, 3)$ and $(r, Y, 60)$.   □

With key-specific key-range locking, a single-tuple insert or delete action on a relation $r$ with primary key $X$ and secondary keys $Y_1, \dots, Y_n$ requires $n + 1$ commit-duration locks and $n + 1$ short-duration locks of the finest granularity, besides the IX lock on $r$.

Another option is to use *primary-key-only key-range locking* (or *data-only locking*) in which all tuple-level lock names are formed from the primary keys of the tuples.

*Example 11.2*  With primary-key-only locking, the bulk-read action by transaction $T_2$ in Example 11.1 is protected by commit-duration S locks on $(r, 4)$, $(r, 5)$, and $(r, 6)$, and the insert action by transaction $T_3$ is protected by commit-duration X locks on $(r, 2)$ and $(r, 4)$ and by short-duration X locks on $(r, 3)$, $(r, 5)$, and $(r, 6)$.

□

When primary-key-only locking is coupled with the partition-based locking protocol of Sect. 10.2, the number of locks needed can further be reduced.

## 11.2  Secondary B-Tree Indexes

To accelerate actions based on a secondary key, a *secondary index* to the relation $r(\underline{X}, Y, V)$ is constructed. Such an index is a *dense index* in that for every single tuple $(x, y, v)$ in the relation, the secondary index contains an index record $(y, x, p)$ that points to the data page, $p$, that contains (or used to contain) the tuple and identifies (in this case by the primary key $x$) the referenced tuple in the page.

We assume that the secondary index based on secondary key $Y$ is organized as a B-tree on the unique key composed of the pair $(Y, X)$ of the secondary key and the primary key. The leaf pages of this B-tree thus store the records $(y, x, p)$ sorted first by secondary key $y$ and records with a duplicate secondary key sorted by primary key $x$ (Fig. 11.1). For accessing the secondary index, a saved path is maintained as for the primary sparse B-tree.

**Fig. 11.1** Indexes on relation
$r(X, Y, V)$: (**a**) the leaf pages
of the sparse B-tree index on
primary key $X$; (**b**) the leaf
pages of the dense secondary
B-tree index on $Y$

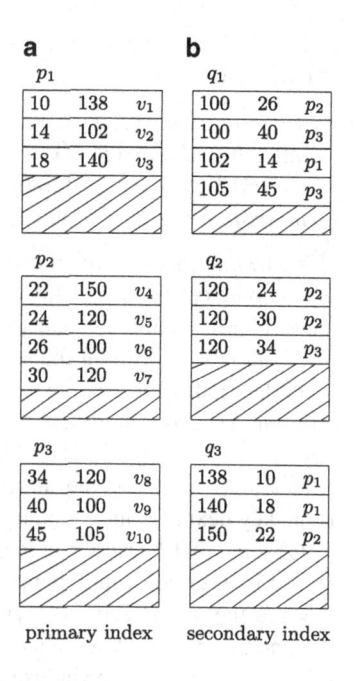

| | | |
|---|---|---|
| **a** | | |
| $p_1$ | | |
| 10 | 138 | $v_1$ |
| 14 | 102 | $v_2$ |
| 18 | 140 | $v_3$ |

| | | |
|---|---|---|
| $p_2$ | | |
| 22 | 150 | $v_4$ |
| 24 | 120 | $v_5$ |
| 26 | 100 | $v_6$ |
| 30 | 120 | $v_7$ |

| | | |
|---|---|---|
| $p_3$ | | |
| 34 | 120 | $v_8$ |
| 40 | 100 | $v_9$ |
| 45 | 105 | $v_{10}$ |

| | | |
|---|---|---|
| **b** | | |
| $q_1$ | | |
| 100 | 26 | $p_2$ |
| 100 | 40 | $p_3$ |
| 102 | 14 | $p_1$ |
| 105 | 45 | $p_3$ |

| | | |
|---|---|---|
| $q_2$ | | |
| 120 | 24 | $p_2$ |
| 120 | 30 | $p_2$ |
| 120 | 34 | $p_3$ |

| | | |
|---|---|---|
| $q_3$ | | |
| 138 | 10 | $p_1$ |
| 140 | 18 | $p_1$ |
| 150 | 22 | $p_2$ |

primary index          secondary index

As noted earlier, some database management systems store the tuples of a relation in an unordered file (heap), where the tuples, once inserted, remain in their place until deleted or until the entire file structure is reorganized. In that case it is customary that the index records in the secondary indexes are "hard pointers" to the tuples, that is, of the form $(y, (p, i))$, where $(p, i)$ is the tuple identifier of the referenced tuple.

We assume here, as before, that the tuples are stored in the leaf pages of a sparse B-tree index, which is a highly dynamic structure, with tuples moving from a page to another in page splits and merges. With index records of the form $(y, (p, i))$, this would mean constantly updating references in the secondary indexes. With index records of the form $(y, x, p)$, on the contrary, the primary-key component $x$ remains valid when the tuple moves, and only the page-id component $p$ turns invalid.

If the page-id components of index records are not updated when the referenced tuples move, the page-id component can only be used as a guess of the current location of the referenced tuple. In locating a tuple given a secondary-index record $(y, x, p)$, we latch page $p$ and inspect its contents so as to find out if the page is still a data page that contains a tuple with primary key $x$. If not, the tuple is located from the primary B-tree index by traversing it in search for key $x$.

The secondary indexes are updated at the end of the calls that implement the forward-rolling insert and delete (and write) actions (Algorithms 6.7 and 6.8) or undo such actions (Algorithms 4.7 and 4.9). For example, in the call $insert(T, x, y, v)$ that implements the insert action $I[x, y, v]$, the secondary index is updated after inserting the tuple $(x, y, v)$ into data page $p$, unlatching the page and the page $p'$ that contains the key $x'$ next to $x$, and releasing the short-duration X lock on $x'$:

   *insert-into-page*$(T, p, x, y, v)$
   *unlatch-and-unfix*$(p, p')$
   *unlock*$(T, x', \mathrm{X}, \text{short-duration})$
   *insert-index-record*$(T, y, x, p)$

Here the call *insert-index-record*$(T, y, x, p)$ inserts the index record $(y, x, p)$ into the secondary index and logs the insertion with a redo-undo log record in the name of transaction $T$. The logic of the call is similar to that of the plain call *insert*$(T, x, y, v)$ without the secondary-index update; it includes a traversal of the secondary index for locating the index page that will hold the index record $(y, x, p)$, with any structure modifications needed to accommodate the insertion, and, with key-specific locking, the acquisition of the secondary-key-specific commit-duration X lock on $(r, Y, y)$ and the short-duration X lock (or IX lock) on $(r, Y, y')$, where $y'$ is the key next to $y$ in secondary-key order.

Observe that because the latch on the data page $p$ is released before the index record is inserted, it may happen that the tuple $(x, y, v)$ no longer resides in page $p$ when the index record carrying that page-id arrives at the secondary index. We have no other option than releasing the latch on $p$, because, if key-specific locking is used, the insertion of the index record involves acquiring the secondary-key-specific locks, which may cause lock waits, which are only allowed if no latches are held.

Inserting a tuple $(x, y, v)$ by transaction $T$ generates the pair of redo-undo log records:

   $n_1$: $\langle T, I, p, x, y, v, n_0 \rangle$
   $n_2$: $\langle T, ix\text{-}I, q, y, x, p, n_1 \rangle$

where $p$ is the page-id of the data page that received the tuple $(x, y, v)$, and $q$ is the page-id of the leaf page of the secondary index that received the index record $(y, x, p)$. The action name *ix-I* denotes the action of inserting an index record into the secondary index. If the transaction rolls back, the following pair of redo-only log records is written:

   $n_3$: $\langle T, ix\text{-}I^{-1}, q', y, x, n_1 \rangle$
   $n_4$: $\langle T, I^{-1}, p', x, n_0 \rangle$

Here $q'$ is the page-id of the leaf page that holds the index record $(y, x, p)$ and $p'$ the page-id of the data page that holds the tuple $(x, y, v)$, at the time the undo actions are performed.

## 11.3   Index Construction Without Quiescing Updates

We now tackle the problem of constructing a secondary index for a relation $r(\underline{X}, Y, V)$ during normal transaction processing when transactions are allowed concurrent access to the relation. First we note that a trivial solution is obtained

by quiescing updates by transactions accessing the relation during index construction:

1. Begin a system transaction $T$.
2. S-lock relation $r$ for $T$ for commit duration.
3. Scan the tuples in $r$ in primary-key order, extract the index records $(y, x, p)$, and
   write them to a file.
4. Sort the file on the unique key $(Y, X)$.
5. Build the secondary index from the sorted file of index records, logging all index-
   record insertions with redo-undo log records and structure modifications with
   redo-only log records.
6. Update the system catalog with information about the new index.
7. Commit $T$.

With the above solution, the index is created on a transaction-consistent state
of the relation $r$, because of the commit-duration S lock. A major problem is that
while the index construction is in progress, updates are *quiesced*: the relation $r$
is only accessible for reading, while transactions that want to update the relation
must wait for the release of the S lock on $r$, which only happens at the end of the
index construction, meaning a very long wait in the case of a large relation. Another
problem is related to *restartability* of a failed index construction: in the event of
a system crash, the entire system transaction $T$ is rolled back, after which index
construction must be restarted from the beginning.

We present a solution with which updates are not quiesced during index
construction: transactions can update the relation while the index construction is in
progress (Fig. 11.2). Naturally, transactions may not use the index being constructed
until it is completed. Thus the tuples in the relation must be accessed via access paths
that exist and are declared available in the system catalog.

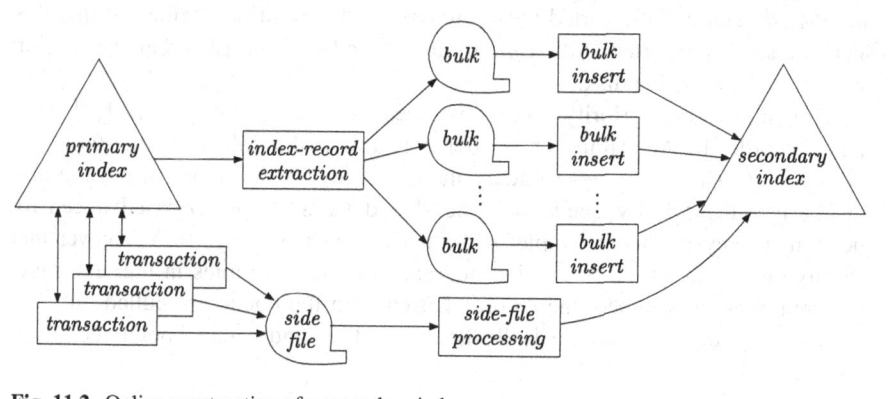

**Fig. 11.2**  Online construction of a secondary index

## 11.4   Extraction of the Index Records

Algorithm 11.1 is used to scan, in primary-key order, the sparse B-tree primary index of the relation, to extract the index records for the secondary index being constructed, and to store the records onto disk. This is done using repeated calls of *find-page-for-bulk-read*($p, p', Cursor, \infty$) (see Sect. 10.3), where the shared variable *Cursor* holds the primary key of the next tuple to be scanned.

We assume that the system-catalog page that stores index information for the relation contains a variable that tells the state of the secondary index, that is, whether the index is "under construction," "available," or "does not exist." The updating of this information is protected by a short-duration write latch on the page containing the information. The updating of the shared variable *Cursor* is protected by a write latch kept just for the instant of setting its value.

At the start of the scan, the state of the index in the system catalog is updated from "does not exist" to "under construction." For the short duration of this update, the relation is S-locked so as to prevent an update action by a transaction seeing the index state as "does not exist" and later another update action (or an undo action) by the same transaction seeing the state as "under construction." No locks other than the short-duration S lock are acquired on the relation or on the tuples read during the scan.

The initial value of *Cursor* is $-\infty$, the lowest primary-key value. During the traversal of the B-tree, pages are read-latched in the usual way, as explained earlier in Chaps. 7 and 10. The scan ends when *Cursor* reaches the highest primary-key value $\infty$.

To address the problem of restartability, the relation is scanned in bulks of tuples that fit in main memory, each such bulk is sorted and forced to disk, and a checkpoint record is written to the log that contains the current value of *Cursor* and the addresses of the sorted bulks on disk. In the event of a failure during the scan, the scan is restarted at the *Cursor* value found in the most recent checkpoint record found on the log disk.

A main-memory priority queue is used to create the sorted bulks of index records. In Algorithm 11.1, after performing the call *find-page-for-bulk-read*($p, p', Cursor, \infty$), the secondary-index records for the tuples in page $p$ are pushed into the priority queue, and the shared variable *Cursor* is advanced to the primary key of the first tuple in the page $p'$ next to page $p$. Whenever the priority queue has no room for the index records for the tuples in the leaf page $p$ currently being scanned, the records currently in the queue are pulled out and written to a new file, the file is flushed onto disk, and a checkpoint record is logged.

---

**Algorithm 11.1** Scan the relation and extract the index records

allocate space for a main-memory priority queue indexed by the secondary key
$F \leftarrow \emptyset$ {set of identifiers of files created to store sorted bulks}
acquire a short-duration S lock on the relation
declare in the system catalog that the secondary index is under construction
$Cursor \leftarrow -\infty$ {a shared variable for scanning the primary-key space}
release the short-duration S lock on the relation
**while** $Cursor < \infty$ **do**
   *find-page-for-bulk-read*$(p, p', Cursor, \infty)$
   $V \leftarrow \{(y, x, p) \mid (x, y, v) \text{ in page } p\}$
   **if** the priority queue cannot accommodate $|V|$ new elements **then**
      $f \leftarrow$ create a file for the next bulk
      $F \leftarrow F \cup \{f\}$
      pull all the records from the priority queue and write them to file $f$
      flush file $f$
      $log(n, \langle checkpoint\text{-}index\text{-}extract, F, Cursor \rangle)$
   **end if**
   push the records in $V$ to the priority queue
   **if** $p' = p$ **then**
      $Cursor \leftarrow \infty$
   **else**
      $Cursor \leftarrow$ the least primary key $x$ in $p'$
   **end if**
   *unlatch-and-unfix*$(p, p')$
**end while**
$f \leftarrow$ create a file for the last bulk
$F \leftarrow F \cup \{f\}$
pull all the records from the priority queue and write them to file $f$
flush file $f$
$log(n, \langle checkpoint\text{-}index\text{-}extract, F, Cursor \rangle)$

---

## 11.5 Concurrent Updates by Transactions

During index-record extraction, it may happen that transactions insert, delete, or write tuples in the already scanned portion of the relation, that is, tuples with primary key less than *Cursor*. Those updates are missed by the scan, while inserts, deletes, and writes done on the not-yet-scanned portion of the relation (containing tuples with primary key $\geq$ *Cursor*) are captured by the scan.

The updates done on the scanned portion of the relation could be captured from the log, by remembering the LSN of the begin-transaction log record written for the index constructor. However, an easier and more efficient solution is obtained by using a *side file* to store the updates by transactions done after the start of the index construction on the scanned portion of the relation. After the relation has been scanned and the index records thus extracted have been inserted into the index being constructed, the side file is processed and the updates stored there are applied to the index.

When the secondary index is ready and declared available for use in query-execution plans, a transaction $T$ inserting a tuple $(x, y, v)$ into data page $p$ must also insert the index record $(y, x, p)$ into the secondary index. This is done by the call *insert-index-record*$(T, y, x, p)$ at the end of the call *insert*$(T, x, y, v)$, as explained above in Sect. 11.2. Similarly, a transaction $T$ deleting a tuple $(x, y, v)$ from data page $p$ must use, at the end of the call *delete*$(T, x, y, v)$, the call *delete-index-record*$(T, y, x, p)$ to delete the associated index record $(y, x, p)$ from the secondary index (cf. Problem 11.5).

When the secondary index is under construction but not yet available, a transaction inserting a tuple cannot insert the associated index record into the secondary index, nor can a transaction deleting a tuple delete the associated index record from the secondary index even if such a record already appeared in the index. When the index construction is in progress, transactions can only append to the side file records that describe secondary-index updates.

In the case of an insert action $I[x, y, v]$ by transaction $T$ on data page $p$, if the index-record extraction process has not yet advanced past the primary key $x$ of the inserted tuple, $T$ must not append anything to the side file, because in that case the extraction process will capture the inserted tuple in its then current page $p'$, create the index record $(y, x, p')$, and store it on disk for insertion into the index. Otherwise, if the extraction process has advanced past $x$, $T$ must append a record $(ix\text{-}I, y, x, p)$ to the side file and log the append with a log record

$$n_2 : \langle T, append\text{-}to\text{-}side\text{-}file, q, ix\text{-}I, y, x, p, n_1 \rangle, \tag{11.1}$$

where $q$ is the page-id of the side-file page that received the appended record. Moreover, to make it possible for a later undo action by $T$ to observe the index record, the log record

$$n_2 : \langle T, no\text{-}append\text{-}to\text{-}side\text{-}file, ix\text{-}I, y, x, p, n_1 \rangle \tag{11.2}$$

must be written in the case that no record is appended to the side file.

---

**Algorithm 11.2** Secondary-index update in procedure *insert*$(T, x, y, v)$

---
*insert-into-page*$(T, p, x, y, v)$
**if** the secondary index is under construction **then**
    $n_1 \leftarrow$ UNDO-NEXT-LSN$(T)$
    **if** $x <$ *Cursor* **then**
        *unlatch-and-unfix*$(p, p')$
        *unlock*$(T, x', $X$,$ short-duration$)$
        $q \leftarrow$ the last page of the side file (allocating one if needed)
        *fix-and-write-latch*$(q)$
        append the record $(ix\text{-}I, y, x, p)$ into page $q$
        $log(n_2, \langle T, append\text{-}to\text{-}side\text{-}file, q, ix\text{-}I, y, x, p, n_1 \rangle)$
        PAGE-LSN$(q) \leftarrow n_2$
        *unlatch-and-unfix*$(q)$
    **else** $\{x >$ *Cursor*$\}$
        *unlatch-and-unfix*$(p, p')$
        *unlock*$(T, x', $X$,$ short-duration$)$
        $log(n_2, \langle T, no\text{-}append\text{-}to\text{-}side\text{-}file, ix\text{-}I, y, x, p, n_1 \rangle)$
    **end if**
    UNDO-NEXT-LSN$(T) \leftarrow n_2$
**else if** the secondary index is available **then**
    *unlatch-and-unfix*$(p, p')$
    *unlock*$(T, x', $X$,$ short-duration$)$
    *insert-index-record*$(T, y, x, p)$
**else** {the secondary index is neither available nor under construction}
    *unlatch-and-unfix*$(p, p')$
    *unlock*$(T, x', $X$,$ short-duration$)$
**end if**

---

To accomplish the above, the sequence of calls

*insert-into-page*$(T, p, x, y, v)$
*unlatch-and-unfix*$(p, p')$
*unlock*$(T, x', $X$,$ short-duration$)$

at the end of the call *insert*$(T, x, y, v)$ must be replaced by the statements given as Algorithm 11.2. Here the short-duration X lock on the key $x'$ next to $x$ can be released immediately after unlatching the page $p$ and the next page $p'$, before performing the actions specific to the secondary index.

In Algorithm 11.2, it is essential that the data page $p$ that receives the inserted tuple is kept write-latched during the reading of the current value of *Cursor*, because then the extraction process cannot advance its scan past page $p$, if it has not done so before the page was write-latched for the insert action by $T$. If $T$ observes that $x >$ *Cursor*, it unlatches $p$ and writes a *no-append-to-side-file* log record, after which the extraction process will in due time get a read latch on $p$ when the scan has advanced at that page and hence will capture the insertion. On the other hand, it is also correct to unlatch $p$ immediately after observing that $x <$ *Cursor* before appending the record to the side file, because in that case the scan has already advanced past the page. We also note that the case $x =$ *Cursor* is impossible.

The reading of the value of *Cursor* is protected by a read latch kept just for the instant of reading. The querying of the state of the index is protected by a short-duration read latch on the system-catalog page that contains the index information.

After inserting the tuple $(x, y, v)$ into data page $p$ and logging the insertion, Algorithm 11.2 performs the actions specific to the secondary index. Depending on the state of the index and the position of *Cursor*, there are four cases:

Case (1). The index is under construction and the extraction process has advanced past primary key $x$. In this case the record $(ix\text{-}I, y, x, p)$ is appended into the side file and the log records

$n_1$: $\langle T, I, p, x, y, v, n_0 \rangle$
$n_2$: $\langle T, append\text{-}to\text{-}side\text{-}file, q, ix\text{-}I, y, x, p, n_1 \rangle$

are written, where $q$ is the page-id of the side-file page that receives $(ix\text{-}I, y, x, p)$.

Case (2). The index is under construction but the extraction process has not advanced past $x$. Then the log records

$n_1$: $\langle T, I, p, x, y, v, n_0 \rangle$
$n_2$: $\langle T, no\text{-}append\text{-}to\text{-}side\text{-}file, ix\text{-}I, y, x, p, n_1 \rangle$

are written.

Case (3). The index is available. Then the index record $(y, x, p)$ is inserted into the secondary index and the log records

$n_1$: $\langle T, I, p, x, y, v, n_0 \rangle$
$n_2$: $\langle T, ix\text{-}I, q', y, x, p, n_1 \rangle$

are written, where $q'$ is the page-id of the index leaf page that receives $(y, x, p)$.

Case (4). The index is neither available nor under construction, that is, does not exist at all. Then only the log record

$n_1$: $\langle T, I, p, x, y, v, n_0 \rangle$

is written.

Next we turn to the question of what is involved in undoing the above secondary-index-specific actions should the transaction $T$ perform a total or partial rollback. In cases (1) and (2), we must check whether the index has become available in the interim and, in case (2), if the index is not yet available, whether the extraction process has advanced past key $x$.

In case (1), if the index has become available, the index constructor has processed the side file and inserted the record $(y, x, p)$ to the index; therefore, the append-to-side-file action is undone by traversing the index in search for $(y, x, p)$, deleting it, and logging the deletion with the redo-only log record

$n_3$: $\langle T, q', ix\text{-}I^{-1}, y, x, p, n_1 \rangle$, where $q'$ is the page-id of the index leaf page in which the record resided. Otherwise, if the index is not yet available so that the side file is still open for transactions to append records, the append-to-side-file action is undone by appending the record $(ix\text{-}\Gamma^{-1}, y, x, p)$ to the side file and logging the append with the redo-only log record

$n_3$: $\langle T, append\text{-}to\text{-}side\text{-}file, q'', ix\text{-}\Gamma^{-1}, y, x, p, n_1 \rangle$,

where $q''$ is the page-id of the side-file page that received the record. Observe that all updates (appends) to the side file are redo-only actions, no matter whether they result from forward-rolling or backward-rolling actions by transactions.

In case (2), if the index has become available, the index constructor has captured the inserted tuple $(x, y, v)$ in its then current data page $p'$ and inserted the record $(y, x, p')$ to the index; therefore, the no-append-to-side-file action is undone by traversing the index in search for record $(y, x, p')$, deleting it and logging the deletion with the redo-only log record

$n_3$: $\langle T, q', ix\text{-}\Gamma^{-1}, y, x, p', n_1 \rangle$,

where $q'$ is the page-id of the index leaf page in which the record resided.

Otherwise, in case (2), when the index is not yet available, we have to check if the extraction process has now advanced past $x$. To ensure consistency of this check, we must write-latch the primary-index leaf page that currently holds the tuple $(x, y, v)$. Of course, we guess that the page is still $p$ and therefore write-latch it and check if the page still holds the record and also that the page will not underflow by the deletion; if not, we must use the call *find-page-for-undo-insert*$(p, x)$ (see Sect. 7.6) to traverse the primary index so as to locate and write-latch the page $p'$ that currently holds the tuple (recall that this call also performs any structure modifications needed to accommodate the undo action).

Then we compare $x$ and *Cursor*. If $x >$ *Cursor*, the extraction process still has not advanced past $x$, so no undoing of an index-record insertion needs to be done, and we just set UNDO-NEXT-LSN back to $n_1$ and proceed to undo the insertion of the tuple. Otherwise, when $x <$ *Cursor*, the extraction process has advanced past $x$, capturing the inserted tuple $(x, y, v)$ in its then current page $p'$ and inserting the record $(y, x, p')$ to the index. Therefore, as the side file is still open for transactions to append records, the no-append-to-side-file action is undone by appending the record $(ix\text{-}\Gamma^{-1}, y, x, p)$ to the side file and logging the append with the redo-only log record

$n_3$: $\langle T, append\text{-}to\text{-}side\text{-}file, q'', ix\text{-}\Gamma^{-1}, y, x, p, n_1 \rangle$,

where $q''$ is the page-id of the side-file page that receives the record.

In case (3), the index-record insertion is undone in the obvious way from the index and logged with the redo-only log record

$n_3$: $\langle T, q'', ix\text{-}\Gamma^{-1}, y, x, p, n_1 \rangle$,

where $q''$ is the page-id of the index leaf page in which the record currently resided.

In case (4), because the short-duration S lock acquired by the index constructor at the start of the extraction process ensures that the construction of the index cannot begin while transactions that have updated the relation are still active, the index must still be unavailable and not in construction when the transaction rolls back the insertion. Therefore, no index-specific actions need be done, and the transaction proceeds to undo the insertion of the tuple.

The index-specific action to be done in the case of deleting a tuple from the relation are analogous to those for inserting a tuple. We leave the details as an exercise (see the Problems section).

## 11.6 Populating the Index

When the index constructor has completed the index-record extraction process, it begins populating the index with the records written into the files $f \in F$. A straightforward solution is to use a bulk insertion algorithm similar to but simpler than that used in the procedure *bulk-insert* (Algorithm 10.2). Now that the records in each file $f$ are already in sorted order, no sorting needs to be done. Moreover, no locks need to be acquired for protecting the index, because it is only visible to the index constructor.

We assume that a sorted bulk $f$ of index records is inserted to the dense B-tree index with the call *index-bulk-insert*$(T, f)$ (Algorithm 11.3), where $T$ is the identifier of the system transaction created in whose name the bulk insert is logged, using redo-undo log records.

We use one transaction per bulk so as to make a failed index population restartable. The call *find-page-for-index-bulk-insert*$(p, y', x', p', y, x, p)$, which is very similar to the call *find-page-for-bulk-insert*$(p, h, p', x, p)$ in Algorithm 10.2, is used to determine the leaf page $p$ of the secondary index that covers key $(y, x)$, the high key $(y', x')$ of $p$, and the page $p'$ next to $p$ (if any), write-latching $p$, and read-latching $p'$. The call also arranges, with appropriate structure modifications, that page $p$ has room at least for the record $(y, x, p)$. If no next page $p'$ exists (i.e., $(y', x') = (\infty, \infty)$), then $p' = p$. The call *insert-index-bulk-into-page*$(T, p, S)$, similar to *insert-bulk-into-page*$(T, p, S)$ (Algorithm 10.4), is used to insert the records in $S$ into the write-latched leaf page $p$ of the secondary index and to log the insertions with redo-undo log records.

Algorithm 11.4 first populates the index by inserting the bulks $f \in F$ and then processes the side file by scanning it from the beginning to the end. For restartability, at the start of the population, the set $F$ of file identifiers is logged, and after the commit of each bulk insertion of a bulk $f$, the file identifier $f$ is logged. The insertion of the bulks in $f \in F$ is performed by $k$ concurrent process threads each inserting a subset $F_i$ of the bulks in $F$, $i = 1, \ldots, n$, where $k$ is a chosen degree of concurrency.

Algorithm 11.5 processes the records in the side file in their original appending order and, for each record, performs on the index the update action represented by the record. The processing is done in the name of a single transaction, whose identifier is logged for restartability. Transactions may continue appending to the side file while it is being scanned by the index constructor. For each side-file record that represents an index-record insertion (i.e., *ix-I* or *ix-D*$^{-1}$), an index-record insertion is performed, and for each side-file record that represents an index-record deletion (i.e., *ix-D* or *ix-I*$^{-1}$), an index-record deletion is performed. Each such insertion or deletion is logged with a redo-undo log record.

---

**Algorithm 11.3** Procedure *index-bulk-insert*$(T, f)$

$(y, x, p) \leftarrow$ the first record in $f$
**while** $(y, x) \neq (\infty, \infty)$ **do**
    *find-page-for-index-bulk-insert*$(p, y', x', p', y, x, p)$
    $S \leftarrow$ the records $(y'', x'', p'')$ in $f$ with $(y, x) \leq (y'', x'') < (y', x')$ fitting in $p$
    *insert-index-bulk-into-page*$(T, p, S)$
    *unlatch-and-unfix*$(p, p')$
    **if** the last record in $f$ is in $S$ **then**
        $(y, x, p) \leftarrow (\infty, \infty, 0)$
    **else**
        $(y, x, p) \leftarrow$ the first record in $f$ not in $S$
    **end if**
**end while**

---

**Algorithm 11.4** Populating the index with sorted bulks $f \in F$

$log(n, \langle begin\text{-}populate\text{-}index, F \rangle)$
partition $F$ into disjoint sets $F_1, \ldots, F_k$ for some $k$
**for all** $i = 1, \ldots, k$ concurrently **do**
    **for all** files $f \in F_i$ **do**
        $T \leftarrow$ new transaction identifier
        $log(n, \langle T, B \rangle)$
        *index-bulk-insert*$(T, f)$
        *commit*$(T)$
        $log(n, \langle populated\text{-}index\text{-}with, f \rangle)$
    **end for**
**end for**
process the side file with Algorithm 11.5
$log(n, \langle end\text{-}populate\text{-}index, F \rangle)$

---

**Algorithm 11.5** Side-file processing

---

$T \leftarrow$ new transaction identifier
$log(n, \langle begin\text{-}process\text{-}side\text{-}file\text{-}with, T \rangle)$
**for** all pages $q'$ in the side file of the index **do**
   $fix\text{-}and\text{-}read\text{-}latch(q')$
   **for** all records $t$ in page $q'$ **do**
      **if** $t$ is of the form $(ix\text{-}I, y, x, p)$ **then**
         $insert\text{-}index\text{-}record(T, y, x, p)$
      **else if** $t$ is of the form $(ix\text{-}D, y, x, p)$ **then**
         $delete\text{-}index\text{-}record(T, y, x, p)$
      **else if** $t$ is of the form $(ix\text{-}I^{-1}, y, x, p)$ **then**
         $delete\text{-}index\text{-}record(T, y, x, p)$
      **else if** $t$ is of the form $(ix\text{-}D^{-1}, y, x, p)$ **then**
         $insert\text{-}index\text{-}record(T, y, x, p)$
      **end if**
   **end for**
   **if** $q'$ is the last page of the side file **then**
      $commit(T)$
      set the index as available in the system catalog
      $unlatch\text{-}and\text{-}unfix(q')$
      deallocate the side file
   **else**
      $unlatch\text{-}and\text{-}unfix(q')$
   **end if**
**end for**

---

## 11.7  Restoration of Clustering

An index structure such as the B-tree maintains its balance under insertions, updates, and deletions, so that the search-path length for a single key is logarithmic in the number of indexed tuples (as stated in Lemma 7.3). However, as new pages are allocated in page splits and emptied pages are deallocated in page merges, the structure gradually becomes *physically fragmented*, so that the B-tree pages are scattered all over the disk, even if all the pages were initially located close to each other in consecutive disk blocks of adjacent cylinders.

In traditional *offline reorganization* of a primary index structure (sparse B-tree), the database system administrator runs periodically a system transaction that S-locks the relation, thus quiescing all updates by transactions; traverses the entire structure in key order and builds a new structure in a newly allocated disk area, thus removing physical fragmentation; and finally upgrades the S lock on the relation to an X lock, write-latches the system-catalog page containing the index information, and substitutes the address of the new structure for the old one. During the reorganization, which may take a long time, other transactions cannot update the relation.

*Online reorganization* attempts to circumvent the drawback of offline reorganization. Other transactions are allowed to access the old structure when the reorganization is in progress; when the new structure is ready, it is atomically substituted for the old structure, and the space occupied by the old structure is deallocated. The problem here, as in online index construction, is how to capture the updates made by transactions to the old structure when the construction of the new structure is in progress.

One solution is to use a technique reminiscent to maintaining a *remote backup copy* of a database (to be discussed in Sect. 13.8). The solution outlined below is based on unloading the data from the structure to be reorganized, reloading the data in reorganized form into a *shadow copy*, applying the log to bring the shadow copy up to date so as to reflect the updates by transactions into the original copy, and substituting the shadow copy atomically for the original copy. The method applies to the reorganization of both sparse primary B-trees and dense secondary B-tree indexes.

The online reorganization of the primary B-tree for relation $r(\underline{X}, V)$ consists of the following steps (assuming that no secondary indexes exist on $r$):

1. Record the current tail (LSN) of the log.
2. Initialize an empty sparse B-tree structure, the shadow copy.
3. Scan the tuples in the leaf pages of the original B-tree of relation $r$ and upload the tuples into the shadow copy using several system transactions with bulk-insert actions; no locks on the tuples are acquired by these transactions.
4. Perform a system transaction that scans the log forward, starting from the recorded LSN, and, at every logged update action $(I, D, W, I^{-1}, D^{-1}, W^{-1})$ on $r$ encountered, applies the logical action to the shadow copy, skipping any log records for structure modifications encountered.
5. Record the current tail (LSN) of the log.
6. U-lock $r$.
7. Repeat step 4.
8. X-lock $r$.
9. Write-latch the system-catalog page $q$ that contains the primary-index information for $r$.
10. Abort all transactions that are waiting for a lock on $r$.
11. Make the primary-index information in page $q$ to point to the shadow copy.
12. Unlatch $q$. Unlock $r$.

The update-mode lock (U lock) on $r$ acquired in step 6 quiesces updates and thus ends the accumulation of further log records; so step 7 will capture any updates that arrive after performing step 4 for the first time. The U lock also suspends the granting of all further IS or S locks on $r$. As each transaction must protect its updates by commit-duration X locks, the U lock on $r$ is granted only after all the transactions that have updated the original B-tree of $r$ have committed (or completed their rollback). Finally X-locking $r$ in step 8 quiesces all scans of the original B-tree of $r$ by transactions that run at an isolation level higher than $1°$.

It is necessary to abort every transaction that is waiting for a lock on $r$ in step 10, even though the locks requested are logical and hence readily appropriate to the new structure of $r$ as well. If a lock on $r$ were granted to a transaction that has located the target of its action by traversing the old structure of $r$ before being forced to wait for the lock, then, after the lock is granted, the transaction would use the page-id saved from the access, latch the page, and inspect its contents to find out if the page is still the correct target for the action, as explained in Sect. 6.7 and Chap. 7. If the PAGE-LSN of the page has not changed, the transaction would operate on that page and not traverse the new structure of the relation.

## Problems

**11.1** Identify the different kinds of dirty writes, dirty reads, and unrepeatable read that can appear in histories of the transactions that can contain both primary-key-based and secondary-key-based actions, as defined in Sect. 11.1.

**11.2** What locks must be acquired for a secondary-key-based bulk-delete action $D_Y[s_Y, s_{XYV}]$ when using (a) key-specific key-range locking, and (b) partition-based key-range locking?

**11.3** Assume that the secondary index shown in Fig. 11.1 does not yet exist and that the following sequence of actions is executed:

1. The index constructor begins the extraction of the index records for the secondary index; it write-latches page $p_1$, scans the tuples, and unlatches page $p_1$.
2. Transaction $T$ starts and performs the insert actions $I[27, 101, v_{11}]$ and $I[12, 150, v_{12}]$.
3. The index constructor write-latches page $p_2$, scans the tuples, and unlatches $p_2$.
4. $T$ performs the actions $I[15, 103, v_{13}]$ and $I[41, 104, v_{14}]$, the abort action $A$, and the undo action $I^{-1}[41, 104, v_{14}]$.
5. The index constructor write-latches page $p_3$, scans the tuples, and unlatches $p_3$.
6. $T$ performs the undo actions $I^{-1}[15, 103, v_{13}]$ and $I^{-1}[12, 150, v_{12}]$.
7. The index constructor populates the index, processes the side file, and completes the index construction.
8. $T$ performs the undo action $I^{-1}[27, 101, v_{11}]$ and the complete-rollback action $C$.

Give the contents of the side file at each step. Give the set of index records in the leaf pages of the secondary index at the end of steps 7 and 8. What log records are generated for transaction $T$?

**11.4** What actions are needed to restart the index-construction process when a failure occurs during the population of the index in Algorithm 11.4 or during the side-file processing in Algorithm 11.5?

**11.5** Give the secondary-index-specific actions needed in performing a delete action $D[x, y, v]$ and an undo-delete action $D^{-1}[x, y, v]$ by transaction $T$ on relation $r(\underline{X}, Y, V)$ when the secondary index may be in construction at the same time.

**11.6** In the case of our model of transactions with primary-key-based and secondary-key-based actions on the relation $r(\underline{X}, Y, V)$, the log contains all the necessary information for completing the construction of the secondary index with updates done by transactions during the key-extraction process, so that maintaining a separate side file might seem unnecessary. Evaluate the pros and cons of these two solutions.

**11.7** We cannot allow the secondary index to be dropped while transactions are accessing it. Explain why. What actions are involved in dropping an index? What locks and latches are needed to protect index dropping?

**11.8** In the index-construction algorithm, it is possible to get rid of the short-duration S lock acquired on the relation at the beginning of index-record extraction if we include in the log record written for a tuple insertion or deletion a field that indicates the current state of the secondary index. What changes are needed in the algorithms? Also note that the no-append-to-side-file log records are no longer needed.

**11.9** Generalize the online index-construction algorithm to the case in which several secondary indexes can exist on a relation.

**11.10** With a B-tree secondary index having records of the form $(y, x, p)$ in its leaf pages, the page-id $p$ of the data page holding the indexed tuple $(x, y, v)$ is not maintained when the tuple migrates to another data page $p'$. Outline an online reorganization algorithm for the secondary index that updates the page-ids left outdated due to tuple migration.

**11.11** Assume that relation $r$ is reorganized using the online reorganization algorithm outlined in Sect. 11.7. When can the old structure of $r$ be deallocated?

**11.12** The online reorganization algorithm outlined in Sect. 11.7 does not observe secondary indexes that may exist on the relation whose primary B-tree index is being reorganized. Modify the algorithm so as to observe secondary indexes.

## Bibliographical Notes

The organization of secondary indexes presented in this chapter follows that used for index-organized tables in Oracle8i [Srinivasan et al., 2000]. The key-specific locking protocol is our adaption of the index-specific locking protocol ARIES/KVL by Mohan [1990a, 1996a]. We prefer the attribute key-specific to index-specific so as to emphasize that the locks are logical, not physical.

The problem of online index construction is addressed by Stonebraker [1989], who proposes partial indexes as a solution. A partial index is an index under construction with an attached tuple-id-range or primary-key-range predicate stating the subset of tuples it currently indexes. Other methods for online index construction are proposed by Srinivasan and Carey [1991, 1992] and Mohan and Narang [1992], among others. These algorithms are surveyed by Sockut and Iyer [2009].

The index-construction algorithm in this chapter has been adapted from that presented by Mohan and Narang [1992] for their method named "bottom-up index build with side file (algorithm SF)." Mohan and Narang [1992] suggest using a restartable merge-sort algorithm to merge the sorted runs created from tuples pulled from the priority queue, so that the B-tree can be constructed bottom-up, filling each page with a prespecified amount of records from the single sorted sequence of records.

Sockut and Iyer [1996] survey the different methods for the restoration of clustering in DB2 databases. Sockut et al. [1997] present an algorithm for online reorganization that scans the data from the old structure and stores it into a new clustered structure, removing overflow and distributing free space evenly, and finally brings the new structure up-to-date by capturing from the log transactions' updates missed by the scan. The algorithm outlined in Sect. 11.7 is an adaption of one of the methods of Sockut and Iyer [1996]. Sockut and Iyer [2009] present an extensive survey of methods for online reorganization of databases. Salzberg and Dimock [1992] address the problem of updating the references to tuples relocated in online reorganization. Zou and Salzberg [1996a,b] give online algorithms for relocating and compacting the pages of sparsely populated B-trees. Sun et al. [2005] present an algorithm for online merging of two B-trees.

# Chapter 12
# Concurrency Control by Versioning

All the concepts related to transactional isolation and concurrency control discussed in the previous chapters pertain to a *single-version database* model in which for each data item (identified by a unique key) in the logical database, only a single version, namely, the most recent or the *current version*, of the data item is available at any time. When a transaction is permitted, at the specified isolation level, to read or update a data item, the database management system always provides the transaction with the current version of the data item.

In this chapter we discuss an alternative model of a logical database, called the *multiversion database*, in which besides the current version also older versions of data items are kept available for transactions. The data-item versions are stamped by the transaction identifiers of the transactions that created them, and the versions of a data item are ordered by the commit times of those transactions. The main motivation for this is that these older versions, which are all committed, can be given to other transactions for reading without locks. Most of the major database management systems now offer the option of running transactions under *snapshot isolation*, an isolation level achieved by transient versioning of data items.

When the old versions of data items are stored permanently, a multiversion database provides access to prior states of the database via historical "as of" queries. Such accountability and traceability are needed in applications such as queryable online database backup, banking-account auditing, trend analysis, and moving-objects databases.

## 12.1 Transaction-Time Databases

The multiversion databases considered in this chapter are *transaction-time databases*, meaning that each version of a data item is stamped with the commit time of the transaction that created the version, or, actually, with the transaction

© Springer International Publishing Switzerland 2014
S. Sippu, E. Soisalon-Soininen, *Transaction Processing*, Data-Centric Systems and Applications, DOI 10.1007/978-3-319-12292-2_12

identifier of the transaction, which is then, if the transaction does commit, mapped to a *commit timestamp* that reflects the time when the transaction committed.

Thus, instead of tuples $(x, v)$ of a single-version relation $r(\underline{X}, V)$, in a transaction-time relation of the same schema, we have *versioned tuples* of the form $(x, T, v)$, where $T$ is the transaction identifier of the transaction that created the tuple. Here the pair $(x, T)$ forms a unique key of the versioned tuple among all versioned tuples.

When a transaction $T'$ updates a tuple with key $x$ for the first time in the transaction, it is given the most recent versioned tuple $(x, T, v)$, if one exists. This tuple must be committed, before $T'$ can update it. The first update by $T'$ on the tuple creates a new versioned tuple $(x, T', v')$. Any further updates by $T'$ on $x$ occur "in place," that is, they just change the value component $v'$ of the tuple. If $T'$ commits, it leaves behind the newest version of the tuple containing the last update by $T'$ on it. Otherwise, if $T'$ does not commit but aborts and rolls back, no new version created by $T'$ is left behind.

The deletion by $T'$ of a tuple with key $x$ is represented by the new versioned tuple $(x, T', \perp)$. The insertion by $T'$ of a tuple $(x, v')$ is legal only if there is not yet any versioned tuple with key $x$ or if the most recent versioned tuple with key $x$ is $(x, T, \perp)$.

The above rules mean that we only allow *linear version histories*: for any key $x$, the versioned tuples with key $x$ form a sequence

$$(x, T_1, v_1), (x, T_2, v_2), \ldots, (x, T_n, v_n),$$

where the transactions have committed in the order $T_1, T_2, \ldots, T_n$ and, for all $i = 2, \ldots, n$, transaction $T_i$ has only seen version $(x, v_{i-1})$ of the tuple with key $x$ when creating version $(x, v_i)$.

The mapping from transaction identifiers to commit timestamps is maintained in a relation of schema

*commit-time*(*transaction-id, timestamp*),

called the *commit-time table*. This table is organized as a sparse B-tree index with transaction identifiers as keys. Concurrent operations on the table are controlled by a single *tree latch* on the B-tree, so that only one transaction at a time is allowed to insert to the table.

When transaction $T$ is about to commit, it write-latches the commit-time table, reads the system clock, creates a new timestamp $\tau$, inserts the tuple $(T, \tau)$ into the commit-time table, logs the insertion with a redo-undo log record in the normal way, unlatches the table, and then commits, that is, performs the normal commit processing. This procedure ensures that the commit timestamps are unique and respect the true chronological commit order, that is, the order of the commit log records in the log.

As insertions into the commit-time table are logged with usual redo-undo log records, if it happens that $T$, after inserting into it, does not commit after all, then in the rollback of $T$, the inserted tuple is deleted from the commit-time table, as are all the versioned tuples created by $T$ deleted from the versioned relation.

For a committed transaction $T$, the unique timestamp $\tau$ with $(T, \tau)$ in the commit-time table is denoted by *commit-time(T)*.

Consider a transaction-time database $r$ and let $x$ be a key such that there exists at least one tuple with key $x$ created by a committed transaction that committed at time $\tau$ or earlier. Thus, there exist versioned tuples $(x, T, v)$ with *commit-time(T)* $\leq \tau$. Let $(x, T, v)$ be the one among those tuples with the greatest *commit-time(T)*. Then the tuple $(x, v)$ is called the *committed version* or *snapshot* of the tuple with key $x$ in $r$ *at time $\tau$* or *tuple of version $\tau$*.

We say that key $x$ is (1) *deleted at time $\tau$*, if the committed version of the tuple with key $x$ at time $\tau$ is $(x, \perp)$; (2) *present* (or *alive*) *at time $\tau$*, if the committed version of the tuple with key $x$ at time $\tau$ is $(x, v)$ with $v \neq \perp$; and (3) *absent at time $\tau$*, if either it is deleted at $\tau$ or there is no versioned tuple with key $x$. The *committed version* or *snapshot* of the transaction-time database *at time $\tau$* (or *database of version $\tau$*) is the set of tuples $(x, v)$ with keys $x$ present at time $\tau$.

The SQL language for a transaction-time database might include an extension that allows *historical as of queries*, also called *time-travel queries*, such as

**select** $*$ **from** $r$ **where** $C(X, V)$ **as of** $\tau$.

This query is supposed to return the set of tuples $(x, v)$ that satisfy $C(x, v)$ and are alive at time $\tau$.

## 12.2   Read-Only and Update Transactions

We assume two kinds of transactions on a transaction-time database: read-only transactions and update transactions. A *read-only transaction* is one that contains only read actions and is specifically declared so, such as with the SQL statement **set transaction read-only**. All other transactions are *update transactions*.

When a read-only transaction $T$ begins, or when it is about to perform its first read action, it is given a *start timestamp*, denoted by *start-time(T)*. The start timestamp is created by reading the system clock. The start timestamps need not be unique.

Usually, a read-only transaction $T$ reads from the version of the database at time *start-time(T)*, and these readings are performed without acquiring any locks. Thus, a read action $R[x, \theta z, v]$ by $T$ reads from the database of version *start-time(T)* the tuple $(x, v)$ where $x$ is the least key with $x \, \theta \, z$.

*Example 12.1* Assume that the transaction-time database contains initially the versioned tuples $(x, T_0, u_0)$ and $(y, T_0, v_0)$ and that the commit-time table contains the tuple $(T_0, \tau_0)$, indicating that transaction $T_0$ created the tuples and committed at time $\tau_0$. Consider the read-only transaction

$$T_1 = BR[x, u_0] R[y, v_0] R[y, v_0] C,$$

which reads from the version $start\text{-}time(T_1)$, runs concurrently with the update transaction

$$T_2 = BW[x, u_0, u_2]W[y, v_0, v_2]C.$$

The following history is possible in a transaction-time database:

$T_1$:                          $BR[x, u_0]R[y, v_0]$                          $R[y, v_0]C$
$T_2$:    $BW[x, u_0, u_2]$                          $W[y, v_0, v_2]C$

Here the read action $R[x, u_0]$ by $T_1$ must not be termed a dirty read nor must the first $R[y, v_0]$ be termed an unrepeatable read, because they read the earlier, committed versions of the tuples with keys $x$ and $y$ and the write actions by $T_2$ create new versions. The history is equivalent to the serial history

$T_1$:    $BR[x, u_0]R[y, v_0]R[y, v_0]C$
$T_2$:                          $BW[x, u_0, u_2]W[y, v_0, v_2]C$

At the end, the database contains the versioned tuples $(x, T_0, u_0)$, $(x, T_2, u_2)$, $(y, T_0, v_0)$, and $(y, T_2, v_2)$, and the commit-time table contains the tuples $(T_0, \tau_0)$ and $(T_2, \tau_2)$, where $\tau_2$ is the commit timestamp of $T_2$. Note that the reversed serial history

$T_1$:                          $BR[x, u_0]R[y, v_0]R[y, v_0]C$
$T_2$:    $BW[x, u_0, u_2]W[y, v_0, v_2]C$

where the commit order of the transactions is the same as in the concurrent history is not possible at all, because here $start\text{-}time(T_1) \geq commit\text{-}time(T_2)$ and hence $T_1$ must read the updates of $T_2$.                                                               □

When versioning is only used for concurrency control, not all versions of the database are maintained forever. To prevent unlimited growth of the database, older versions are periodically purged from the database. Thus it may happen that in the case of some very old read-only transaction $T$, the system can no longer provide the transaction with tuple versions at time $start\text{-}time(T)$ for reading. Such a transaction must be aborted.

Defining the meaning of read actions performed by update transactions is not so straightforward. An update transaction $T$ may read from the version at $start\text{-}time(T)$, or from some later version, depending on the isolation level chosen; besides that $T$ is assumed to see its own updates. Under the isolation levels to be discussed in this chapter, a read action $R[x, \theta z, v]$ always reads either from the version at $start\text{-}time(T)$ or from the version of the time when the read action started, except that tuples inserted or written by $T$ itself always take precedence.

Thus, in executing a read action $R[x, \theta z, v]$ on version $start\text{-}time(T)$, if it is found that the least key $x$ with $x \, \theta \, z$ in the database version of $start\text{-}time(T)$ is greater than a key $x'$ that has been inserted or written by $T$ itself and also satisfies $x' \, \theta \, z$, then $T$ must read $x'$ instead of $x$.

*Example 12.2* Assume that the transaction-time database contains initially the versioned tuples $(1, T_0, u_0)$ and $(4, T_0, v_0)$. Consider different interpretations of the following history of two update transactions:

$T_1$:   $BI[2, w_1]R[x, > 1, w]$                     $R[y, > 2, w']C$
$T_2$:                              $BI[3, w_2]C$

As it is natural that transactions see their own updates, the read action $R[x, > 1, w]$ must retrieve the tuple $(2, w_1)$. But should the read action $R[y, > 2, w']$ retrieve the tuple $(3, w_3)$, whose key is the least key greater than 2 in the most recent committed database version (*commit-time*$(T_2)$), or the tuple $(4, v_0)$, whose key is the least key greater than 2 in the database of version *start-time*$(T_1)$? In the latter case, the history is equivalent to the serial history

$T_1$:   $BI[2, w_1]R[2, > 1, w_1]R[4, > 2, v_0]C$
$T_2$:                              $BI[3, w_2]C$

where the commit order of the transactions is reversed. In the former case, the history is equivalent to the reversed serial history

$T_1$:                      $BI[2, w_1]R[2, > 1, w_1]R[3, > 2, w_2]C$
$T_2$:   $BI[3, w_2]C$

that respects the original commit order. Note that in this case *start-time*$(T_1) \geq$ *commit-time*$(T_2)$, so that $T_1$ must also see the update of $T_2$.                     □

Obviously, read actions by an update transaction that read anything other than the most recent data are unrepeatable and can cause the integrity of the database to be lost, as shown in the examples given in Sect. 5.4.

An update action, that is, a write action $W[x, u, v]$, an insert action $I[x, v]$, or a delete action $D[x, v]$, is always based on the most recent committed database version, that is, version $\tau$, where $\tau$ is the greatest timestamp in the commit-time table at the time the action is executed. Moreover, no update actions by other transactions on key $x$ must be ongoing at the same time. This is necessary to maintain the constraint that only linear version histories are allowed.

Accordingly, to perform a write action $W[x, u, v]$ in update transaction $T$, if this is the first update by $T$ on key $x$, the tuple $(x, u)$ with key $x$ in the most recent committed database version is retrieved, the versioned tuple $(x, T, v)$ is inserted into the database, and the insertion is redo-undo logged in the usual way. In this case the most recent committed database version must contain a versioned tuple $(x, T', v')$ with $v' \neq \perp$. Otherwise, if $T$ has earlier inserted or written the tuple with key $x$, so that a versioned tuple $(x, T, u)$ exists, then that tuple is replaced by the tuple $(x, T, v)$, and the replacement is redo-undo logged.

To perform an insert action $I[x, v]$, if this is the first update on key $x$ by $T$, the versioned tuple $(x, T, v)$ is inserted into the database, and the insertion is redo-undo logged. In this case the most recent committed database version must not contain any tuple with key $x$ other than one of the form $(x, T', \perp)$. Otherwise, if $T$ has earlier deleted the tuple, so that a versioned tuple $(x, T, \perp)$ exists, then that tuple is replaced by the tuple $(x, T, v)$ and the replacement is redo-undo logged.

To perform a delete action $D[x, v]$, if this is the first update on key $x$ by $T$, the version $(x, v)$ of the tuple with key $x$ is retrieved from the most recent committed database version, the versioned tuple $(x, T, \bot)$ is inserted into the database and the insertion is redo-undo logged. The tuple $(x, v)$ must exist and satisfy $v \neq \bot$. Otherwise, if $T$ has earlier inserted or written the tuple, so that a versioned tuple $(x, T, v)$ with $v \neq \bot$ exists, then that tuple is replaced by the tuple $(x, T, \bot)$ and the replacement is redo-undo logged.

The undo actions $W^{-1}[x, u, v]$, $I^{-1}[x, v]$, and $D^{-1}[x, v]$ operate as expected, using the log record for the corresponding forward-rolling action to undo the action and logging the undoing with a redo-only log record.

## 12.3  Transaction-Level Read Consistency

Under *transaction-level read consistency*, the read actions of every read-only transaction $T$ read from the same database version, usually that of *start-time*$(T)$, while every update transaction runs at the serializable isolation level. Accordingly, the read and update (insert, delete, write) actions of every update transaction operate on the most recent tuple versions.

*Example 12.3*  The history

$T_1$:                        $BR[x, \geq x, u_0] R[y, \geq y, v_0]$                        $R[y, \geq y, v_0]C$
$T_2$:  $BW[x, u_0, u_2]$                                        $W[y, v_0, v_2]C$

(cf. Example 12.1) is possible under transaction-level read consistency and so is the history

$T_1$:  $BI[2, w_1] R[2, > 1, w_1]$                        $R[3, > 2, w_2]C$
$T_2$:                                $BI[3, w_2]C$

of Example 12.2.                                                                                    $\square$

Isolation anomalies that may appear in this setting are the usual dirty writes, dirty reads, and unrepeatable reads between two update transactions as defined in Sect. 5.3 for single-version databases. Thus, it is a dirty write if a transaction writes, inserts, or deletes a tuple with key $x$ when a tuple with key $x$ has been written, inserted, or deleted by another, still active transaction. As before, dirty writes must be prevented under all circumstances. When using a pessimistic concurrency-control protocol such as locking, this means that the simultaneous existence of two versioned tuples $(x, T, v)$ and $(x, T', v')$, where both $T$ and $T'$ are active, is not allowed.

A read action $R[x, \theta z, v]$ by an update transaction $T_2$ is a dirty read if some key $y$ with $x \geq y \, \theta \, z$ has an uncommitted update by some other transaction $T_1$, for example, there is a versioned tuple $(y, T_1, w)$, where $T_1$ is still active. A read action $R[x, \theta z, v]$ by an update transaction $T_1$ is an unrepeatable read if some other transaction $T_2$ later updates a tuple with key $y$ with $x \geq y \ \theta \ z$ while $T_1$ is still forward-rolling.

A read-only transaction cannot do any dirty read, because under transaction-level read consistency, by definition, all the read actions read committed data. No read action by such a transaction can be termed as an unrepeatable read either, even though some update transaction might later, while the read-only transaction is still active, update a tuple read by the read-only transaction. This is because read-only transactions do no updates and, again by definition, the same read action repeated would still read the same old version of the tuple. Thus no isolation anomalies can occur between read-only and update transactions.

We have:

**Theorem 12.4** *Assume a history H of any number of forward-rolling, backward-rolling, committed, and rolled-back transactions run at transaction-level read consistency. If in H no update transaction does any dirty writes or dirty or unrepeatable reads, then H is equivalent to a serial history H' in which the transactions run at transaction-level read consistency, the committed and rolled-back update transactions commit in the same order as in H, and every read-only transaction T is ordered before any update transaction T' with start-time(T) < commit-time(T') and after any update transaction T'' with start-time(T) $\geq$ commit-time(T'').* ☐

To ensure transaction-level read consistency, all the actions by update transactions are protected by commit-duration locks. However, unlike with a single-version database, a next-key X lock is needed only for insert actions but not for delete actions, because deleted keys are explicitly represented as versioned tuples $(x, T, \perp)$. When read actions are implemented so that they observe these tuples, it suffices that a single X lock is acquired, namely, a commit-duration X lock on the key to be deleted. An X lock on key $x$ prevents other update transactions from reading any versioned tuple with key $x$ or creating a new version of a tuple with key $x$ but does not prevent a read-only transaction from reading a committed versioned tuple with key $x$, because read-only transactions do not observe locks.

*Example 12.5* Assume that the database contains initially the versioned tuples $(1, T_0, v_1)$ and $(3, T_0, v_3)$. The following history of two update transactions $T_1$ and $T_2$ is not possible at transaction-level read consistency because of the dirty read by $T_2$:

$T_1$:   $BD[1, v_1]$                                                                  $D[2, v_2]C$
$T_2$:                          $BR[x, \geq 1, v]I[2, v_2]C$

The dirty read $R[x, \geq 1, v]$ is prevented if $T_1$ protects the delete action $D[1, v_1]$ by acquiring a commit-duration X lock on key 1, and $T_2$ observes the uncommitted versioned tuple $(1, T_1, \perp)$ and acquires an S lock on key 1.

The following history of two update transactions $T_3$ and $T_4$ is not possible at transaction-level read consistency because of the unrepeatable read by $T_3$:

$T_3$:   $BR[3, > 1, v]$                                                              $D[2, v_2]C$
$T_4$:                          $BI[2, v_2]C$

Here $T_3$ protects the read action $R[3, > 1, v]$ by acquiring a commit-duration S lock on key 3, but this does not prevent $T_4$ from getting an X lock on key 2 for the insert action $I[2, v_2]$. Thus, to prevent the unrepeatable read, we still have to obey the standard key-range locking protocol and acquire a short-duration X lock on the next key 3.                                                                                                □

## 12.4  Statement-Level Read Consistency

Under *statement-level read consistency*, the read actions of every read-only transaction read from the start-time version, and the update actions of every update transaction operate on the most recent tuple versions. Unlike with transaction-level read consistency, the read actions by update transactions may be unrepeatable and read from the most recent committed version as of the time when the SQL statement that gave rise to the read action was started. Thus, all the tuples read by a single SQL statement are read from the same committed version (except that, again, read actions operating on tuples updated by the transaction itself read from the current version). In a way, the start timestamp of an update transaction is advanced every time a new SQL statement is invoked by the database application program that generates the transaction.

Obviously, transactions run at statement-level read consistency, do neither dirty writes nor dirty reads.

*Example 12.6* Assume that the database contains initially the versioned tuples $(1, T_0, v_1)$, $(2, T_0, v_2)$, and $(3, T_0, v_3)$. In a single-version database, the history

$T_1$:    $BD[2, v_2]$                         $\ldots C$
$T_2$:                    $BR[x, > 1, v]C$

would exhibit a phantom dirty read, because then the read action $R[x, > 1, v]$ is $R[3, > 1, v_3]$.

However, in a transaction-time database at statement-level read consistency the read action is $R[2, > 1, v_2]$. While searching the database for the tuple with the least key $x$ greater than 1, the read action $R[x, > 1, v]$ encounters the tuples $(2, T_0, v_2)$ and $(2, T_1, \perp)$. As the version at time *commit-time*$(T_0)$ is the most recent committed version, the read action must ignore $(2, T_1, \perp)$ and read the tuple $(2, T_0, v_2)$. Thus the history produced is equivalent to the serial history

$T_1$:                    $BD[2, v_2] \ldots C$
$T_2$:    $BR[2, > 1, v_2]C$

                                                                                                □

**Theorem 12.7** *Statement-level read consistency prevents dirty writes and dirty reads of all kinds and hence implies SQL isolation level 2° (read committed).*     □

The unrepeatable reads that may occur at statement-level read consistency include lost updates, as defined in Sect. 9.5.

*Example 12.8*  Consider again the banking-database transactions

$$T_1 = BR[x, v_1]W[x, u_1, v_1 - 100]C$$
$$T_2 = BR[x, v_2]W[x, u_2, v_2 - 200]C$$

("withdraw 100/200 euro from account $x$") of Example 9.6. The history

$T_1$:    $BR[x, v]$                                                $W[x, v - 200, v - 100]C$
$T_2$:                        $BR[x, v]W[x, v, v - 200]C$

is possible at statement-level read consistency when the read action is generated from a **select** query and the write action from a separate **update** statement.

In the above history, the read and write actions of both transactions operate on the most recent committed version of the tuple with key $x$. However, because of the unrepeatable read by $T_1$, the history is not possible at transaction-level read consistency: both transactions are update transactions and are therefore, by definition, required to run at the serializable isolation level.                □

In implementing statement-level read consistency by locking, no S locks are used, because the read actions by both read-only and update transactions operate on committed versions of the (advanced) start times of the transactions. Commit-duration X locks are used to protect the update actions.

If read actions and update actions are generated from the same SQL statement, then the system may protect both actions by X locks (or U locks upgraded later to X locks), in which case no unrepeatable reads may occur after all.

*Example 12.9*  If both the read and write actions in the transactions of Example 12.8 are generated from the single SQL statements

**update** *r* **set** $V = V - 100$ **where** $X = :x$
**update** *r* **set** $V = V - 200$ **where** $X = :x$

then the key $x$ is X-locked for commit duration before reading and updating the tuple with key $x$, so no lost update will occur.                                                    □

## 12.5  Snapshot Isolation

Several well-known database management systems that use versioning for concurrency control enforce an isolation level called *snapshot isolation*. With snapshot isolation, no distinction is made between read-only and update transactions. All read actions by any transaction read from the start-time version of the transaction, except that, again, read actions on data updated by the transaction itself read the current data. All update actions operate on the most recent tuple versions.

Under snapshot isolation, if two committed transactions $T_1$ and $T_2$ are concurrent, that is, at some timepoint both are active, then they are required to have the *disjoint-write property* defined as follows: the *write sets* of $T_1$ of $T_2$, that is, the sets of identifiers (unique keys) of the data items updated by the transactions, must be disjoint:

$$write\text{-}set(T_1) \cap write\text{-}set(T_2) = \emptyset.$$

*Example 12.10* The two banking-database transactions given in Example 12.8 are concurrent but do not have the disjoint-write property, because

$$write\text{-}set(T_1) = \{x\} = write\text{-}set(T_2).$$

Thus, the transactions are not snapshot isolated.                                                □

The requirement of the disjoint-write property only pertains to pairs of two committed transactions, not for pairs of transactions of which one is rolled back.
It can be shown:

**Theorem 12.11** *In a snapshot-isolated history, no committed transaction contains any dirty writes or dirty reads or lost updates.*                                                □

Unrepeatable reads other than of the lost-update type can still appear in a snapshot-isolated history, and hence the integrity of the database may be lost.

*Example 12.12* Consider the two transactions

$$T_1 = BR[y, v] W[x, u, -v]C$$
$$T_2 = BR[x, u] W[y, v, -u]C$$

of Example 5.19, where transaction $T_1$ (resp. $T_2$) can be interpreted as "a withdrawal of the sum of balances $u + v$ of accounts $x$ and $y$ from account $x$ (resp. $y$)." Both transactions bring the sum of the balances of accounts $x$ and $y$ down to zero, thus preserving the integrity constraint that states that the sum of the balances must be nonnegative. The history

$$T_1: \quad BR[y, v] \qquad\qquad\qquad\qquad\qquad W[x, u, -v]C$$
$$T_2: \qquad\qquad\qquad BR[x, u] W[y, v, -u]C$$

is possible at snapshot isolation because (1) both read actions read from the start-time versions, (2) both write actions operate on the most recent committed versions, and (3) the write sets of the transactions do not intersect:

$$write\text{-}set(T_1) \cap write\text{-}set(T_2) = \{x\} \cap \{y\} = \emptyset.$$

As in Example 5.19 for a single-version database, the unrepeatable read action $R[y, v]$ causes the integrity constraint on a transaction-time database to be broken whenever the sum of the balances is positive initially.                                                □

## 12.6  Enforcing the Disjoint-Write Property

The disjoint-write property can be enforced with the use of commit-duration X locks in the following way.

When transaction $T$ wants to perform its first forward-rolling update action on a tuple with key $x$ (i.e., one of the actions $I[x, v]$, $D[x, v]$, or $W[x, u, v]$), the lock table is examined in order to find out if some other transaction currently holds an X lock on $x$.

If no other transaction holds an X lock on $x$, the versioned tuple $(x, T', v')$, if any, is determined where $T'$ is the committed transaction with the greatest *commit-time*$(T')$. If

$$commit\text{-}time(T') > start\text{-}time(T),$$

then $T'$ and $T$ are concurrent, and since $T'$ has updated $x$ and committed, to maintain the disjoint-write property, $T$ must be prevented from updating $x$ and committing; thus $T$ must be aborted and rolled back. Otherwise, if

$$commit\text{-}time(T') < start\text{-}time(T),$$

$T$ is allowed to read and update data items update by $T'$, so a commit-duration X lock on $x$ is granted to $T$. The lock is also granted in the case the database does not yet contain a versioned tuple $(x, T', v')$.

If, on the contrary, a versioned tuple $(x, T', v')$ exists and transaction $T'$ holds an X lock on $x$, then $T'$ and $T$ are concurrent and $T$ must wait for the outcome of $T'$, that is, whether $T'$ commits or rolls back. If $T'$ commits, then, to maintain the disjoint-write property, $T$ and all other transactions that are also waiting for an X lock on $x$ must be prevented from updating $x$ and committing; hence $T$ and the other waiting transactions are aborted and rolled back. Otherwise, if $T'$ aborts and rolls back, $T$ and the other transactions are allowed to wait for a lock on $x$, and if $T$ is the first in the queue, it is granted a commit-duration X lock on $x$.

As with conventional locking protocols, the undo actions $I^{-1}[x, v]$, $D^{-1}[x, v]$, and $W^{-1}[x, u, v]$ are performed under the protection of the commit-duration X lock acquired for the corresponding forward-rolling action.

Because no S locks are used to protect read actions, no read action ever waits for an updating transaction to commit, nor does any update action wait for a read action to complete, but update actions need to wait for an updating transaction to commit or roll back. Transactions are aborted when their updates come too late or after waiting for a transaction that commits.

## 12.7   Version Management with B-Trees

The basic sparse B-tree index discussed in Chaps. 7 and 8 offers as such a simple, although not the most efficient, physical database structure for managing a transaction-time database. We just regard versioned tuples $(x, T, v)$ of a logical database $r(\underline{X}, V)$ as tuples in a relation $r(\underline{X}, \underline{T}, V)$, where the ordered pair $XT$ is the key of the relation. We call such a B-tree a *versioned B-tree*.

In the leaf pages of the versioned B-tree for $r$, the different versions

$$(x, T_1, v_1), (x, T_2, v_2), \dots, (x, T_n, v_n)$$

of the tuple with key $x$ are stored side by side, ordered by the transaction identifiers, so that the first versioned tuple $(x, T_1, v_1)$, that is, the one with the least transaction identifier $T_1$, resides in the leaf page covering key $(x, T_1)$, and the last versioned tuple $(x, T_n, v_n)$, that is, the one with the greatest transaction identifier $T_n$, resides in the leaf page covering key $(x, T_n)$.

Assume now that transaction $T$ wants to execute a read action $R[x, \theta z, v]$ on the database of version $\tau$, where $\tau$ is some timestamp greater than or equal to *start-time*$(T)$ and less than or equal to the start time of the action $R[x, \theta z, v]$. Such reads appear in read-only transactions running at transaction-level read consistency and in all transactions at statement-level read consistency and snapshot isolation.

We must determine the least key $x$ with $x \theta z$ among the keys $x$ present at time $\tau$ or inserted or written by $T$ itself. To that end, we scan keys $x$ in ascending order starting at the least key $x$ with $x \theta z$ until we find one that is present at time $\tau$ or has been inserted or written by $T$.

First, a procedure call *find-pages-for-read*$(P, \theta z)$ is used to locate and read-latch the set $P$ of leaf pages that contain the versioned tuples with the least key $x$ with $x \theta z$. The commit-time table is read-latched, the set of records for the transactions that created the versioned tuples with key $x$ is retrieved from the commit-time table, and the commit-time table is unlatched. Among the versioned tuples $(x, T', v')$ with *commit-time*$(T') \leq \tau$, if such tuples exist, the one with the greatest commit time is determined. Also the tuple $(x, T, v)$ created by $T$, if one exists, is determined.

We have to consider the following cases (of which only cases 3, 4, and 5 can appear in a read-only transaction):

1. $(x, T, v)$ exists and $v \neq \perp$. Then $T$ itself has inserted or written a tuple with key $x$; the tuple $(x, v)$ is returned.
2. $(x, T, v)$ exists and $v = \perp$. Then $T$ itself has deleted the tuple with key $x$; we move to the key next to $x$.
3. $(x, T, v)$ does not exist but $(x, T', v')$ with $v' \neq \perp$ exists. The tuple $(x, v')$ is returned.
4. $(x, T, v)$ does not exist but $(x, T', v')$ with $v' = \perp$ exists. Then we move to the key next to $x$.
5. Neither $(x, T, v)$ nor $(x, T', v')$ exists. The tuple $(\infty, 0)$ is returned.

If the tuple with key $x$ does not qualify for reading (cases 2 and 4), we unlatch the pages in $P$ and move to the key next to $x$, that is, to the versioned tuples with the least key greater than $x$, and repeat the checking for that key, and so on. The move to the key next to $x$ is performed by the procedure call *find-pages-for-read*$(P, > x)$. Note that even if new versioned tuples with keys already scanned may appear, those tuples must be of versions created by transactions other than $T$ committing after the start of the read action and hence do not qualify for reading. If no key $x$ is found with a non-deleted tuple, then by our convention, the tuple $(\infty, 0)$ is returned (case 5).

The call *find-pages-for-read*$(P, \theta z)$ first performs an optimistic traversal of the versioned B-tree, so that only the leaf page that covers the key $z$ is accessed and read-latched. If the page does not contain all of the versioned tuples with the least key $x$ with $x \ \theta \ z$, then the page is unlatched and a second traversal is performed that read-latches at each level a set of pages whose subtrees together contain all the versioned tuples with the least key $x$ with $x \ \theta \ z$. Saved paths are used as in the B-tree algorithms of Chaps. 7 and 8. The second traversal saves a path that leads to the leaf page containing the last versioned tuple with key $x$, that is, the one with the greatest transaction identifier.

In analogy with the complexity result stated in Theorem 10.14 for bulk-read actions, we have:

**Theorem 12.13** *Assume that a read-only transaction $T$ reading from the database of version start-time$(T)$ performs a bulk-read action $R[s_X, s_{XV}]$ on a versioned B-tree with n leaf pages. Let $P$ be the set of leaf pages that contain tuples of any version with keys in some of the ranges in $s_X$. Then, in the absence of structure modifications, the complexity bound (10.1) proved in Sect. 10.7 holds for the total number of page latchings performed.*

*Proof* The proof is similar to that of Theorem 10.14. Now for each key $x$ in one of the scanned ranges in $s_X$, we have to read the entire set of versioned tuples with key $x$ in order to find out which version to read and to skip over to the next key. Unlike in Theorem 10.14, the result holds even if update transactions run concurrently with the read-only transaction, if only no structure modifications are triggered. Because a read-only transaction does not acquire or observe any locks, no waits other than those caused by latching occur between the read-only transaction and the update transactions. The absence of structure modifications in turn implies that the saved path maintained for the read-only transaction is not invalidated between two read actions, except that the changes in leaf-page PAGE-LSNs due to updates may at key-range borders cause a traversal to be started at one page higher than necessary on the saved path.                                                                                □

To the complexity bound mentioned in the above theorem, we must also add the number of latchings of the commit-time table and the number of fixings of its pages. Obviously, as we must eventually purge oldest versions from the B-tree, we must not allow the commit-time table to grow indefinitely either; we address this question in the Problems section.

There exist more efficient index structures for versioned data than the basic B-tree organization discussed here. One of the more sophisticated index structures, called the *multiversion B-tree*, can be considered optimal in that a bulk-read action as in Theorem 12.13 can be performed for any database version $\tau$ asymptotically as efficiently as with a B-tree that contains only the tuples of version $\tau$. Here $\tau$, the version to be read, can be any version equal to or earlier than the start time of the transaction.

## 12.8   Update Transactions on Versioned B-Trees

In this section we discuss the implementation of read, insert, and delete actions in update transactions running at transaction-level read consistency, when the physical structure of the database is a versioned B-tree. As explained in Sect. 12.3, commit-duration locks are used to protect these actions. The locking protocol used is an adaption of key-range locking to a transaction-time database.

In executing a read action $R[x, \theta z, v]$ in an update transaction $T$, we scan keys $x$ in ascending order starting at the least key $x$ with $x\ \theta\ z$. First, the procedure call *find-pages-for-read*$(P, \theta z)$ is used to locate and read-latch the set $P$ of leaf pages that contain the versioned tuples with the least key $x$ with $x\ \theta\ z$. The commit-time table is read-latched, the set of records for the transactions that created the versioned tuples with key $x$ is retrieved from the commit-time table, and the table is unlatched. Among the committed versioned tuples $(x, T', v')$, if such tuples exist, the one with the greatest commit time is determined. Also the tuple $(x, T'', v'')$ created by some active transaction, if one exists, is determined. Note that because of X-locking, at most one such tuple $(x, T'', v'')$ may exist at a time. Here $T''$ may be $T$ or some other update transaction.

We have to consider the following cases:

1. $(x, T'', v'')$ exists and $T'' \neq T$. Then key $x$ is currently X-locked by $T''$. All latches are released, $x$ is unconditionally S-locked, and the search for the correct key is restarted when the lock is granted.
2. $(x, T'', v'')$ exists and $T'' = T$ and $v'' \neq \bot$. Then $T$ itself holds an X lock on $x$; the tuple $(x, v'')$ is returned.
3. $(x, T'', v'')$ exists and $T'' = T$ and $v'' = \bot$. Then $T$ itself has deleted $x$; we release all latches and move to the key next to $x$.
4. $(x, T'', v'')$ does not exist but $(x, T', v')$ with $v' \neq \bot$ exists. If $T$ holds a lock on $x$, then the tuple $(x, v')$ is returned. Otherwise, key $x$ is conditionally S-locked; if the lock is granted, the tuple $(x, v')$ is returned—otherwise, all latches are released, $x$ is unconditionally S-locked, and the search for the correct key is restarted when the lock is granted.
5. $(x, T'', v'')$ does not exist but $(x, T', v')$ with $v' = \bot$ exists. Then the current version of the tuple with key $x$ is committed and deleted. If $T$ holds a lock on $x$, we release all latches and move to the key next to $x$; otherwise, all latches are

released and $x$ is unconditionally S-locked, and the search for the correct key is restarted when the lock is granted.

6. Neither $(x, T'', v'')$ nor $(x, T', v')$ exists. If $T$ holds a lock on the key $\infty$, then the tuple $(\infty, 0)$ is returned. Otherwise, all latches are released, $\infty$ is unconditionally S-locked, and the search for the correct key is restarted when the lock is granted.

If the current version of the tuple with key $x$ is deleted (cases 3 and 5), we unlatch the pages in $P$ and move to the key next to $x$, that is, to the versioned tuples with the least key greater than $x$, and repeat the checking for that key, and so on. The move to the key next to $x$ is performed by the procedure call *find-pages-for-read*$(P, > x)$. If no key $x$ is found with a non-deleted tuple, then by our convention, the key $\infty$ is S-locked and the tuple $(\infty, 0)$ is returned (case 6).

In case 5 it is necessary to lock the key $x$ of the currently deleted tuple, because otherwise some transaction that commits before $T$ can insert a tuple with key $x$. This is possible because the pages in $P$ are unlatched before moving to the key next to $x$. It is necessary to unlatch the pages before the call *find-pages-for-read*$(P, > x)$, because we must never latch an ancestor of a latched page.

In executing a delete action $D[x, v]$ in an update transaction $T$, we first unconditionally X-lock key $x$ for commit duration and then use the procedure call *find-pages-for-delete*$(P, x, T)$ to access and write-latch the set $P$ of leaf pages that contain the versioned tuples with key $x$. The call also ensures, by appropriate structure modifications, that the leaf page that covers the key $(x, T)$ has room for the versioned tuple $(x, T, \perp)$. The commit-time table is read-latched, the set of records for the transactions that created the versioned tuples with key $x$ is retrieved from the commit-time table, and the table is unlatched. Among the committed versioned tuples with key $x$, if such tuples exist, let $(x, T', v')$ be the one with the greatest commit time, and let $(x, T, v'')$ be the one created by $T$ itself, if one exists. Note that because of commit-duration X-locking, there cannot exist any tuple $(x, T'', v'')$ where $T'' \neq T$ is an active transaction.

We have the following cases:

1. $(x, T, v'')$ with $v'' \neq \perp$ exists. Then the last update on key $x$ is an insert or a write action by $T$ itself. The tuple $(x, T, v'')$ is replaced by the tuple $(x, T, \perp)$, the replacement is redo-undo logged, and the tuple $(x, v'')$ is returned.
2. $(x, T, v'')$ with $v'' = \perp$ exists. Then $T$ is logically inconsistent and must be aborted.
3. $(x, T, v'')$ does not exist but $(x, T', v')$ with $v' \neq \perp$ exists. Then the last update on key $x$ is an insert or a write action by committed transaction $T'$. The tuple $(x, T, \perp)$ is inserted, the insertion is redo-undo logged, and the tuple $(x, v')$ is returned.
4. $(x, T, v'')$ does not exist but $(x, T', v')$ with $v' = \perp$ exists. Then the delete action is illegal in this context and $T$ must be aborted.
5. Neither $(x, T, v'')$ nor $(x, T', v')$ exists. Then the delete action is illegal in this context and $T$ must be aborted.

In executing an insert action $I[x, v]$ in an update transaction $T$, we first unconditionally X-lock key $x$ for commit duration and then use the procedure call *find-pages-for-insert*$(P, x, T, v, y)$ to access and write-latch the set $P$ of leaf pages that contain the versioned tuples with key $x$ and the key $y$ next to $x$. The call also ensures, by appropriate structure modifications, that the leaf page that covers the key $(x, T)$ has room for the versioned tuple $(x, T, v)$. The key $y$ is X-locked for short duration in the usual way, first with a conditional lock request, and if this does not succeed, with an unconditional request. The key $y$ need not be one that is currently present, because read actions observe deleted keys and S-lock them before moving to the next key.

The commit-time table is read-latched, the set of records for the transactions that created the versioned tuples with key $x$ is retrieved from the commit-time table, and the table is unlatched. Among the committed versioned tuples with key $x$, if such tuples exist, let $(x, T', v')$ be the one with the greatest commit time, and let $(x, T, v'')$ be the one created by $T$ itself, if one exists. Again note that because of commit-duration X-locking, there cannot exist any tuple $(x, T'', v'')$ where $T'' \neq T$ is an active transaction.

We have the following cases:

1. $(x, T, v'')$ with $v'' = \perp$ exists. Then the last update on key $x$ is a delete action by $T$ itself. The tuple $(x, T, v'')$ is replaced by the tuple $(x, T, v)$ and the replacement is redo-undo logged.
2. $(x, T, v'')$ with $v'' \neq \perp$ exists. Then $T$ is logically inconsistent and must be aborted.
3. $(x, T, v'')$ does not exist but $(x, T', v')$ with $v' = \perp$ exists. Then the last update on key $x$ is a delete action by committed transaction $T'$. The tuple $(x, T, v)$ is inserted and the insertion is redo-undo logged.
4. $(x, T, v'')$ does not exist but $(x, T', v')$ with $v' \neq \perp$ exists. Then the insert action is illegal in this context and $T$ must be aborted.
5. Neither $(x, T, v'')$ nor $(x, T', v')$ exists. The tuple $(x, T, v)$ is inserted and the insertion is redo-undo logged.

## Problems

**12.1** When can old versions be purged from a transaction-time database, when maintaining (a) transaction-level read consistency, (b) statement-level read consistency, and (c) snapshot isolation?

**12.2** Give log records for the forward-rolling update actions $I[x, v]$, $D[x, v]$, and $W[x, u, v]$ and for their undo actions $I^{-1}[x, v]$, $D^{-1}[x, v]$, and $W^{-1}[x, u, v]$, performed on a transaction-time database.

**12.3** When commit time is used as the basis of stamping and ordering the versioned tuples, we need the separate commit-time table for mapping transaction identifiers of committed transactions to commit timestamps. Such a table would not be needed if the versioned tuples created by a transaction were stamped with the start time of the transaction. But then we have to force the transactions to commit in their start-time order. Explain why. Also explain how such a commit order can be enforced.

**12.4** Design an efficient method for maintaining the write sets needed in enforcing the disjoint-write property of transactions. Observe that unlike locks, which only exist while the owner transaction is active, write sets need be stored for committed transactions. When can the write set of a committed transaction be purged?

**12.5** Give an algorithm for the procedure $find\text{-}pages\text{-}for\text{-}read(P, \theta z)$ discussed in Sect. 12.7.

**12.6** Give an algorithm for the procedure $find\text{-}pages\text{-}for\text{-}insert(P, x, T, v, y)$ discussed in Sect. 12.8.

**12.7** The logic of the procedures $find\text{-}pages\text{-}for\text{-}read(P, \theta z)$, $find\text{-}pages\text{-}for\text{-}insert(P, x, T, v, y)$, and $find\text{-}pages\text{-}for\text{-}delete(P, x, T)$ can obviously be simplified, and the number of page latchings reduced, by linking the leaf pages of the versioned B-tree from left to right, so that each leaf page $p$ contains the record $(high\text{-}key(p), p')$, where $p'$ is the page-id of the leaf page next to page $p$ (if $p$ is not the last leaf page) or the null page-id (otherwise). Work out this simplification.

**12.8** Analyze the complexity of insert, delete, and write actions $I[x, v]$, $D[x, v]$, and $W[x, u, v]$ on a versioned B-tree.

**12.9** Explain how the actions in the history

$T_1$:   $BI[2, w_1] R[2, > 1, w_1]$                                  $R[3, > 2, w_2]C$
$T_2$:                                     $BI[3, w_2]C$

of Example 12.3 are executed on a versioned B-tree using the algorithms outlined in Sect. 12.8. What locks are acquired?

**12.10** Explain how a write action $W[x, u, v]$ is executed in a transaction running at transaction-level read consistency on a versioned B-tree. Explain how the actions in the history

$T_1$:                       $BR[x, \geq x, u_0] R[y, \geq y, v_0]$                          $R[y, \geq y, v_0]C$
$T_2$:  $BW[x, u_0, u_2]$                                  $W[y, v_0, v_2]C$

of Example 12.3 are executed. What locks are acquired?

**12.11** The commit-time table may become a bottleneck because in order to access the correct versioned tuple in read and update actions, the commit-time table must be consulted for the mapping from transaction identifiers to the commit timestamps of the transactions. Also, the insertion of the commit-time tuple into the table within the transaction, when the transaction is committing, means an overhead. Performance can be enhanced (1) by allowing a transaction to commit without updating the commit-time table, by just logging the commit timestamp and updating the commit-time table later, and (2) by substituting the commit timestamps for the transaction identifiers in data-item versions when they are first needed in subsequent accesses to the versions. Elaborate on these ideas. What changes would be needed in the algorithms outlined in Sect. 12.8?

## Bibliographical Notes

POSTGRES was the first transaction-time database management system implemented. The system maintains the entire history of data items, and its query language, POSTQUEL, provides for historical "as of" queries [Stonebraker and Rowe, 1986, Stonebraker, 1987, Stonebraker et al. 1990]. The system combines the use of historical database states with the force buffering policy to obtain instantaneous recovery from system crashes, while strict two-phase locking is used for concurrency control [Stonebraker, 1987].

In POSTGRES, when a transaction updates a data item, the transaction's identifier is associated with the new version of the item, and when the transaction commits, its commit time is stored in the persistent commit-time table, as explained in Sect. 12.1. The new data-item versions may later be revisited and the commit times be substituted for the transaction identifiers (cf. Problem 12.11). This revisitation is done lazily or not at all [Stonebraker, 1987].

To achieve a transaction-consistent view of historical database states, it is necessary to use the same timestamp for all data-item versions created by a transaction, and these timestamps must respect the serialization order of the transactions, as do the commit timestamps [Salzberg, 1994]. However, SQL statements within a transaction may request "current time." To address this problem, Jensen and Lomet [2001] and Lomet et al. [2005b] present methods for earlier choice of transaction timestamps while respecting the serialization order.

The theory of multiversion concurrency control was developed by Bayer et al. [1980] and by Bernstein and Goodman [1983], among others. Algorithms can be found in the textbooks by Bernstein et al. [1987], Papadimitriou [1986], and Weikum and Vossen [2002], to name a few. Mohan et al. [1992b] present an algorithm that uses transient versioning of data items to permit read-only transactions to run without acquiring locks. Snapshot isolation was defined and analyzed by Berenson et al. [1995].

The isolation anomalies possible at snapshot isolation are discussed by Berenson et al. [1995], Fekete et al. [2005], Kifer et al. [2006], and Cahill et al. [2009], among others. Cahill et al. [2009] present a concurrency-control algorithm based on snapshot isolation that permits only serializable histories. This algorithm has been implemented in the BERKELEY DB embedded database management system and in the INNODB transactional backend for MYSQL [Cahill et al. 2009] and also in POSTGRESQL [Ports and Grittner, 2012].

The IMMORTAL DB system provides both snapshot isolation and full transaction-time functionality, with permanently stored versions, built into the SQL Server database engine [Lomet et al. 2005a, 2006, Lomet and Li, 2009]. Transactions commit without updating the persistent commit-time table, carrying the commit timestamps in their commit log records. The timestamps are later inserted as group updates into the commit-time table, from where they are propagated lazily to the data-item versions when the versions are needed in subsequent accesses [Lomet and Li, 2009] (cf. Problem 12.11).

When the data-item versions carry their final commit timestamps, the two-dimensional space of keys and versions can be indexed using some multidimensional access method. In POSTGRES, the R-trees of Guttman [1984] are used to index historical versions moved to archival storage, while recent data is indexed by a B-tree [Stonebraker, 1987]. In IMMORTAL DB, the time-split B-tree (TSB-tree) of Lomet and Salzberg [1989] is used to index both current and historical data [Lomet et al. 2008, Lomet and Li, 2009]. The multiversion B-tree (MVBT) of Becker et al. [1996] is adapted to a transaction-time environment by Haapasalo [2010] and Haapasalo et al. [2009a,b]. Lomet and Salzberg [1993] show how a TSB-tree-indexed transaction-time database can also provide the database-backup function of media recovery.

# Chapter 13
# Distributed Transactions

In the preceding chapters, we have considered transaction processing in a central-ized database environment: we have assumed that each transaction accesses data items of a single database only and is thus run entirely on a single database server. However, a transaction may need access to data distributed across several databases governed by a single organization. This gives rise to the concept of a *distributed transaction*, that is, a transaction that contains actions on several intraorganization databases connected via a computer network.

A distributed transaction is actually a set of two or more local transactions each running at a single server and operating on a single database. A major task in the management of such a transaction is to coordinate the execution of the local transactions such that the basic (ACID) properties are ensured for the entire distributed transaction. The basic method for ensuring atomic commitment for a distributed transaction is the *two-phase commit protocol*, which we present in the context of our transaction model and ARIES-based transaction processing.

An important type of a distributed database is a *replicated database*, in which copies (called replicas) of a data item are stored at several servers, in order to increase data-access speed and data availability in events of failure. For protecting a database from natural disasters, a special kind of a replicated database, called a *remote backup database*, is often maintained, for duplicating the updates performed on the primary database.

Many modern web applications that involve millions of users all over the Internet require high availability, so that every request should succeed and receive a response. The applications are also expected to be tolerant to failures of sites and communication links on the network. These requirements have given rise to the development of highly distributed and replicated systems called *key-value stores* or NoSQL *data stores* that scale to huge workloads of relatively simple update actions better than full-fledged SQL database systems, at the expense of relaxed consistency.

© Springer International Publishing Switzerland 2014
S. Sippu, E. Soisalon-Soininen, *Transaction Processing*, Data-Centric Systems and Applications, DOI 10.1007/978-3-319-12292-2_13

## 13.1   Distributed Databases

In a centralized database environment all data in the database are stored on the disks controlled by a single database server, where several server processes and their threads, usually applying the transaction-shipping (query-shipping) paradigm, service requests coming from client application processes. These client processes may run on machines different from the server machine and communicate with the server-process threads over a network, but, in any case, in this setting, we have only a single database server and a single database which is shared by all the applications, and every transaction operates on the data items of that database only.

A *distributed database system* is a collection of database systems within a single organization in which applications can run transactions spanning several databases, so that a transaction can contain read and update actions on data items stored in different databases. Thus we assume that there are database servers $s_0, s_1, \ldots, s_k$ that can communicate between each other in such a way that it is possible to execute a transaction that, say, reads from a relation $r_1$ stored at server $s_1$ and updates a relation $r_2$ stored at server $s_2$.

The database servers of a distributed database system usually run on different machines, some of which may be located near to each other and connected via a fast interconnection network, while some may be located at distant sites on the Internet. The database at a machine can be accessed only via the server running at that machine or, in the case of a shared-disks system (to be discussed in Chap. 14), only via the servers belonging to the same cluster. In what follows we assume that each single database in the distributed database system is managed by a single server and that each server has the functionality of a transaction server, as explained in Sect. 1.1.

A fully integrated distributed database system runs the same database management software on every server and is capable of optimizing and executing SQL queries and update operations that span data items stored at different servers. The distribution of data across the servers may be transparent to the application, so that the application programmer need not know exactly at which server a particular data item resides. Obviously, the system catalog or data dictionary at each server must then store the schema of the entire distributed database, that is, descriptions of all the relations at all servers, together with information about the placement of different relations (or their parts) across different servers.

The distribution of the data of the database does not need to follow the decomposition of the data into logical units such as relations and tuples. A logical relation may be *partitioned* into *fragments*, either horizontally or vertically.

In *horizontal partitioning*, or *sharding*, a logical relation $r$ is partitioned into *horizontal fragments* or *shards*, that is, disjoint subrelations $r_1, \ldots, r_k$ with

$$r = r_1 \cup \ldots \cup r_k$$

and each fragment $r_i$ having the same relation schema as $r$.

In *vertical partitioning*, a logical relation is partitioned into *vertical fragments* $r_1, \ldots, r_k$ with

$$r = r_1 \bowtie \ldots \bowtie r_k,$$

where the intersection of the schemas of the fragments contains the primary key of $r$. Typically, the schema of a vertical fragment contains the attributes that are often queried together.

In both horizontal and vertical partitioning, the fragments $r_i$ may be placed at different servers, but the application programmer still sees only a single logical relation $r$. A large relation may be partitioned in both ways. The system catalog at each server must, naturally, store a description of the partitioning of the relation.

Sometimes an individual database server wants to retain sole access to its local database schema and hence withholds it from other servers. Such a server may still allow distributed transactions that access its database via stored procedures. A *stored procedure* is a program stored in the database that accesses data items of the database. Distributed transactions invoked at a remote server may be granted permission to execute a stored procedure with suitable input arguments and to receive results via output arguments.

With stored procedures, it is not possible to do global SQL query optimization. Instead, each SQL query or update statement issued on the local database during the execution of the stored procedure is only optimized locally. The system catalog at the remote servers only contains the name of the stored procedure, a description of its arguments, and the identifier of the server that stores the procedure, not the schemas of the relations the procedure accesses.

## 13.2 Transactions on a Distributed Database

Any of the database servers of a distributed database system can execute *local transactions* that operate only on the data items stored at the database managed by that server. The servers can also participate in *global transactions*, also called *distributed transactions*, that read or update data items stored at more than one server.

A distributed transaction that operates on data items stored at servers $s_0, s_1, \ldots, s_k$ consists of *subtransactions* $T_0, T_1, \ldots, T_k$, one for each participating server, where $T_i$ is a local transaction that operates only on data items stored at server $s_i$, for $i = 0, \ldots, k$. Thus $T_i$ consists of all the read and update actions of the distributed transaction that operate on data items stored at $s_i$. Server $s_i$ is called a *cohort* or *participant* of the distributed transaction.

The execution of a distributed transaction consisting of subtransactions $T_0, T_1, \ldots, T_k$ is *coordinated* by one of the participating servers $s_0, s_1, \ldots, s_k$. That server is called the *coordinator* of the distributed transaction. Often, the coordinator is the server at which the transaction was started. In the following we assume this is always the case and that $s_0$ is the identifier of the coordinator. The subtransaction $T_0$ at $s_0$, which hence is the first of the subtransactions created, is called the *parent transaction*, and the other subtransactions $T_1, \ldots, T_k$ are called *child transactions*. We identify a distributed transaction with its parent transaction.

Assume that an application starts a transaction at server $s_0$. It is given a transaction identifier $T_0$, the start is logged at $s_0$, and an entry for the forward-rolling transaction $T_0$ is inserted into the active-transaction table at $s_0$. Any read and update actions on data items stored at $s_0$ are executed in the name of $T_0$ as in a centralized database system. That is, any locks needed to protect the action are registered in the lock table at $s_0$ as locks owned by $T_0$, and, in the case of an update action, a log record for $T_0$ is written to the log at $s_0$.

The first action on a remote data item, that is, one stored at server $s_i$ different from $s_0$, starts a new transaction at $s_i$, where it is given a transaction identifier $T_i$ and marked as a subtransaction of $T_0$ at $s_0$; the start of $T_i$ is logged at $s_i$, and an entry for $T_i$ is inserted into the active-transaction table at $s_i$. Server $s_0$ is informed about this new subtransaction. The action is now executed and (if an update) logged at $s_i$ in the name of $T_i$, with appropriate locks registered in the lock table at $s_i$. Any further actions requested to be executed for the distributed transaction $T_0$ on data items at $s_i$ are executed there in the name of the already existent subtransaction $T_i$.

In Fig. 13.1, a request to execute an SQL query or update statement $R$ starts a transaction $T_0$ at server $s_0$. As $R$ is found to touch remote data, SQL statements $R_1$ and $R_2$ are formed and send to the servers, $s_1$ and $s_2$, that store the remote data. Statement $R_1$ is executed at $s_1$ in the name of a new subtransaction, $T_1$, and statement $R_2$ is executed at $s_2$ in the name of a new subtransaction, $T_2$. The result for $R$ is formed at $s_0$ from the results returned from $s_1$ and $s_2$ for $R_1$ and $R_2$.

We assume that requests to execute SQL statements sent by client application processes are processed in the coordinating server $s_0$ by the call *process-SQL-request*$(T_0, R)$ (Algorithm 13.1), where $R$ is the SQL statement to be executed and $T_0$ is the identifier of the transaction in whose name the statement is to be executed. The transaction identifier $T_0$ is not given as input when this is the first request for a new transaction, because it is not yet known. In this case the coordinating server $s_0$ creates a new transaction identifier $T_0$, logs the begin-transaction log record, and inserts a record for the new transaction in the active-transaction table. The transaction identifier $T_0$ is returned to the requesting client and must be included in subsequent requests for the same transaction.

If the SQL statement $R$ touches remotely stored data, then in the procedure *process-SQL-request*, requests of the form $(s_0, T_0, R_i)$ are sent to remote servers $s_i$ to execute SQL statements $R_i$ that only touch data stored at $s_i$. Such a request is serviced at server $s_i$ with the procedure call *service-remote-SQL-request*$(s, T, R')$ (Algorithm 13.2). The result of the original statement $R$ is computed from the results returned for statements $R_i$ by these calls.

A distributed transaction $T_0$ is represented in the active-transaction table at the coordinator with a transaction record

$$(T_0, L, S, n),$$

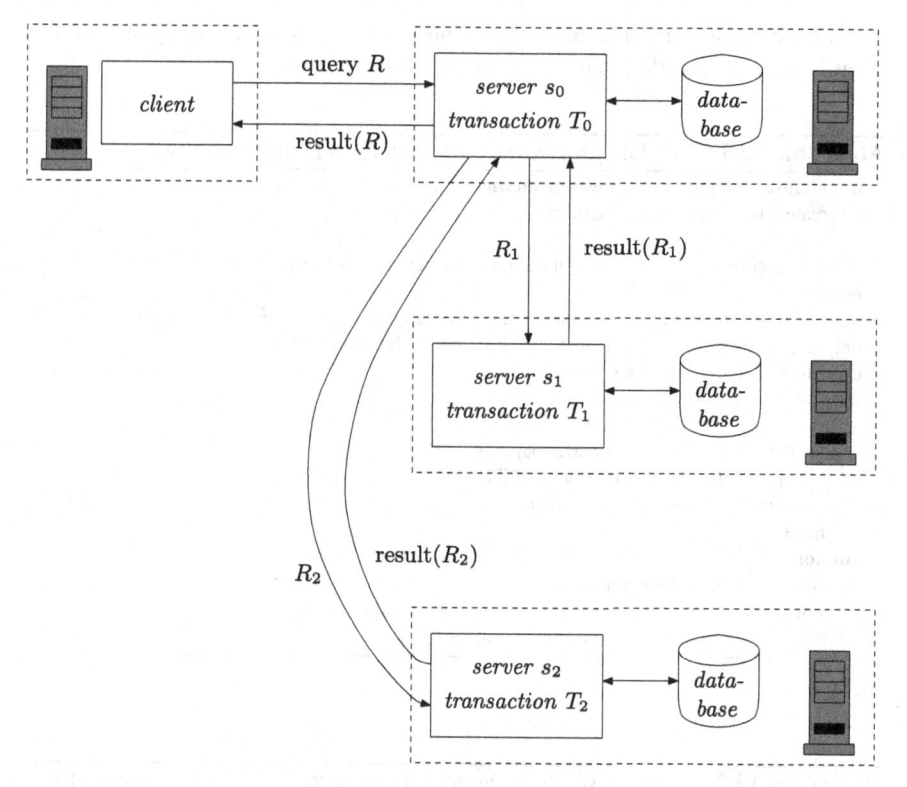

**Fig. 13.1** A distributed transaction consisting of subtransactions $T_0$ (the parent) and $T_1$ and $T_2$ (the children). Server $s_0$ is the coordinator and servers $s_1$ and $s_2$ are participants

where $S$ is the state of $T_0$, $n$ is the UNDO-NEXT-LSN, and $L$ is a list of identifiers of the participants (excluding the coordinator) of $T_0$. The addition of server $s_i$ to $L$ is logged with the log record

$$\langle T_0, \textit{add-participant}, s_i \rangle. \tag{13.1}$$

The subtransaction $T_i$ of $T_0$ at remote server $s_i$ is represented with a transaction record

$$(T_i, s_0, T_0, L, S, n)$$

in the active-transaction table at $s_i$, where the list $L$, initially empty, will eventually contain the identifiers of all the participants of $T_0$ other than $s_0$ and $s_i$, at the time $T_0$ is about to commit. The start of $T_i$ is logged at $s_i$ with

$$\langle T_i, s_0, T_0, B \rangle. \tag{13.2}$$

Thus, each participant of a distributed transaction stores permanently at least the identifiers of the coordinating server and the parent transaction.

---

**Algorithm 13.1** Procedure *process*-SQL-*request*($T_0$, $R$), executed at $s_0$

---
**if** no transaction identifier $T_0$ was given **then**
    create a new transaction identifier $T_0$
    $log(n, \langle T_0, B \rangle)$
    insert record ($T_0$, $\emptyset$, forward-rolling, $n$) to the active-transaction table
**end if**
transform $R$ into an equivalent statement with arguments $R_0, \ldots, R_k$, where each $R_i$ touches only data stored at server $s_i$, $i = 1, \ldots, k$ ($s_0$ being the coordinator)
execute $R_0$ locally in the name of $T_0$
**for** $i = 1, \ldots, k$ **do**
    send a request ($s_0$, $T_0$, $R_i$) to $s_i$ to execute $R_i$ for $T_0$
    **if** $s_i$ is not yet in the list of participants of $T_0$ **then**
        add $s_i$ to the list of participants of $T_0$
        $log(n, \langle T_0, add\text{-}participant, s_i \rangle)$
    **end if**
**end for**
wait for the requests to be serviced
compute the result of $R$ from the results received for $R_0, \ldots, R_k$
return the result of $R$ with the transaction identifier $T_0$

---

**Algorithm 13.2** Procedure *service-remote*-SQL-*request*($s_0$, $T_0$, $R_i$), executed at $s_i$

---
**if** the local active-transaction table contains no record for a subtransaction of $T_0$ coordinated by server $s_0$ **then**
    create a new transaction identifier $T_i$
    $log(n, \langle T_i, s_0, T_0, B \rangle)$
    insert record ($T_i$, $s_0$, $T_0$, $\emptyset$, forward-rolling, $n$) to the active-transaction table
**else**
    let $T_i$ be the active subtransaction of $T_0$ coordinated by $s_0$
**end if**
execute the SQL statement $R_i$ in the name of $T_i$
return the result of $R_i$

---

*Example 13.1* Consider the transaction generated by the following fragment of an application program:

**exec sql update** $r$ **set** $V = 2 * V$;
**exec sql select sum**($V$) **into** :new_sum **from** $r$;
**exec sql commit**.

The execution of this transaction on a centralized database system was considered in Example 1.1.

Assume now that the relation $r$ has been horizontally partitioned across servers $s_0, s_1, \ldots, s_k$, so that $r = r_0 \cup r_1 \cup \ldots \cup r_k$, where $r_i$ denotes the fragment of $r$ stored at server $s_i$. Obviously, the transaction must be executed as a distributed transaction having a subtransaction, $T_i$, running at each participating server $s_i$, $i = 0, \ldots, k$.

Assume that the application has opened a connection to server $s_0$, which, we assume, will also be the coordinator of the distributed transaction. Processing the request to execute the **update** statement in Algorithm 13.1 causes the coordinator to start a new transaction, $T_0$. Finding out from the system catalog that relation $r$ is partitioned across servers $s_0, \ldots, s_k$, the coordinator sends the **update** statement to be executed at servers $s_1, \ldots, s_k$ as part of subtransactions of distributed transaction $T_0$. These requests are represented as $(s_0, T_0, R)$, where $R$ is the **update** statement. The coordinator also executes the statement locally in the name of $T_0$, with the fragment of $r$ at $s_0$ being X-locked for commit duration at $s_0$ for $T_0$ and redo-undo log records for the tuples updated by $T_0$ being written to the log at $s_0$. The log records $\langle T_0, add\text{-}participant, s_i \rangle$, $i = 1, \ldots, k$, are written to the log at $s_0$. The list $L$ of participants of $T_0$ in the transaction record for $T_0$ is now $\{s_1, \ldots, s_k\}$.

At server $s_i$, $i = 1, \ldots, k$, when executing the procedure call *service-remote-SQL-request*$(s_0, T_0, R)$, it is found that there is not yet any subtransaction for $T_0$; so a new transaction, $T_i$, is started and marked as a subtransaction of $T_0$ at $s_0$, the **update** statement $R$ is executed in the name of $T_i$, with the fragment of $r$ at $s_i$ being X-locked for commit duration at $s_i$ for $T_i$ and redo-undo log records for the tuples updated by $T_i$ being written to the log at $s_i$, and a result code indicating successful execution is returned to the coordinator.

When the statement has been executed successfully at all the servers, control returns to the application program, which now requests the execution of the **select** query in the name of $T_0$. Again finding out that $r$ is partitioned across servers $s_0, \ldots, s_k$, the coordinator sends the **select** query to be executed at servers $s_1, \ldots, s_k$ as part of subtransactions of $T_0$ and also executes the query locally in the name of $T_0$.

At server $s_i$, $i = 1, \ldots, k$, it is found that there is already a subtransaction, $T_i$, active for $T_0$; so the **select** query is executed in the name of $T_i$. The result (the computed sum) is returned to the coordinator.

When the coordinator has received the results of all the **select** queries, it sums up the results and returns the result to the application, which then assigns it to the program variable new_sum.

The last request, **commit**, should now lead to the commit of all the transactions $T_0, T_1, \ldots, T_k$, or, if some of these fail after all, to the abort and rollback of all of them. The case in which the transaction commits is depicted in Fig. 13.2 (to be discussed in the next two sections).                                                          $\square$

Coordinator $s_0$:                                        Participant $s_i$:

log $\langle T_0, B \rangle$
perform **update** $r$ **set** $V = 2 * V$
for $T_0$ on the local fragment of $r$
send **update** $r$ **set** $V = 2 * V$ to each $s_i$
to be performed as part of $T_0$                $\longrightarrow$
log $\langle T_0, add\text{-}participant, s_i \rangle$ for each $s_i$        log $\langle T_i, s_0, T_0, B \rangle$
$L \leftarrow \{s_1, \ldots, s_k\}$                                perform **update** $r$ **set** $V = 2 * V$
wait for the results                                  for $T_i$ on the local fragment of $r$
                                                 $\longleftarrow$ return result code

after receiving all the result codes:
perform **select** **sum**$(V)$ **from** $r$
on the local fragment of $r$
send **select** **sum**$(V)$ **from** $r$ to each $s_i$
to be performed as part of $T_0$                $\longrightarrow$
wait for the results                                  perform **select** **sum**$(V)$ **from** $r$
                                                 on the local fragment of $r$
                                                 $\longleftarrow$ return sum

after receiving all the sums:
add up the sums
log $\langle T_0, prepare \rangle$
flush the log
$state(T_0) \leftarrow$ "prepared-to-commit"
send $prepare\text{-}to\text{-}commit(s_0, T_0, L)$ to each $s_i$ $\longrightarrow$
wait for the responses                                log $\langle T_i, prepare, s_0, T_0, L \rangle$
                                                 flush the log
                                                 $state(T_i) \leftarrow$ "prepared-to-commit"
                                                 $\longleftarrow$ return "ready" vote

after receiving a "ready" vote from all $s_i$:
log $\langle T_0, C \rangle$
flush the log
send $terminate(s_0, T_0, \text{"commit"})$ to each $s_i$ $\longrightarrow$
release all locks held by $T_0$                          log $\langle T_i, C \rangle$
remove $T_0$                                          flush the log
                                                 release all locks held by $T_i$
                                                 remove $T_i$

**Fig. 13.2** Actions performed at the coordinator and at the participants when executing the transaction of Examples 13.1 and 13.2 under the two-phase commit protocol

## 13.3   Atomic Commitment

The requirement of atomicity of transactions on a centralized database generalizes naturally to the following requirement of atomicity of transactions on a distributed database: the coordination of a distributed transaction must lead either to the commit of all its subtransactions or to the abort and rollback of all its subtransactions. This requirement is sometimes called *global atomicity*, to emphasize the difference from the *local atomicity* naturally required of each subtransaction.

For example, in the case of the distributed transaction of Example 13.1, the update on relation $r$ must either be accomplished and persist at all the servers $s_i$, $i = 0, \ldots, k$, or be rolled back at all these servers.

To ensure global atomicity, a subtransaction of a distributed transaction can be allowed to commit only if we are sure that all the other subtransactions can also be committed. Even if a subtransaction had successfully completed all the database actions it was requested to do, it cannot unilaterally to decide to commit. This is because some other subtransaction may still fail and abort or has done so already. If some subtransaction aborts, then all the other subtransactions must abort, too.

We also note that a subtransaction cannot tell whether or not it has completed all the actions it is expected to do, until the coordinator tells it so. Thus, after performing a requested action, the subtransaction must wait for the next request, which may be a request for a new database action or a request to abort or to *prepare to commit*. Only if all the subtransactions are prepared to commit, can all of them commit.

When the application process requests to commit a distributed transaction, the coordinator performs an *atomic commit protocol* in order to find out whether or not the transaction can be committed, that is, whether or not all its subtransactions are prepared to commit, and if so, to tell the participants to commit their subtransactions.

We say that a forward-rolling subtransaction $T_i$ of a distributed transaction is a *prepared-to-commit* transaction or a transaction in the *prepared-to-commit* state, if it has written a special *prepare-to-commit* log record and forced the log onto disk. In the active-transaction table, the prepared-to-commit state of $T_i$ is represented as

$$state(T_i) = \text{``prepared-to-commit.''}$$

Both commit and rollback must be made possible for every prepared-to-commit subtransaction. When the requirements imposed by the basic ARIES algorithm are satisfied, so that, among other things, all transactions run at isolation level 1 or higher, we know that it is possible to abort and roll back any forward-rolling transaction (assuming, of course, that there exist sufficient computational and storage resources for the execution of the undo actions). That also commit is possible for a prepared-to-commit transaction may not be immediately obvious, because the system may have to abort and roll back transactions due to a deadlock. With lock-based concurrency control, a transaction involved in a deadlock may have to be aborted and rolled back. But such a transaction is waiting for a lock and hence has not yet performed all its database actions; thus, it cannot be a prepared-to-commit transaction. By similar reasoning, with optimistic concurrency control, a transaction that is aborted and rolled back due to an action coming too late cannot be a prepared-to-commit transaction.

## 13.4 Two-Phase Commit Protocol

The most common atomic commit protocol in use is called the *two-phase commit protocol*, or *2PC*, for short. This protocol comes in several variants, of which we present here the simplest one in which two rounds of message interchange occur between the coordinator and the other participants of a distributed transaction. The two phases of the protocol are the voting phase followed by the decision phase.

When the application requests to commit transaction $T_0$, the coordinator starts the first phase, or the *voting phase*, of the commit protocol. The actions in this phase are included in the procedure call *coordinate-voting-phase*($T_0, L$) (Algorithm 13.3). A prepare-to-commit log record

$$\langle T_0, prepare \rangle \tag{13.3}$$

is written to the log at the coordinator. The coordinator then flushes its log onto disk. Next, the coordinator requests all the other participants $s_i$ of $T_0$ to prepare to commit their subtransaction $T_i$ of $T_0$. The request to be sent includes the identifiers $s_0$ and $T_0$ of the coordinator and the transaction, respectively, and the list $L$ of the identifiers of the other participants.

Upon receiving a prepare-to-commit request, each participant performs the procedure call *prepare-to-commit*($s_0, T_0, L$) (Algorithm 13.4): the active-transaction table is consulted so as to find out if a subtransaction $T_i$ of transaction $T_0$ coordinated by $s_0$ is active and forward-rolling, and if so, the prepare-to-commit log record

$$\langle T_i, prepare, s_0, T_0, L \rangle \tag{13.4}$$

is written and the log is flushed onto disk, after which the call returns with a "ready" vote. Otherwise, the subtransaction of $T_0$ must have aborted and either is rolling back or has rolled back already. In this case the call *prepare-to-commit*($s_0, T_0, L$) returns with an "aborting" vote.

When all the participants have voted or a preset time limit has expired, the coordinator performs the *decision phase* by calling *coordinate-decision-phase*($T_0, L, V$) (Algorithm 13.5) with a set $V$ of votes received. If all the participants have voted "ready," the coordinator decides to commit the transaction and sends to each participant the request *terminate*($s_0, T_0,$ "commit") so as to request the participant to commit its subtransaction of $T_0$. Otherwise, the coordinator sends the request *terminate*($s_0, T_0,$ "abort") to each participant that voted "ready," thus requesting the participant to abort and roll back the subtransaction. To implement the coordinator's decision $D$, participants perform the procedure call *terminate*($s_0, T_0, D$) (Algorithm 13.6).

---

**Algorithm 13.3** Procedure *coordinate-voting-phase*($T_0, L$), executed at $s_0$

---

$L \leftarrow$ list of all participants of $T_0$ (except the coordinator $s_0$)
$log(n, \langle T_0, prepare \rangle)$
*flush-the-log*()
$state(T_0) \leftarrow$ "prepared-to-commit"
**for** all participants $s_i \in L$ **do**
    send the request *prepare-to-commit*($s_0, T_0, L \setminus \{s_i\}$) to $s_i$
**end for**
wait for the votes to arrive within a preset time limit
$V \leftarrow$ the set of votes arrived within the time limit
return $V$

---

---

**Algorithm 13.4** Procedure *prepare-to-commit*($s_0$, $T_0$, $L$), executed at $s_i$

---

**if** the active-transaction table contains a record for a forward-rolling subtransaction $T_i$ of transaction $T_0$ coordinated by $s_0$ **then**
    $log(n, \langle T_i, prepare, s_0, T_0, L \rangle)$
    *flush-the-log*()
    *other-participants*($T_i$) $\leftarrow L$
    *state*($T_i$) $\leftarrow$ "prepared-to-commit"
    return with vote "ready"
**else**
    return with vote "aborting"
**end if**

---

**Algorithm 13.5** Procedure *coordinate-decision-phase*($T_0$, $L$, $V$), exec. at $s_0$

---

**if** $V$ contains "ready" votes from all the participants in $L$ **then**
    *commit*($T_0$)
    **for** all participants $s_i \in L$ **do**
        send the request *terminate*($s_0$, $T_0$, "commit") to $s_i$
    **end for**
**else**
    *rollback*($T_0$)
    **for** all participants $s_i \in L$ with a "ready" vote in $V$ **do**
        send the request *terminate*($s_0$, $T_0$, "abort") to $s_i$
    **end for**
**end if**

---

**Algorithm 13.6** Procedure *terminate*($s_0$, $T_0$, $D$), executed at $s_i$

---

**if** the active-transaction table contains a record for a forward-rolling subtransaction $T_i$ of transaction $T_0$ coordinated by $s_0$ **then**
    **if** $D = $ "commit" **then**
        *commit*($T_i$)
    **else**
        *rollback*($T_i$)
    **end if**
**end if**

---

*Example 13.2* Consider the actions involved in executing the **commit** statement in Example 13.1 when distributed transactions are coordinated using the two-phase commit protocol.

In the voting phase, the coordinator $s_0$ first prepares to commit its own subtransaction $T_0$ by writing the prepare-to-commit log record $\langle T_0, prepare \rangle$, by flushing the log, and by setting *state*($T_0$) $\leftarrow$ "prepared-to-commit." Then the coordinator sends the request *prepare-to-commit*($s_0$, $T_0$, $L_i$) to each participant $s_i$, $i = 1, \ldots, k$, where $L_i = \{s_1, \ldots, s_{i-1}, s_{i+1}, \ldots, s_k\}$. Participant $s_i$ responds by writing the log record $\langle T_i, prepare, s_0, T_0, L_i \rangle$, flushing the log, setting $state(T_i) \leftarrow$ "prepared-to-commit," and returning with a "ready" vote.

In the decision phase, the coordinator finds out that all participants have voted "ready"; so the coordinator decides to commit the distributed transaction by executing $commit(T_0)$ (Algorithm 3.3) and requesting each participant to commit its own subtransaction of $T_0$. Upon receiving the request, participant $s_i$ executes $commit(T_i)$.

In conclusion, at the end of the execution of the two-phase commit protocol, the tail of the log on disk at coordinator $s_0$ contains the log records

$\langle T_0, prepare \rangle$,
$\langle T_0, C \rangle$,

and the tail of the log on disk at participant $s_i$ contains the log records

$\langle T_i, prepare, s_0, T_0, L_i \rangle$,
$\langle T_i, C \rangle$,

for $i = 1, \ldots, k$ (Fig. 13.2).                                               □

The prepare-to-commit log records written at the participating servers carry the lists $L_i$ of other participants, because this information may be needed to find out the fate of a distributed transaction in the case of coordinator failures (see Sect. 13.5).

When the application itself wants to roll back the transaction or if the coordinator decides unilaterally to abort it for some reason, the coordinator uses procedure call $distributed\text{-}rollback(T_0)$ (Algorithm 13.7) to abort and roll back the entire distributed transaction. Thus, the coordinator and the other participants all use the call $rollback(T_i)$ to abort and rollback their subtransactions $T_i$ of $T_0$ (recall Algorithm 4.1).

---

**Algorithm 13.7** Procedure $distributed\text{-}rollback(T_0)$, executed at $s_0$

---

$L \leftarrow$ list of all participants of $T_0$ (except the coordinator $s_0$)
$rollback(T_0)$
**for** all participants $s_i \in L$ **do**
    send the request $terminate(s_0, T_0, \text{"abort"})$ to $s_i$
**end for**

---

*Example 13.3* Assume that the application program fragment of Example 13.1 ends at rollback rather than commit:

**exec sql update** $r$ **set** $V = 2 * V$;
**exec sql select sum**($V$) **into** :new_sum **from** $r$;
**exec sql rollback**.

In executing the **rollback** request, the coordinator first aborts and rolls back $T_0$ and then sends to every participant $s_i$ the request $terminate(s_i, T_0, \text{"abort"})$. As a response to this request, $s_i$ aborts and rolls back $T_i$.                              □

## 13.5 Recovering a Server from a Failure

Server $s_i$ recovers from a failure by executing the analysis, redo, and undo passes of the ARIES recovery algorithm. Only minor modifications are needed to the basic (centralized-database) recovery algorithm so as to handle distributed transactions.

In the analysis pass, the log at server $s_i$ is scanned from the begin-checkpoint log record for the last completely taken checkpoint to the end of the log, reconstructing the active-transaction and modified-page tables. The actions performed during the scan are as in the analysis pass for centralized database recovery (Algorithm 4.12), except that the procedure *analyze-log-record*() (Algorithm 4.13) must also observe log records of the forms (13.1)–(13.4). For (13.1), the server identifier $s_i$ is inserted to the list $L$ of participants in the active-transaction table being reconstructed. For (13.2) with LSN $n$, the record $(T_i, s_0, T_0, n)$ is inserted into the active-transaction table, and for (13.3) and (13.4), the state of the transaction is changed to "prepared-to-commit."

The redo pass of ARIES recovery at server $s_i$ is performed exactly in the same way as in a centralized database system (Algorithm 4.14).

In the undo pass of ARIES recovery (Algorithm 4.16) at server $s_i$, the set of active transactions is now partitioned into five groups:

1. Not-prepared-to-commit forward-rolling transactions coordinated by the server $s_i$
2. Not-prepared-to-commit forward-rolling subtransactions of transactions coordinated by servers other than $s_i$
3. Prepared-to-commit transactions coordinated by server $s_i$
4. Prepared-to-commit subtransactions of transactions coordinated by servers other than $s_i$
5. Aborted transactions

The transactions in groups 1–3 are aborted and rolled back, and the rollback of the transactions in group 5 is run into completion. Note that the transactions in groups 1 and 2 are not yet prepared to commit; thus server $s_i$ can abort them by a unilateral decision. As regards transactions in group 3 (or in group 1), $s_i$ is their coordinator server and thus is entitled to decide about the fate of the transactions.

The transactions in group 4 are termed *in-doubt transactions*: their fate cannot be deduced from the information surviving in the log at server $s_i$. The parent of an in-doubt transaction may have committed or aborted at the coordinator, or another subtransaction of the same distributed transaction may have aborted at some other participant. Only when the commit/abort information becomes available can an in-doubt transaction be either committed or aborted at $s_i$.

For each transaction in group 2, the coordinator of the transaction is informed about the abort of the transaction. Every participant of each transaction in groups 1 and 3 is informed about the abort of the transaction. Note that the record for a subtransaction in group 2 in the reconstructed active-transaction table contains the identifiers of the distributed transaction and its coordinator and that the record for a transaction in groups 1 and 3 contains a list of the identifiers of all the participants of the transaction. A group 1 transaction has no prepare-to-commit log record in the log, but the record for the transaction in the active-transaction table reconstructed in the analysis pass contains the list of participants, because each addition to the list was logged.

As long as the fates of the in-doubt (group 4) transactions are not known, those transactions must remain in the prepared-to-commit state at $s_i$, implying that their uncommitted updates must be protected from dirty reads and dirty writes by new transactions that access server $s_i$ after its recovery. With lock-based concurrency control, appropriate commit-duration locks must be acquired. This is done by scanning the log at $s_i$ backwards along the UNDO-NEXT-LSN chains in search for logged updates by in-doubt transactions. To make possible the acquiring of appropriate locks in this way, the log records for the updates must carry sufficient information about the lock names required. For example, with the key-range locking protocol, the lock name for the key $y$ next to the key $x$ of a deleted tuple $(x, v)$ must be included in the log record for the deletion, so that a commit-duration X lock can be acquired on $y$.

For each in-doubt (or group 4) subtransaction $T_i$, the fate of the distributed transaction $T_0$ must be inquired from the coordinator of $T_0$ or from one of the other participants. Recall that the reconstructed transaction record for such a subtransaction $T_i$ contains the identifiers of the parent transaction $T_0$, the coordinator $s_0$, and of the other participants.

In our simple variant of the two-phase commit protocol, a log scan is possibly needed to respond to such an inquiry. This is because both of the calls $commit(T_i)$ and $rollback(T_i)$ performed in Algorithms 13.5 and 13.6 remove the transaction record for $T_i$ from the active-transaction table, as well as the locks held by $T_i$ from the lock table, so that no trace of $T_i$ remains in the main-memory data structures. In the Problems section, we consider a practical variant, called *presumed abort*, of the two-phase commit protocol, in which the nonexistence of the transaction record of $T_i$ in the active-transaction table of the coordinator implies that $T_i$ has aborted.

## 13.6  Isolation and Concurrency Control

The isolation concepts considered in Chap. 5 for a centralized database system readily generalize to a distributed database system. For example, an update action $W[x]$ by subtransaction $T_i$ of a distributed transaction $T_0$ is a dirty write if the data item $x$ has an uncommitted update by some subtransaction $T_j'$ of another distributed transaction $T_0'$. Similarly, we can generalize the isolation anomalies dirty read and unrepeatable read. Using these generalized definitions for the isolation anomalies, we can define SQL isolation levels for distributed transactions in the obvious way.

We assume that each server controls the concurrency of actions on its own data items with one of the concurrency-control mechanisms defined previously for a centralized database system. Accordingly, each server $s_i$ ensures isolation for its local transactions and for any subtransactions of distributed transactions running at $s_i$.

As is seen from Algorithms 13.5 to 13.7, when a locking protocol is used to ensure isolation, all commit-duration locks held at server $s_i$ by a subtransaction $T_i$ of a distributed transaction $T_0$ are only released after $T_i$ has committed at $s_i$ (which occurs after $T_0$ has committed at the coordinating server $s_0$) or after $T_i$ has rolled back at $s_i$ (which implies that $T_0$ will roll back at $s_0$). No isolation anomalies are caused even though subtransactions $T_j$ of $T_0$ at servers other than $s_i$ may still be forward-rolling when $T_i$ rolls back and releases its locks. Recall that once $T_i$ has rolled back, all the data items updated by it have been returned to a clean state and can hence be read and written by other transactions at $s_i$. Also note that, when the call *prepare-to-commit*$(s_0, T_0, L)$ (Algorithm 13.4) is performed at server $s_i$ in response to a prepare-to-commit request, if the subtransaction $T_i$ of $T_0$ has rolled back and hence its transaction record is no longer found in the active-transaction table at $s_i$, then the call returns with an "aborting" vote.

*Example 13.4* Consider again the transaction of Example 13.1. Assume that concurrency is controlled by the multi-granular locking protocol described in Chap. 9, with a three-level hierarchy of lock names: database, relation, and tuple. At each server $s_i$, the database at $s_i$ is IX-locked for commit duration, and the entire fragment of the relation $r$ at $s_i$ is X-locked for commit duration. These locks are released at the end of the *commit*$(T_i)$ call.

If, on the contrary, the application requests the transaction to be rolled back (Example 13.3), then the coordinator $s_0$ requests each server $s_i$ to abort and roll back the subtransaction $T_i$ (Algorithm 13.7). At server $s_i$ the locks held by $T_i$ are released at the end of the *rollback*$(T_i)$ call (Algorithms 13.6 and 4.1).                □

Besides deadlocks between local transactions at one server, in a distributed database system deadlocks between subtransactions at different servers are possible. Such deadlocks are called *distributed deadlocks*.

*Example 13.5* The following scenario is possible in a distributed database system with distributed transactions $T_0$ and $T_0'$ on servers $s_1$ and $s_2$:

1. At server $s_1$, subtransaction $T_1$ of $T_0$ acquires an X lock on data item $x_1$.
2. At server $s_2$, subtransaction $T_2'$ of $T_0'$ acquires an X lock on data item $x_2$.
3. At server $s_1$, subtransaction $T_1'$ of $T_0'$ requests a lock on $x_1$ and hence must wait.
4. At server $s_2$, subtransaction $T_2$ of $T_0$ requests a lock on $x_2$ and hence must wait.

Neither of the two distributed transactions can proceed, due to a distributed deadlock: the wait by $T_1'$ for $T_1$ at $s_1$ implies that $T_0'$ waits for $T_0$, and the wait by $T_2$ for $T_2'$ at $s_2$ implies that $T_0$ waits for $T_0'$. But these waits are not seen locally: the wait-for graph at $s_1$ contains only the edge $T_1' \rightarrow T_1$ and the wait-for graph at $s_2$ contains only the edge $T_2 \rightarrow T_2'$, so that neither graph contains any cycles.

However, interpreting the edges between subtransactions as edges between the corresponding distributed transactions, that is, the edge $T_1' \rightarrow T_1$ as $T_0' \rightarrow T_0$ and the edge $T_2 \rightarrow T_2'$ as $T_0 \rightarrow T_0'$, and uniting the two graphs, we would get a graph with a cycle: $T_0' \rightarrow T_0 \rightarrow T_0'$.                □

As suggested by the above example, in order to detect distributed deadlocks, the servers must exchange information about waits between their transactions. Instead of explicitly constructing a union of the local wait-for graphs, cycles in the union graph may also be detected by generating sequences of messages telling about single wait-for relationships $T_0 \rightarrow T_0'$ forwarded from a server to another. A cycle exists in the union graph if such a message is received by the originator of the message sequence (see the Problems section).

## 13.7   Replicated Databases

A *replicated data item* is a data item that comes in two or more copies, or *replicas*, stored at different servers of a distributed database system. A *replicated database* is a distributed database that stores replicated data items. Such a database is *fully replicated*, if each server stores a replica of every data item, and *partially replicated* otherwise.

Actually, every distributed database is partially replicated because it necessarily replicates at least some metadata, namely, schemas of relations or names of stored procedures that must be accessible from servers other than the ones that store them. When actual data such as whole relations or fragments of relations are replicated, the system catalog at each server must also replicate metadata that tells which data items are replicated and which servers store the replicas.

An important motivation for replication is to increase *data availability* in events of failure: when a server crashes or loses connection to the other servers, replicas of data items stored in the crashed server may still be accessible at other servers. Replication can also decrease data-access time and hence improve transaction throughput and response time, because actions of transactions can be directed to the closest available replicas; in the best case, local replicas may be available.

Downsides of replication are the increased storage space needed and, most importantly, the additional burden of maintaining the consistency of the replicas under updates. If data item $(x, v_1)$ stored at server $s_1$ and data item $(x, v_2)$ stored at server $s_2$ are replicas of the same data item (i.e., the one with key $x$), the system must maintain the integrity constraint $v_1 = v_2$.

Maintaining *strong mutual consistency* of replicas means that all committed data-item replicas are to be kept consistent. In a transaction-processing environment, the only way to ensure this is that every transaction that updates a replicated data item $x$ also updates all the other replicas of $x$. In this setting, an update action on a replicated data item automatically makes the transaction a distributed transaction whose execution is coordinated by an atomic commit protocol. Such replication is called *transactional replication*.

Maintaining strong mutual consistency is often called *synchronous replication* or *eager replication*. Such replication also represents *read-one-write-all replication*, or ROWA, for short: a read action $R[x]$ can be satisfied by accessing a single replica of $x$ only (and any can do), while an update action $W[x]$, $I[x]$ or $D[x]$ has to be performed at all servers that can hold a replica of $x$.

Read-one-write-all replication is a special case of a more general *quorum consensus replication*, in which a read action on a data item $x$ replicated at $n$ servers can be satisfied by accessing $p$ replicas, and a write action on $x$ can be satisfied by accessing and updating $q$ replicas, where $p$ and $q$ are fixed integers with $p + q > n$ and $q > n/2$. Here $q$ can be less than $n$, implying that mutual consistency is not maintained. Each replica must carry the commit timestamp of the updating transaction, so that a read action can select the most recent replica from the set of $p$ replicas accessed.

In *symmetric update propagation*, the same SQL statement that specifies the update in the service request coming from the application program is performed at all the servers that are designated to hold a replica of a data item to be updated.

*Example 13.6* Assume that relation $r(\underline{X}, Y, V)$ is replicated at all the servers, $s_0, s_1, \ldots, s_k$, and that a transaction is generated by performing the following program fragment at server $s_0$:

**exec sql update** $r$ **set** $V = f(Y, V)$ **where** $Y = z$;
**exec sql select sum**($V$) **into** :new_sum **from** $r$;
**exec sql commit**.

Here $f$ is some arithmetic expression over arguments $Y$ and $V$, and $z$ is constant in the domain of attribute $Y$. We assume that there is no index to $r$ on $Y$, so that the entire relation must be scanned so as to find the tuples that must be updated.

With symmetric propagation of updates, for all $i = 0, \ldots, k$, the replica of $r$ at $s_i$ is SIX-locked for commit duration for subtransaction $T_i$ started at $s_i$, and the **update** statement is performed at $s_i$, X-locking for commit duration the tuples to be updated and writing redo-undo log records to the log at $s_i$ for the updated tuples. The **select** statement is only performed at the coordinator server $s_0$. The commit of the distributed transaction consisting of subtransactions $T_0, T_1, \ldots, T_k$ is coordinated by $s_0$ with the two-phase commit protocol. ☐

Symmetric propagation of updates often repeats work unnecessarily, as is demonstrated in the above example, where the entire relation $r$ is scanned $k + 1$ times for the update. In *asymmetric update propagation*, the original update statement is performed at one server only, where the update triggers the generation of the exact set of tuplewise updates to be performed at the other servers.

*Example 13.7* With asymmetric update propagation, if the statistics available in the system catalog suggest that the condition $Y = z$ is expected to select only a few tuples, a temporary trigger on $r$ is specified that—upon update of tuple $(x, y, v)$ of $r$ by this server-process thread—sends to servers $s_1, \ldots, s_k$ a request to perform the statement

**exec sql update** $r$ **set** $V = v'$ **where** $X = x$,

where $v' = f(y, v)$. Then the original update statement

**exec sql update** $r$ **set** $V = f(Y, V)$ **where** $Y = z$

is performed at $s_0$, firing the trigger, after which the trigger is removed. Assuming that there is at $s_i$ an index to $r$ on the key attribute $X$, for all $i = 1, \ldots, k$, the update is far more efficient than with the symmetric propagation in Example 13.6.         □

With read-one-write-all replication, transactions block in the event of a failure at one of the servers that hold a replica of a data item to be updated. A relaxed protocol, called *read-one-write-all-available*, or ROWAA, allows a transaction to proceed if all available replicas are updated; separate transactions are run to update the unavailable replicas when the replica sites are working again. Obviously, with this protocol, strong mutual consistency of the replicas is lost.

Some relaxed protocols, such as ROWAA, maintain *weak mutual consistency*, also called *eventual consistency*, among the replicas: the updates on a data item are guaranteed to be propagated to all the replicas eventually.


## 13.8   Remote Backup Databases

For protecting a database from natural disasters such as fires, floods, and earthquakes, a *remote backup database* is often maintained. Such a database resides at a site that is at a "safe" distance from the site of the corresponding *primary database*. The remote backup database system propagates in a continuous fashion all updates performed on the primary database, trying to keep the backup database up to date, that is, identical or nearly identical to the primary database, so that in the event of a disaster, the backup database system can assume the role of the primary database system with as little delay as possible.

A primary database server and its remote backup database server thus together form a kind of a fully replicated database system, with the difference however that transactions are only run on the primary database. We assume that each backup-database server propagates updates from a single database server only, so that in the case of a distributed database system there is one backup server for each primary server. We also assume that the primary database server and its backup database server run exactly the same database management system software and that both database structures are configured identically, having the same page sizes and formats, etc.

The work done for maintaining the remote backup database should be minimized, because it is all pure overhead if no disaster occurs. A most cost-effective solution is based on *log-record shipping* and *physical redoing*: updates on the primary database are propagated to the backup database system by shipping log records and redoing the updates on each page in the original LSN order, as in the redo pass of ARIES recovery. Thus, the state of each page in the physical backup database will be kept identical to a recent state of the same page in the physical primary database.

The log records coming from the primary server are stored on the log of the backup server in their original LSN order, and the usual write-ahead logging and steal-and-no-force policies are applied in the redo process. However, buffering-specific log records are omitted. These include log records containing a copy of the modified-page table written when taking a checkpoint at the primary server. Such log records are of no use at the backup server, because the set of pages buffered at the backup server at any time is different from that at the primary server. For one thing, the backup server never fetches from disk a page that was only read but not updated at the primary. Also, as explained below, the updates on different pages may not be applied in exactly the same order in which they were performed at the primary server.

The log records are stored in the log of the backup server with new LSNs, retaining also the LSNs used at the primary server. A mapping from primary-server LSNs to backup-server LSNs is maintained in a main-memory *primary-to-backup*-LSN mapping table at the backup server. Before storing a log record, its UNDO-NEXT-LSN value, if any, is replaced by the corresponding backup-server LSN obtained from the mapping table. The mapping table need only store records with backup-server LSNs greater than or equal to the backup-server COMMIT-LSN (see Sect. 9.6). As the log records also contain the primary-server LSNs, the insertions into the mapping table need not be logged.

Even though redoing a logged update is faster than performing the update logically, and even though read actions are not repeated, a single server process at the backup server cannot alone manage to handle the flow of all the log records coming from many concurrent transactions running at the primary server. The work at the backup server must be divided between as many concurrent process threads as are needed to keep pace with the flow of log records.

The flow of log records is directed to different *work queues*, each maintained in main memory and processed by a single server process. Since for each database page the log records for updates on that page must still be processed in the original LSN order, the log records for updates on a given page must all go to the same work queue, in the LSN order. If a log record describes a structure modification that involves pages assigned to different work queues, then the log record must go into all these work queue.

We assume that the assignment of updates is defined by a hash function $h$ such that, given page-id $p$, $h(p)$ gives the address of the work queue assigned for the updates on page $p$. For example, all pages residing on a given disk may be assigned to the same work queue, thus avoiding contention for the same disk arm between different server processes.

Algorithms 13.8 and 13.9 outline the process of tracking the updates from the primary server and of their redoing at the backup server. In the case of redoing a structure modification involving pages assigned to different work queues, we must be careful to redo the modification selectively only on those pages that are assigned to the work queue in question. As we have stated previously in Sect. 4.11, selective redoing is possible for all the structure modifications we consider in this book. The active-transaction table is maintained only for the purpose of accelerating the takeover by the backup server of normal transaction processing at a primary-server failure.

---

**Algorithm 13.8** Tracking updates from the primary database system

---

$r \leftarrow$ the next log record from the primary system
**while** $r$ is a log record with LSN $n$ **do**
    $r \leftarrow r$ with its UNDO-NEXT-LSN, if any, replaced with the mapped LSN
    **if** $r$ is other than a buffering-specific log record **then**
        $log(n', r)$
        insert $(n, n')$ to the primary-to-backup LSN mapping
        **if** $r$ describes updates on some pages **then**
            **for** each updated-page page-id $p$ in $r$ **do**
                append the log record $r$ with LSN $n'$ to work queue $h(p)$
            **end for**
        **else if** $r$ is a transaction-control-action log record for transaction $T$ **then**
            **if** $r$ is a commit-transaction log record **then**
                *flush-the-log*()
                remove the record of $T$ from the active-transaction table
            **else** {$r$ is a begin-transaction, abort-transaction, set-savepoint, rollback-to-savepoint or
            complete-rollback-to-savepoint log record}
                update the record of $T$ in the active-transaction table
            **end if**
        **end if**
    **end if**
    $r \leftarrow$ the next log record from the primary system
**end while**

---

**Algorithm 13.9** Processing work queue $w$

---

$r \leftarrow$ the first record on work queue $w$
**while** $r$ is a log record with backup-LSN $n$ **do**
    WORK-QUEUE-REC-LSN$(w) \leftarrow n$
    **if** $r$ describes updates on pages $p_1, \ldots, p_k$ **then**
        **for** all $i = 1, \ldots, k$ **do**
            **if** $h(p_i) = w$ and REC-LSN$(p_i) \leq n$ **then**
                *fix-and-write-latch*$(p_i)$
                **if** page $p_i$ is unmodified **then**
                    REC-LSN$(p_i) \leftarrow$ PAGE-LSN$(p_i) + 1$
                **end if**
                **if** PAGE-LSN$(p_i) < n$ **then**
                    apply the update described by $r$ on page $p_i$
                    PAGE-LSN$(p_i) \leftarrow n$
                **end if**
                *unlatch-and-unfix*$(p_i)$
            **end if**
         **end for**
    **end if**
    remove $r$ from work queue $w$
    $r \leftarrow$ the first record on work queue $w$
**end while**

---

For efficient handling of failures of the backup database system, the backup server takes periodically its own checkpoints. These checkpoints include, besides copies of the modified-page and active-transaction tables, a copy of the primary-to-backup-LSN mapping table and, for each work queue, a value WORK-QUEUE-REC-LSN that is the backup-server LSN of the last processed log record for that work queue. The WORK-QUEUE-REC-LSN values are needed during restart recovery to repopulate the work queues with log records that remained unprocessed due to the failure.

Restart recovery from a failure of the backup database system only includes an analysis pass that starts with reading the tables copied at the most recent completely taken checkpoint. The modified-page and active-transaction tables are initialized as in the normal ARIES analysis pass (Sect. 4.7). Also the primary-to-backup-LSN table is initialized from a copy taken at the checkpoint.

For repopulating the work queues, the forward log scan must be started at the log record with an LSN value BACKUP-REDO-LSN, which is the greatest among the LSN values that are less than or equal to WORK-QUEUE-REC-LSN$(w) + 1$ for every work queue $w$ and less than or equal to REC-LSN$(p)$ for every page $p$ in the modified-page table. The log scan is performed as in the normal ARIES analysis pass, with the addition that log records for updates are distributed into the work queues as in Algorithm 13.8 and that the primary-to-backup-LSN mapping table is reconstructed from the log records (that, we recall, store both the primary and backup LSNs).

In the event of a primary-server failure, the backup server takes over the role of the primary server as follows. First, all the remaining log records that have arrived are distributed into the work queues and processed, that is, Algorithm 13.8 and, for every work queue, Algorithm 13.9, are run into completion. Then the undo pass of ARIES (Sect. 4.9) is performed, aborting and rolling back all forward-rolling transactions and completing the rollback of all backward-rolling transactions. The backup database is now in a transaction-consistent state and ready to process new transactions.

## 13.9 Transactions on Key-Value Stores

A *key-value store* (or a *datastore*) as a logical database is just a single, though usually very big, relation $r(\underline{X}, V)$, where the tuples are versioned. The data in a typical key-value store for Internet applications resides in a big datacenter where it is partitioned among hundreds or thousands of servers. For high data availability and protection against natural disasters, the data is replicated on a small number of datacenters located in geographically distant locations.

A typical key-value store management system offers no transactions, but only single-tuple actions of the following forms that can be executed atomically:

1. *Read actions* of the form $R[x, \tau, v]$ that read the tuple $(x, v)$ current *as of* time $\tau$, that is, the latest version of the tuple with key $x$ created at time $\tau$ or before. If no timestamp $\tau$ is specified, the action returns the most recent version.

2. *Write actions* of the form $W[x, \tau, u, v]$ that retrieve the current version $(x, u)$ of
   the tuple with key $x$ (if any) and create a new version $(x, v)$ with timestamp $\tau$.
   If no timestamp $\tau$ is specified, a timestamp is generated (and returned) that is
   greater than the timestamp for any existing version. If no version $(x, u)$ exists,
   $u = \perp$.
3. *Check-and-write actions* of the form $CAW[x, u, v]$ that check whether or not
   $(x, u)$ is the current version of the tuple with key $x$ and, if so, performs the write
   action $W[x, u, v]$; otherwise, the write action is not performed and an error is
   returned.

The check-and-write actions are used to eliminate inconsistencies arising from
unrepeatable reads: if an application first reads the latest version $(x, u)$ of a tuple
and later wants to create a new version $(x, v)$ where the new value $v$ depends on the
read value $u$, the atomic action $CAW[x, u, v]$ is used to create the new version. If the
action returns with error, the application knows that $(x, u)$ is no longer the current
version and hence must reconsider the update.

Some recent key-value-store management systems allow transactions with the
restriction that a transaction can only update a group of co-located tuples residing on
a single server. At the logical database level, this setting can be portrayed as follows.
We assume that our key-value store consists of a single relation $r(\underline{X}, \underline{Y}, V)$, where
each tuple $(x, y, v)$, identified by key $(x, y)$, represents an *entity* belonging to an
*entity group* (also called a *key group*) identified by $x$. The values $x$ of the attribute
set $X$ thus divide the entire entity set into disjoint entity groups.

Such a key-value store is horizontally partitioned into many horizontal fragments
$r_i(\underline{X}, \underline{Y}, V)$, each of which is composed of whole entity groups, so that for any
value $x$ of $X$ only one fragment contains tuples $(x, y, v)$ for some values $y$ of $Y$
and $v$ of $V$. Each fragment is placed in its entirety on a single server in a datacenter.
The tuples in the fragments stored on a server are organized as a clustered index
structure such as a sparse B-tree on key $XY$, which thus places the entities of an
entity group near each other.

A transaction on a key-value store must be declared either as a *read-only transaction*, in which case it cannot contain update actions, or an *updating transaction
on* a given *entity group* $x$, in which case it can contain read actions (on any entities)
and update actions on entities in entity group $x$ only. A set of updating transactions
on an entity group is called a *transaction group*.

As each entity group resides on a single server, each updating transaction on
that entity group updates only entities stored on that server, so that any transaction
either runs entirely on one server or is a distributed transaction consisting of a
single updating subtransaction (on the updating site) and one or more read-only
subtransactions (on the other sites). When no log records are written for read-only
transactions, all the log records generated by any transaction are found in the log
of a single server, assuming that the transaction is coordinated by the updating site,
and not some other, remote site. In any case, transaction coordination is expected to
be much more light-weight and efficient than in the case in which a transaction were
allowed to update entities in different entity groups residing on different servers.

As many applications of key-value stores need versioning of entities, a key-value store with entity groups is most conveniently organized as a transaction-time database with snapshot-isolation-based concurrency control. As explained in Sect. 12.5, under snapshot isolation all read actions read from the start-time version of the database (except that updating transactions see their updates on data they read), without doing any locking, while locks (namely, X locks) are only needed on updated entities in the entity group in question. Thus, no distributed deadlocks can appear.

The above applies to a non-replicated key-value store. When the key-value store is fully replicated on a few, say three, datacenters, each update must be performed on three distantly located servers. Most key-value-store management systems apply a replication protocol less heavy than two-phase commit to maintain the consistency of the replicas across the datacenters. Thus, instead of strong consistency, only eventual consistency of the replicas is pursued or achieved.

## Problems

**13.1** Relation $r(\underline{A}, B)$ resides at server $s_1$, and relation $s(\underline{C}, D)$ at server $s_2$, of a distributed database system. At $s_0$, a transaction is started by executing the statements:

    **exec sql update** $r$ **set** $B = B + 1$ **where** $A = a$;
    **exec sql update** $s$ **set** $D = D - 1$ **where** $C = c$;
    **exec sql commit**.

The commitment of the transaction is coordinated using the two-phase commit protocol. We assume that all the operations succeed and that the transaction commits. What log records are written to the log at each server? When is the log forced onto disk? What locks are acquired at each server and when are those locks released, when the transaction is run at the serializable isolation level?

**13.2** Repeat the previous exercise assuming that server $s_2$ crashes just before answering to the prepare-to-commit message.

**13.3** Server $s_1$ of a distributed database system crashes. When $s_1$ recovers, the log disk is found to contain the following log records:

  101: $\langle begin\text{-}checkpoint \rangle$
  102: $\langle active\text{-}transaction\text{-}table, \{\} \rangle$
  103: $\langle modified\text{-}page\text{-}table, \{\} \rangle$
  104: $\langle end\text{-}checkpoint \rangle$
  105: $\langle T_1, s_0, T_0, B \rangle$
  106: $\langle T_1, W, p_1, x_1, u_1, v_1, 105 \rangle$
  107: $\langle T_1', s_0, T_0', B \rangle$
  108: $\langle T_1', W, p_2, x_2, u_2, v_2, 107 \rangle$

109: $\langle T_1, prepare, s_0, T_0, \{s_2\}\rangle$
110: $\langle S, B\rangle$
111: $\langle S, W, p_3, x_3, u_3, v_3, 110\rangle$
112: $\langle S, add\text{-}participant, s_1\rangle$
113: $\langle S, add\text{-}participant, s_2\rangle$
114: $\langle S, prepare\rangle$

Restart recovery is performed using the ARIES algorithm. What are the contents of the active-transaction table reconstructed in the analysis pass? What actions are performed in the redo pass? What actions are performed in the undo pass? What log records are written?

**13.4** A practical variant of the two-phase commit protocol makes use of an assumption called *presumed abort*. With this protocol, when a participant needs to inquire the fate of an in-doubt transaction $T$ of its coordinator, the coordinator can respond to the inquiry by consulting its active-transaction table only: the nonexistence of the transaction record of $T$ in the active-transaction table implies that $T$ has aborted. Make necessary changes in the procedures presented in this chapter so as to implement this protocol. *Hint:* Ensure that the coordinator deletes the transaction record of $T$ only if it aborts $T$ or if no inquiry of the fate of $T$ is to be expected.

**13.5** Another variant of the two-phase commit protocol includes a further round of message exchange after the participants have committed their subtransactions: each participant is required to send an acknowledgement back to the coordinator after committing its subtransaction, and when the coordinator has received such an acknowledgement from each participant, it writes an additional log record representing the final termination of the protocol. What advantage can be gained by this variant of the protocol?

**13.6** Work out the idea of detecting a cycle in the union graph of the local wait-for graphs outlined in Sect. 13.6.

**13.7** A distributed transaction defined in this chapter is composed of a parent transaction with a number of child transactions. In a more general model, the child transactions can be parents for their own children, so that an entire distributed transaction consists of a tree of local transactions. For each subtransaction $T_i$ at a non-leaf node in the tree, the server executing $T_i$ coordinates the parent $T_i$ and its children, that is, the transactions at the child nodes of $T_i$. Such a transaction $T_i$ commits only if all its children commit. Thus, the root of the tree commits only if all the transactions in the tree commit.

Discuss applications in which this kind of *multi-level distributed transactions* might be useful. Generalize the two-phase commit protocol to work with this generalized model of transactions.

**13.8** An atomic commit protocol can be used to coordinate a distributed transaction only if each participating server allows its transactions to participate. In many cases, a server may very well allow one of its stored procedures to be executed as a local transaction at that server, upon a request from a remote server, but, for security or other reasons, the stored procedure is not allowed to be executed as part of a distributed transaction coordinated by a remote server governed by an unfamiliar organization.

A prime example is an online store, where a customer first selects items to a shopping basket and then places the order, which includes paying for the order in advance. Ideally, the placement of the order and the payment for it should together form a single atomic transaction. However, this is not possible to achieve, because the financial institution through which the payment is transferred certainly does not allow its transactions to be coordinated by other organizations.

Assume that the financial institution only allows remote servers governed by other organizations to request the execution of a stored procedure that takes as arguments credit-card information and the sum to be withdrawn, so that this procedure is executed as a local transaction on the server of the financial institution. Discuss possible means of ensuring consistency across the order information stored in the database of the online store and the account balances stored in the database of the financial institution.

**13.9** Relation $r(\underline{X}, V)$ is totally replicated, so that each server of the distributed database system stores a replica of the relation. At one site, a transaction $T$ is started by the following statements:

**exec sql update** $r$ **set** $V = 2 * V$;
**exec sql select sum**($V$) **into** :new_sum **from** $r$;
**exec sql commit**.

Explain the execution of the transaction (including locking and logging), when replicas are managed using (a) the eager read-one-write-all protocol and (b) the primary-copy protocol.

**13.10** Define the three isolation anomalies (dirty writes, dirty reads, unrepeatable reads) for transactions on a replicated database. For ensuring a given level of isolation, are any changes needed in the locking protocols we have presented for transactions on a centralized database?

**13.11** Show that in a replicated database, a deadlock can occur even if each transaction operates on a single data item only and no locks are upgraded.

**13.12** An alternative to temporary triggers in implementing asymmetric replication is to use a *tentative query* on the relation to be updated in order to find out the unique keys of the tuples that are touched by the update, and then, using this set of keys, to generate update statements to be shipped to the replica servers. Work out this method in the case of Example 13.7.

**13.13** Updates from a primary database are propagated to its remote backup database using log-record shipping, as explained in Sect. 13.8. Could this method also be used in implementing asymmetric update propagation in an ordinary replicated database? If so, compare the pros and cons of this method against the trigger-based method. Can updates be applied to a replica in a physical fashion, as is done in the case of a backup database?

**13.14** As explained in Sect. 13.8, when the primary database server crashes, its backup database server first completes the processing of all remaining log records that have arrived before the crash, then performs the undo pass of ARIES, and only then is ready to start normal transaction processing. Using techniques similar to those used in centralized database recovery (Sect. 9.7), new transactions can be accepted to the system before the undo pass has been completed and in fact already before the processing of the remaining log records in the work queues has been completed. Explain how these optimizations can be accomplished.

**13.15** Assume that some subtransaction of a distributed transaction does no updates. If the begin-transaction log record is only written at the time the transaction's first update is logged, no log records appear for a read-only subtransaction, except for the prepare-to-commit and commit log records written in the execution of the two-phase commit protocol. But are those log records really necessary for a read-only subtransaction?

**13.16** Sketch database schemas and transactions for key-value stores with entity groups for the following Internet applications: (a) maintenance of incoming and outgoing e-mail messages of different users and the user profiles; (b) a blogging application in which different users can start a topic and to contribute to topics started by others.

## Bibliographical Notes

The two-phase commit protocol is attributed to Gray [1978] and Lampson and Sturgis [1976]. Many of the basic techniques and theoretical results pertaining to transaction processing in distributed and replicated databases are presented in the textbook by Bernstein et al. [1987]. The variant of the two-phase commit protocol based on the presumed-abort assumption (Problem 13.4) and implemented in R* distributed database management system was defined by Mohan et al. [1986]. Gray and Lamport [2006] present a commit protocol that uses multiple coordinators, ensuring that progress is made if a majority of the coordinators are working.

The read-one-write-all and read-one-write-all-available replication protocols are described by Bernstein et al. [1980], Rothnie et al. [1980], and Bernstein et al. [1987]. The scalability problems with transactional replication are discussed by Gray et al. [1996], among others. The notions of symmetric and asymmetric replication are discussed by Jiménez-Peris et al. [2003], who also advocate read-one-write-all-available replication against quorum consensus replication. The discussion of remote backup databases in Sect. 13.8 is based on an article by Mohan et al. [1993]. Lin et al. [2009] discuss snapshot isolation in the context of replicated databases.

An argument (called the "CAP theorem") has been made that it is impossible for a web service to provide all of the following three guarantees: consistency, availability, and partition tolerance (i.e., tolerance to failures of communication links between sites on the network) [Gilbert and Lynch, 2002, Stonebraker, 2010a]. It is the first of these, consistency, that is sacrificed with the key-value stores or NoSQL data stores. The solutions and performance arguments in favor of such systems are criticized by Stonebraker [2010a,b], Stonebraker and Cattell [2011], and Mohan [2013], among others. Stonebraker et al. [2010] compare the capabilities offered by modern parallel database management systems with the map-reduce parallel computation paradigm adopted by many NoSQL data stores.

Cattell [2010] and Stonebraker and Cattell [2011] survey a number of NoSQL data stores, also covering SQL database systems that scale to the huge workloads of simple read and write actions typical in web applications. They group the systems into key-value stores, document stores, extensible record stores, and scalable relational systems. The systems in the first three groups have limited transactional capabilities, if any. Some systems provide concurrency control for synchronizing single actions, so that only one user at a time is allowed to update an entity (a tuple, an object, or a document). Updates to local replicas may be propagated synchronously, while update propagation to remote replicas is asynchronous in most systems, so that at most eventual consistency among all replicas is achieved.

The idea of providing transactional consistency only within small entity groups (or key groups) of a big key-value store appears in the design of Google's Megastore [Baker et al. 2011], a system implemented on top of BigTable [Chang et al. 2008]. Megastore provides full replication across multiple datacenters, using the Paxos consensus algorithm [Lamport, 1998]. Patterson et al. [2012] describe an enhanced version of Paxos for a similar environment. G-store [Das et al. 2010] is an example of a system that provides transactions on single entity groups, but no replication across datacenters.

Bernstein [1990], Bernstein and Newcomer [1996, 2009], and Kifer et al. [2006] discuss in detail various important aspects of distributed transaction processing we have omitted, such as transactional middleware (or transaction-processing monitor), which is a layer of software components between transaction-processing applications and the underlying database and operating systems. These authors also explain the software mechanisms used for communications between application processes and server-process threads and between processes running at different servers in a distributed systems. These mechanisms include transactional remote procedure calls and persistent queues. General texts on distributed database management, with extensive treatments of transaction processing, have been written by Özsu and Valduriez [2011] (the first edition published in 1991) and by Rahimi and Haug [2010], among others.

# Chapter 14
# Transactions in Page-Server Systems

Thus far we have assumed that a database server, in a centralized as well as in a distributed environment, operates using the query-shipping (transaction-shipping) paradigm, so that client application processes send SQL queries and update statements to the server, which executes them on behalf of the client transaction on the database stored at the server and returns the results to the client application process. The queries and updates are executed on database pages fetched from the server's disk to the server's buffer. The server has exclusive access to the database pages and the buffer and is responsible for the entire task of query processing, that is, parsing, optimizing, and executing the queries and update statements.

In this chapter we discuss two database architectures, called the *page-server* and *shared-disks* systems, in which several sites can buffer pages from a single shared database in their local buffer pool (called the *client cache*) and perform query processing on the buffered pages. Instead of query shipping, such systems do *page shipping*: the clients request database pages from the server, which then ships the pages from the server buffer to the client cache, to be accessed by transactions running at the client.

The main motivation for page-server and shared-disks systems is to offload transaction and query processing from a single server to a number of (powerful) client machines. This offloading comes with the cost of the need to maintain *cache consistency* across the server buffer and the client caches: while a page can be simultaneously cached at more than one client, updates on the page must still form a single linear history, and each transaction must be given the most recent version of any data item it wants to access.

The discussion in this chapter is mainly about page servers, but most of the issues presented are valid for shared-disks systems as well.

© Springer International Publishing Switzerland 2014
S. Sippu, E. Soisalon-Soininen, *Transaction Processing*, Data-Centric Systems
and Applications, DOI 10.1007/978-3-319-12292-2_14

## 14.1  Data Server

A page server is a special case of a more general type of database server called a *data server*. A data server is said to do *data shipping*: it services requests to ship single data items from the database to the clients. In principle, the requested data items might be logical items such as single tuples or relations of a relational database or single objects or object collections in an object-oriented database, or they may be physical items such as pages or files.

The server has the usual *server buffer* for buffering data items fetched from the database disk and updated data items shipped from the clients. Each client buffers the data items shipped from the server in its private *client buffer*, also called the *client cache*. Each client does query processing and runs transactions on the data buffered in the client cache.

The processing of an SQL query at a client typically involves the following steps:

1. Parse the query and request from the server system-catalog information for relations mentioned in the query.
2. Using the received information, check the syntactic and semantic correctness of the query, optimize the query, and generate an execution plan for it.
3. Execute the query according to the plan on the data buffered at the client, requesting from the server any data items needed in the execution that are missing from the client cache.
4. If the SQL operation includes updating, the updated data items are eventually, when they are needed at the server or at some other client, shipped back to the server, replacing the old versions there.

Depending on the typical structure of the data items shipped between the server and a client, data servers are classified as object servers, page servers, and file servers. In an *object server*, the shipped items are logical items such as tuples or relations or objects or object collections. For example, a client of an object-oriented object server may request from the server an object with a given object identifier, and a client of a relational object server may request from the server a relation with a given unique name. Step 1 of SQL query processing at a client of a relational object server involves requesting (by name) some system-catalog relations, while step 3 involves requesting indexes and indexed tuples from the server, given relation or index names and tuple identifiers as arguments of the requests.

In a *page server*, the shipped items are database pages: a request for a page contains the page identifier of the page. Assuming that relations are indexed by B-trees, step 1 of SQL query processing at a client of a relational page server involves requesting pages in root-to-leaf paths of B-trees for system-catalog relations, starting from the root pages, while step 3 involves requesting pages of B-tree-indexed relations.

In a *file server*, the shipped items are database files: a request for a file contains the unique file name or identifier of the file.

In the case of each type of data server, the data items shipped from the server are processed (read and written) at the clients. In fact the entire query processing (parsing, optimization, execution) of an SQL query or update operation is performed at the clients. In that way only it is possible for a client to request a certain object, page, or file from the server. Every transaction is executed from the start to the end at a single client.

Most of transaction processing is thus the responsibility of the clients: a client starts a transaction (generates a transaction identifier), maintains its own active-transaction and modified-page tables, writes log records for the transaction, executes the actions of the transactions on the data cached at the client, and terminates (commits or aborts and rolls back) the transaction.

*Example 14.1* Consider the transaction generated by the following fragment of an application program:

**exec sql update** $r$ **set** $V = 2 * V$;

**exec sql select sum**($V$) **into** :new_sum **from** $r$;

**exec sql commit.**

The execution of this transaction on a centralized database system was considered in Example 1.1 and on a distributed database system in Example 13.1. The execution of the transaction at a client $c$ of a page-server system involves the following:

1. Client $c$ generates a new transaction identifier $T$, generates the begin-transaction log record for $T$, appends it to the log buffer at $c$, and inserts the transaction record of $T$ into the active-transaction table at $c$.
2. Client $c$ requests from the server system-catalog pages to be cached at $c$ for reading until it finds the catalog information for relation $r$.
3. Using the catalog information cached at $c$, the query optimizer at $c$ determines an execution plan for the **update** statement: table scan of the data pages of $r$.
4. For each data page $p$ of $r$, client $c$ requests from the server the page $p$ to be cached at $c$ for updating, updates the $V$ attribute in every tuple in page $p$, and logs the updates.
5. Using the catalog information cached at $c$, the query optimizer at $c$ determines an execution plan for the **select** statement: again a table scan of the data pages of $r$.
6. For each data page $p$ of $r$, if the page is no longer cached at $c$, client $c$ requests from the server the page to be cached at $c$ for reading and sums up the $V$ values.
7. Client $c$ logs $T$ as committed and ships the log records in the log buffer to the server with a request to flush the log onto disk.
8. Upon receiving an acknowledgment from the server, client $c$ releases all locks held by $T$ and deletes $T$'s transaction record from the active-transaction table.

These steps leave open some important questions: how the system keeps track of pages cached at the clients and the mode of caching (for reading, for updating) and how it ensures compatibility of locks granted by different clients on cached data items.                                                                                                   □

In a transaction-server system, almost all of the functionality of database
management and transaction processing resides at the server, while in a data-server
system, much of the functionality is included in the local systems running at the
clients. In a page server, and especially in a file server, the functionality of the
database server is most stripped down: from the point of view of the database server
in a page-server or file-server system, the database is (almost) only a collection of
pages or files whose shared use and persistence are managed by the server. The local
systems running at the clients are aware of the structure of the logical database and
index structures and access paths to the relations or object collections.

Computer-aided design and manufacturing applications (CAD, CAM) running
on powerful workstations are often built on an object-oriented database system
organized as a data-server system. With a data-server system, the computing
resources of workstations are more effectively used than with a transaction-server
system.

## 14.2   Page-Server Architecture

A page-server system consists of a database server $s$ and a set of client machines or
workstations $c_1, \ldots, c_n$, connected to the server machine via a network (Fig. 14.1).
At the server, there are the disk drives that store the disk version of the database and
the log disk that stores the log file. The clients do not store permanently any part
of the database or the log. The server buffer is used to buffer pages fetched from
disk and updated pages shipped from the clients. The private client buffers (or client
caches) are used to buffer pages shipped from the server.

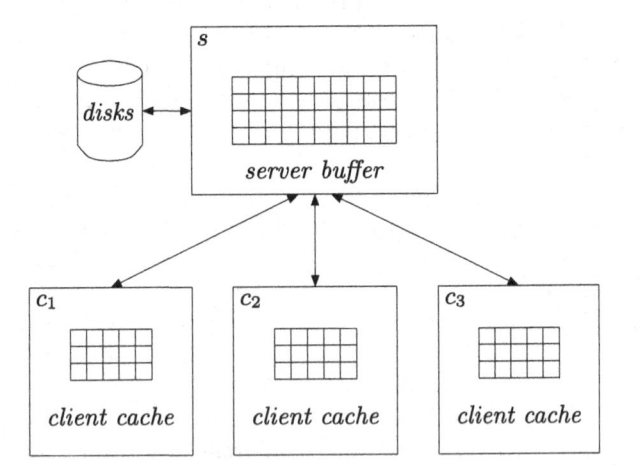

**Fig. 14.1** Page-server architecture

**Fig. 14.2** Processes in a page
server

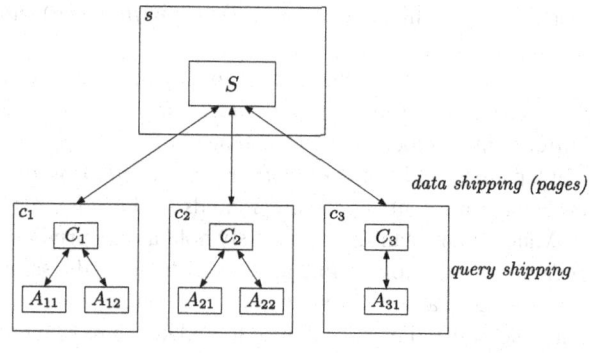

We assume that the following processes appear in a page-server system:

1. The server $s$ runs one or more multi-threaded *database-server processes* $S$ that process the requests coming from the client machines.
2. Each client $c_i$ runs one or more multi-threaded *client database processes* $C_i$ that process the requests coming from the application processes running at $c_i$.
3. Each client $c_i$ runs one or more *client application processes* $A_{ij}$, each of which executes a database application program and sends embedded database requests to the local client database process using the query-shipping (transaction-shipping) paradigm.

From the point of view of an application process $A_{ij}$, the page-server system is a local service that functions using the query-shipping paradigm (Fig. 14.2).

As said previously, every transaction runs from beginning to end at a single client. For simplicity, we might assume that the server runs no transactions. If there is need to run transactions at the server, we can always run an instance of the client database system at the server machine, besides the instance of the database server system.

The client database process $C_i$ parses and optimizes an SQL query or update statement received from its client application processes $A_{ij}$ and executes the corresponding database-action sequence on the pages buffered at client $c_i$ in the name of the application's transaction. From the clients, the log records and the updated pages are eventually shipped to the server.

## 14.3   Client Caching and Page-Server State

To keep track of pages cached at the clients, the server maintains in its main memory a *cached-page table*, that contains, for each page $p$ cached at some client $c$, the page-id of page $p$, the identifier of $c$, and the privilege under which $p$ is cached at $c$. A page is cached under a *caching privilege* that indicates the modes of access allowed on the cached page. A *read privilege* (or *read token*) on a page entitles the client to read the page, while a *write privilege* (or *write token*) entitles the client both to read and update the page.

The caching privilege is stored locally in the cache control block of the page in the client cache. *Cache control blocks* are main-memory structures similar to the buffer control blocks used to manage buffered pages in the server buffer. We say that a page is *cached for reading* (resp. *cached for writing*) at a client, if the page is cached at the client and the client holds read privilege (resp. write privilege) on it.

Many clients can simultaneously hold a read privilege on a page $p$, so that page $p$ can be cached for reading at several clients at the same time. Only one client at a time can hold a write privilege on $p$, so that, even if $p$ is cached at several clients, only one of the clients is allowed to update the page while the other clients can only cache the page for reading. According to the caching protocol considered in this chapter, a page cached for updating at some client is not allowed to be cached at any other client even for reading: granting a write privilege on page $p$ for some client $c$ is possible only if no other client currently caches $p$.

The *fixing* of a page $p$ for a client-database-process thread at client $c$ includes shipping the page from the server if the page is not yet cached at $c$. At the server, the server-process thread that services client $c$ fixes and read-latches $p$, ships a copy of $p$ to $c$, and unlatches and unfixes $p$. Unfixing page $p$ at client $c$ does not cause any communication between the server and the client: page $p$ remains at the client cache, and the cached-page table at the server retains the information about caching $p$ at $c$ with the privilege last granted.

The client calls *fix-and-read-latch*($p$) and *fix-and-write-latch*($p$) are given as Algorithms 14.1 and 14.2. The call *reserve-cache-frame*($p$) reserves a free frame in the client cache for page $p$. If there is currently no free frame, the least-recently-used page $p'$ is evicted from the cache. If $p'$ is a modified page, a copy of it and copies of all log records with LSNs up to and including PAGE-LSN($p'$) written at the client and not yet shipped to the server are now shipped to the server with a message stating that the client gives up caching page $p'$.

---

**Algorithm 14.1** Client procedure *fix-and-read-latch*($p$)

---

  **if** page $p$ is cached at the client **then**
    fix $p$ and acquire a read latch on $p$
  **else**
    *reserve-cache-frame*($p$)
    fix and acquire a read latch on $p$
    send the request *get-page*($p$, "read") to the server
    wait for a response (copy of $p$)
    install page $p$ into the cache frame
    mark $p$ as an unmodified page
    record the read privilege on $p$ in the cache control block
  **end if**
  return the address of the cache frame of $p$ to the caller

---

---

**Algorithm 14.2** Client procedure *fix-and-write-latch*( *p* )

---

**if** page *p* is cached with write privilege at the client **then**
    acquire a write latch on *p*
**else if** page *p* is cached with read privilege at the client **then**
    acquire a write latch on *p*
    send the request *get-write-privilege*( *p* ) to the server
    wait for a response
    record the write privilege on *p* in the cache control block
**else** {page *p* is not cached locally}
    *reserve-cache-frame*( *p* )
    acquire a write latch on *p*
    send the request *get-page*( *p*, "write") to the server
    wait for a response (copy of *p*)
    install page *p* into the cache frame
    mark *p* as an unmodified page
    record the write privilege on *p* in the cache control block
**end if**
return the address of the cache frame of *p* to the caller

---

With the caching protocol considered, the commit of a transaction at client $c$ does not cause any change to the set of pages currently cached at $c$. This is to say that we apply *inter-transaction caching*: a page can reside in a client cache unlatched and unfixed even if there are no transactions currently active at the client.

Inter-transaction caching makes sense because with fine-grained concurrency control we allow reads and writes by several active transactions on tuples in the same page and because a new transaction started at a client may often operate on the same pages as a recently committed transaction. For example, a long computer-aided design session is occasionally interrupted by a "save document" request, which commits the ongoing transaction. The session is then continued with a new transaction that usually operates on tuples that either are the same as or clustered close to the ones accessed by the previous transaction.

For a transaction-server system, we defined the concepts of a disk version, buffer version, and current version of a page and, based on these, a disk version and current version of a database, where the current version of a database is also called the state of the database. In a page-server system, the caching of pages at the clients implies an additional, a third, level for these definitions.

The *disk version* of page $p$ is the version of $p$ on the server's disk. The *server version* of $p$ is the version of $p$ in the server buffer, if $p$ is buffered there, and the disk version of $p$ otherwise. The *server version* of the physical database is the set of server versions of the pages of the database, and the *server version* of the logical database is the set of tuples in the server versions of the data pages.

The *client version* or *current version* of page $p$ is the version of $p$ cached in some client cache, if the page is cached at some client, and the server version of $p$ otherwise. This definition makes sense for the caching protocol discussed in this chapter, because a page can be cached at more than one client at the same time only if each client caches the page only for reading, implying that all the cached copies must also have the same contents. The *current version* of the physical database is the set of current versions of the pages of the database, and the *current version* of the logical database is the set of tuples in the current versions of the data pages.

## 14.4  Update Propagation

We assume that the server applies the usual steal-and-no-force policy in buffering pages in the server buffer. In other words, (1) the server can flush onto disk a page containing uncommitted updates, and (2) at transaction commit, the server does not flush any database page onto disk.

A page in the cache of client $c$ is called a *modified page*, if its contents differ from the server version of the page (i.e., the PAGE-LSN in the client version of the page is greater than that in the server version of the page) and an *unmodified page* otherwise. Recall that a page in the server buffer is called a modified page (from the point of view of the server) if its contents differ from the disk version of the page.

In shipping modified pages from a client back to the server, the *client-side write-ahead logging* or *client-WAL* protocol is applied: whenever a modified page $p$ is shipped from client $c$ to the server, all the log records written at $c$ with LSNs up to and including PAGE-LSN($p$) must also be shipped to the server (if they have not already been shipped). The server naturally follows the ordinary WAL protocol: a modified page $p$ can be flushed from the server buffer onto disk only if all log records with LSNs up to and including PAGE-LSN($p$) have been flushed onto the log disk.

The *client-side commit* protocol states that, to commit a transaction or to complete the rollback of a transaction $T$ at client $c$, all log records with LSNs up to and including that of the commit log record of $T$ must be shipped to the server and flushed onto the log disk.

Within the limits set by the client-side write-ahead-logging and commit protocols, different *update-propagation policies* can be defined. In *immediate update propagation*, an updated page is shipped to the server and installed into the server buffer immediately after every single database action (tuple insert, delete, or update) or structure modification. Thus, after performing a write action $W[x]$ on data page $p$ at client $c$, a copy of the modified page $p$ and the log record written for $W[x]$ are shipped to the server once the write latch on $p$ at $c$ has been released. Similarly, the three B-tree pages (and the space-map page) involved in a B-tree page split and the log record written for it are shipped to the server as soon as the modification is completed. However, the modified pages are not purged from the client cache but remain there with write privilege.

With immediate update propagation, the current version of the database is the same as the server version of the database between any two completely performed actions. A major downside is the high number of page shippings between a client and the server during a long updating transaction.

In the *committed-update propagation*, or *force-to-server-at-commit* policy, whenever a transaction $T$ commits or completes its rollback at client $c$, copies of all modified pages with PAGE-LSNs less than the LSN $n$ of the commit log record of $T$ (together with all log record with LSNs up to and including $n$) are shipped to the server. With this policy, the current version of the database is the same as the server version of the database at times when there are no active transactions in progress at any client.

In the *lazy* or *demand-driven update propagation*, a modified page is shipped from a client to the server only when:

1. The page is needed at some other client for a read or an update action.
2. The page is needed at the server for taking a checkpoint.
3. The client runs out of cache space.

In these cases the client must also give up its write privilege on the cached page.

In what follows we assume that the lazy update-propagation policy, which is the most liberal among the three, is followed.


## 14.5   Cache Consistency

Caching data at clients poses the problem of ensuring *cache consistency*: when a page is latched at a client for reading or updating some tuple in the page, the version of the tuple in the client's cache should be its current version. With the caching protocol considered in this chapter, we even impose the stronger requirement that a page latched at a client for reading or updating must always be the current version of the page.

We need a mechanism by which the server can take away the write privilege held by client $c$ on an unlatched page $p$ cached at $c$ when that page is requested to be cached at another client $c'$ for updating. Also, it should be possible to downgrade the write privilege held by $c$ on $p$ to a read privilege when the page is requested to be cached at $c'$ for reading. Taking away or downgrading a caching privilege held by client $c$ on page $p$ is called a *callback* of $p$ from $c$.

When a server-process thread holds the server version of a page $p$ write-latched, it can call the page back from client $c$ by issuing the call *callback-page*$(p, m, c)$ (Algorithm 14.3), where $m$ is the privilege (read or write) under which some other client wants to cache $p$. The request *callback-page*$(p, m)$ is sent to client $c$. At client $c$, the request is responded by performing the call *respond-to-callback*$(p, m)$ (Algorithm 14.4), which begins with fixing and write-latching the cached page $p$. If $p$ is cached at $c$ for writing, client $c$ ships to the server a copy of $p$, the REC-LSN$(p)$ value and all the log records with LSNs up to and including PAGE-LSN$(p)$ that have not been sent earlier. If $p$ is cached for reading or is requested for writing elsewhere (i.e., $m =$ "write"), $p$ is purged from the cache of $p$; otherwise, $p$ remains cached at $c$ for reading: the write privilege on $p$ held by $c$ is downgraded to a read privilege. Upon receiving the response from $c$, the server-process thread installs $p$ into its buffer, appends the log records to the log buffer, and updates the caching privilege on $p$ at $c$.

If the server does not receive from client $c$ any response to the callback request within a preset time limit, client $c$ is regarded as crashed. In this case all information about pages cached at $c$ is deleted from the cached-page table. This means discarding the current versions of all pages cached for writing at $c$, even if they survive at $c$. For such a page, the next older version, namely, the server version, is now the current version, as it is, by definition, for any page cached for reading at the client. Any further attempts by the client to ship updated pages or log records to the server are turned down.

---

**Algorithm 14.3** Procedure *callback-page*($p, m, c$)

---

send the request *callback-page*($p, m$) to $c$
**if** $c$ does not respond within a preset time limit **then**
    regard client $c$ as crashed
    delete from the cached-page table all information about pages cached at $c$
**else if** client $c$ responded to the callback request with a copy of $p$, REC-LSN($p$) and log records
$L$ **then**
    replace the contents of $p$ in the server buffer by the copy received
    append the log records in $L$ to the server log buffer
    update the LSN-to-address mapping
    mark $p$ as modified in the buffer control block
    acknowledge the receipt of the page and log records
    **if** $m =$ "write **then**
        record $p$ as no longer cached at $c$
    **else**
        record $p$ as cached for reading at $c$
    **end if**
**else** {client $c$ responded, page $p$ was cached for reading at $c$, and $m =$ "write"}
    record $p$ as no longer cached at $c$
**end if**

---

---

**Algorithm 14.4** Procedure *respond-to-callback*($p, m$)

---

*fix-and-write-latch*($p$)
**if** page $p$ is cached for writing **then**
    $L \leftarrow$ all not-yet-shipped log records with LSNs up to PAGE-LSN($p$)
    respond with a copy of $p$, REC-LSN($p$) and $L$
    wait for an acknowledgement
    remember that log records up to PAGE-LSN($p$) have been shipped
    **if** $m =$ "read" **then**
        record $p$ as cached for reading
    **else** {$m =$ "write"}
        purge $p$ from the cache
    **end if**
**else** {page $p$ is cached for reading}
    purge $p$ from the cache
    respond that the client no longer caches $p$
**end if**
*unlatch-and-unfix*($p$)

---

Assume that a client $c$ performs the call *fix-and-read-latch*($p$) (Algorithm 14.1) for a page $p$ that currently is not cached at $c$. Then $c$ sends to the server the request *get-page*($p, m$), where $p$ is the page-id of the requested page and $m$ = "read" is the requested caching privilege. The request is processed at the server by the call *service-get-page*($p, m, c$) (Algorithm 14.5). The server fixes and write-latches the server version of page $p$ and examines the cached-page table to see if the page is cached for writing at some other client $c'$. If so, the server calls the page back from $c'$. Then the server ships a copy of the now current server version of $p$ to client $c$, records $p$ as cached at $c$ for reading, and unlatches and unfixes $p$. Client $c$ installs the page in its cache and continues with the *fix-and-read-latch*($p$) call.

---

**Algorithm 14.5** Procedure *service-get-page*($p, m, c$)

---

*fix-and-write-latch*($p$)
**if** page $p$ is cached for writing at a client $c' \neq c$ **then**
   *callback-page*($p, m, c'$)
**else** {page $p$ is not cached for writing}
   **if** $m$ = "write" **then**
      **for** all clients $c' \neq c$ that cache $p$ **do**
         *callback-page*($p, m, c'$)
      **end for**
   **end if**
**end if**
**if** page $p$ is cached at $c$ and $m$ = "write" **then**
   send write privilege on $p$ to client $c$
**else**
   send a copy of $p$ with $m$ privilege to client $c$
**end if**
record $p$ as cached with $m$ privilege at client $c$
*unlatch-and-unfix*($p$)

---

**Algorithm 14.6** Procedure *service-get-write-privilege*($p, c$)

---

*fix-and-write-latch*($p$)
**for** all clients $c' \neq c$ that cache $p$ **do**
   *callback-page*($p, m, c'$)
**end for**
send write privilege on $p$ to client $c$
record $p$ as cached for writing at client $c$
*unlatch-and-unfix*($p$)

---

Assume then that a client $c$ performs the call *fix-and-write-latch*($p$) (Algorithm 14.2) for a page $p$ that currently is not cached at $c$. Then $c$ sends to the server the request *get-page*($p, m$), where $m$ = "write." When processing the request at the server by the call *service-get-page*($p, m, c$) (Algorithm 14.5), if page $p$ is cached for updating at some other client $c'$, the page is called back, and its current version is installed in the server buffer, as explained above. If, on the contrary, the page is cached for reading at one or more clients, the page is called back from all such clients. Finally the now current server version of $p$ is shipped to $c$ and recorded as cached there for writing.

We still have to consider the case in which client $c$ already caches page $p$ with read privilege at the time it performs the call *fix-and-write-latch*($p$). In this case, with our caching protocol, $c$ already has the current version of $p$ in its cache. Thus $c$ needs only to upgrade its read privilege on $p$ to write privilege. To that end, $c$ sends the request *get-write-privilege*($p$) to the server. This request is processed at the server by the call *service-get-write-privilege*($p$) (Algorithm 14.6).

## 14.6  Log Management

As in a centralized transaction-server system, the permanent log file of a page-server system resides on the log disk at the server. In a transaction-server system, all the log records are also created at the server because all database actions are performed there by some server-process thread upon requests coming from an application process.

In a page-server system, or, more generally, in a data-server system, the log records are created by the client-database-process threads that run at the clients and perform database actions upon requests coming from local application processes. The clients store their log records temporarily in main-memory log buffers and may also use local disks for the purpose, but all the log records created by the clients must be shipped to the server, obeying the client-WAL protocol.

When the server receives log records from a client, it appends them to the log buffer, to be flushed from there to the log file. To be prepared for server crashes, a client must keep a log record until the copy of it shipped to server has been flushed onto the log disk.

Write-ahead logging on the server side requires that all log records for updates on page $p$ with LSNs up to and including PAGE-LSN($p$) go to the server's log disk before page $p$ is flushed onto the server's database disk, while client-side write-ahead logging requires that all log records for updates on page $p$ with LSNs up to and including PAGE-LSN($p$) are appended to the server's log buffer before page $p$ is installed in the server's database buffer. These requirements must be fulfilled in an environment in which log records and updated pages arrive to the server from many clients.

A problem is that the LSNs of the log records created at different clients can no longer form a single sequence that increases with the time of performing the logged action. This is because each client must be allowed to generate LSNs of the log records for updates by its transactions independently of the LSNs generated at the other clients. It would be highly inefficient to enquire a new LSN from some global counter maintained at the server every time an update at some client needs to be logged.

Obviously, a locally created LSN can no longer be the physical address (byte offset) of the log record in the server's log file. Thus, the LSNs created at the clients are purely logical sequence numbers that need explicitly be included in the log records. What is actually needed to enforce write-ahead logging correctly is that, for each page, the LSNs of updates on that page form an increasing sequence that obeys the actual order of the updates performed by different clients.

With our caching protocol, a page can be cached for updating at a single client at a time. To obtain a globally increasing sequence of LSNs, it thus suffices that the LSN value stamped in the PAGE-LSN field of a page cached for updating is always greater than the previous value in the field and, of course, greater than the previous LSN generated locally at that client. For that purpose, the log manager at each client $c$ maintains a counter

LOCAL-MAX-LSN($c$)

that gives the LSN of the last log record written at $c$. An update action or a structure modification that modifies pages $p_1, \ldots, p_k$ is logged with an LSN $n'$ determined as follows:

$$n' = 1 + \max\{\text{PAGE-LSN}(p_1), \ldots, \text{PAGE-LSN}(p_k), \text{LOCAL-MAX-LSN}(c)\}.$$

The new LSN value $n'$ is stamped into the PAGE-LSN field of each page $p_i$, $i = 1, \ldots, k$, and is assigned as the new value of the counter LOCAL-MAX-LSN($c$).

In this way we ensure that (1) for each page, the LSNs of log records for updates on that page form a globally increasing sequence and (2) for each client the, LSNs created by that client form a locally increasing sequence. As each transaction executes entirely at a single client, observation (2) means that also the sequence of LSNs for updates by each transaction forms a sequence that is increasing by time across all transactions at that client.

With purely logical LSNs the traversal of the UNDO-NEXT-LSN chain in transaction rollback is less efficient than with physical LSNs. For accelerating the traversal, and also for other purposes (explained below), the server maintains, for each client $c$, a small LSN-*to-address* mapping consisting of pairs $(n, a)$, where $n$ is an LSN created at client $c$ and $a$ is the log-file address (byte offset) of the log record with LSN $n$ in the log file at the server. New tuples $(n, a)$ are inserted into the mapping when a client ships log records to the server, where they are appended to the server's log file. Not all LSNs are mapped; the location of a given log record can be approximated with the addresses associated with nearest LSNs that are explicitly mapped.

Instead of the REC-LSN values in the modified-page table of a transaction-server system, the *server modified-page table* maintained at the server of a page-server system stores values REC-ADDR, which are physical addresses to the log file. Accordingly, for a page $p$ buffered at the server, the value REC-ADDR($p$) in the server modified-page table indicates that all the updates on page $p$ logged at addresses less than REC-ADDR($p$) are in the disk version of page $p$. In ARIES recovery, the REC-ADDR values are used instead of the REC-LSN values.

Each client $c$ maintains a local *client modified-page table* that records each modified page $p$ cached at $c$ and a value REC-LSN($p$). Recall that a page cached at a client is modified if its contents differ from the server version of the page. Accordingly, the value REC-LSN($p$) indicates that all the updates on page $p$ logged with LSNs less than REC-LSN($p$) are in the server version of $p$.

When a client $c$ ships an updated page $p$ with log records to the server, it also ships the value REC-LSN($p$) (see Algorithm 14.4). If the server version of $p$ is still unmodified, so that either $p$ is not in the server buffer or the modified bit is zero in the buffer control block of the page in the server buffer (indicating that the buffered page has the same contents with the disk version of $p$), the new version of $p$ received from $c$ is recorded in the server modified-page table with a REC-ADDR value $a$ that is the address associated with the greatest LSN $n$ with $n \leq$ REC-LSN($p$) and $(n, a)$ in the LSN-to-address mapping for $c$. Otherwise, page $p$ is already recorded in the server modified-page table, in which case the REC-ADDR($p$) value is left unchanged.

The *client active-transaction table* maintained by each client $c$ contains a transaction record for each transaction active at $c$, where a transaction record contains, as before, the transaction identifier, the transaction state (forward-rolling or backward-rolling), and the UNDO-NEXT-LSN of the transaction. The transaction identifiers are created locally at each client and are thus unique only within that client. A transaction $T$ of client $c$ is identified globally by the pair $(c, T)$; accordingly, both the client identifier $c$ and the transaction identifier $T$ are included in log records written at $c$ for $T$. The client identifier $c$ is also included in log records written for structure modifications performed at $c$.

*Example 14.2* At client $c_1$, a new transaction is started. A new transaction identifier $T_1$ and a new LSN value $n_1$ are generated, the log record

$$n_1 : \langle c_1, T_1, B \rangle$$

is appended to the log buffer at $c_1$, and the transaction record for a forward-rolling transaction $T_1$ with UNDO-NEXT-LSN $= n_1$ is inserted into the active-transaction table at $c_1$.

Transaction $T_1$ then wants to perform action $W[x_1, u_1, v_1]$ on data page $p$, which has not been accessed since the last restart of the page-server system. In the call *fix-and-write-latch*($p$) performed at $c_1$, the request *get-page*($p$, "write") is sent to the server so as to get page $p$ cached for writing at $c_1$.

In processing the request in the call *service-get-page*($p$, "write", $c_1$) (Algorithm 14.5), the server fixes and write-latches $p$, thus bringing $p$ from the server's disk to the server's buffer, records the page as cached with write privilege at $c_1$, sends a copy of the page to $c_1$, and unlatches and unfixes $p$.

Upon receiving page $p$ with write privilege, client $c_1$, continuing with the *fix-and-write-latch*($p$) call, installs the page in its cache as an unmodified page. The action $W[x_1, u_1, v_1]$ is performed on the cached page, and the action is logged with

$$n_2 : \langle c_1, T_1, W, p, x_1, u_1, v_1, n_1 \rangle,$$

where $n_2 = 1 + \max\{$PAGE-LSN($p$), LOCAL-MAX-LSN($c_1$)$\}$, the LSN value $n_2$ is stamped into the PAGE-LSN field of $p$, the UNDO-NEXT-LSN of $T_1$ in the active-transaction table at $c_1$ is advanced to $n_2$, and page $p$ is unlatched and unfixed.

Next a transaction $T_2$ at client $c_2$ wants to perform an action $W[x_2, u_2, v_2]$ on page $p$, where $x_2 \neq x_1$. In the call *fix-and-write-latch*$(p)$ performed at $c_2$, the request *get-page*$(p, \text{"write"})$ is sent to the server so as to get page $p$ cached for writing at $c_2$.

In processing the request in the call *service-get-page*$(p, \text{"write"}, c_2)$, the server fixes and write-latches the server version of $p$ and observes from its cached-page table that page $p$ is currently cached for writing at client $c_1$. Thus, it performs the call *callback-page*$(p, \text{"write"}, c_1)$ so as to call the page back from $c_1$.

Client $c_1$ responds to the callback request by performing the call *respond-to-callback*$(p, \text{"write"})$: client $c_1$ fixes and write-latches $p$; makes a list $L$ of log records not yet shipped to the server (thus including those with LSNs $n_1$ and $n_2$); ships a copy of $p$, the list $L$, and REC-LSN$(p) = n_2$ to the server; waits for an acknowledgement, and purges page $p$ from the cache, remembering that log records up to and including PAGE-LSN$(p) = n_2$ have been shipped.

The server, continuing with the *callback-page*$(p, \text{"write"}, c_1)$ call, replaces the server version of $p$ by the received current version of $p$, appends the log records in $L$ to the log, updating the LSN-to-address mapping, marks $p$ as a modified page in the buffer control block, updates the modified-page table, and acknowledges the receipt of the page and the log records, and records $p$ as no longer cached at $c_1$, and unfixes page $p$. At this point, the log at the server contains the log records

$a_1 : n_1 : \langle c_1, T_1, B \rangle,$

$a_2 : n_2 : \langle c_1, T_1, W, p, x_1, u_1, v_1, n_1 \rangle,$

where $a_1$ and $a_2$ are the addresses of the log records.

Continuing with the *service-get-page*$(p, \text{"write"}, c_2)$ call, the server sends a copy of page $p$ with write privilege to client $c_2$, records $p$ as cached with write privilege at $c_2$, and unlatches and unfixes $p$.

Upon receiving page $p$ with write privilege, client $c_2$, continuing with the *fix-and-write-latch*$(p)$ call, installs the page in its cache as an unmodified page. The action $W[x_2, u_2, v_2]$ is then performed on the cached page, and the action is logged with

$m_2 : \langle c_2, T_2, W, p, x_2, u_2, v_2, m_1 \rangle,$

where $m_2 = 1 + \max\{\text{PAGE-LSN}(p), \text{LOCAL-MAX-LSN}(c_2)\}$ (where PAGE-LSN$(p) = n_2$), the LSN value $m_2$ is stamped into the PAGE-LSN field of $p$, the UNDO-NEXT-LSN of $T_2$ in the active-transaction table at $c_2$ is advanced to $m_2$, and page $p$ is unlatched and unfixed.                                                                                              $\square$

## 14.7  Recovery from a Client Failure

To accelerate recovery from a client crash, the clients take checkpoints periodically, as does the server to accelerate recovery from a server crash. At a *client checkpoint*, the contents of the client active-transaction and modified-page tables are written to the log. When the server receives the log record containing a copy of a client modified-page table, it maps each REC-LSN value there to a corresponding REC-ADDR value and appends the modified-page-table contents with the mapping to the server's log.

As already noted in the discussion on callback processing (Algorithm 14.3), when the server believes that a client $c$ has crashed, all information about pages cached at $c$ is deleted from the cached-page table, so that for every page cached for writing at $c$ the server version becomes the current version of that page. The server then performs the ARIES algorithm on behalf of the failed client $c$. The algorithm is performed selectively, observing only log records originating from $c$.

In the analysis pass, the active-transaction and modified-page tables of the crashed client are reconstructed by a forward scan of the log at the server, starting from the most recent checkpoint of the client found logged at the server, observing only log records written by the client. At the end of the analysis pass a log-address value REDO-ADDR is determined, which is the minimum of the REC-ADDR values corresponding to the REC-LSN values in the reconstructed modified-page table of the client.

The redo pass begins at REDO-ADDR. The only updates that may have to be redone are updates by the failed client on pages whose current versions were cached there for writing at the time of the client crash. Note that if such a page were earlier cached for writing at some other client, then the server version of the page must contain the updates by that client, because the updated page was shipped to the server and from there to the client that then failed.

In the undo pass, the transactions registered as forward-rolling in the client's reconstructed active-transaction table are aborted and rolled back, and the rollback of the backward-rolling transactions is completed, performing all undo actions on the current versions of pages, some of which may need to be called back from other clients.

*Example 14.3* Assume that in Example 14.2, one of the clients crashes while the transaction running there is still active. First consider the case that the crashed client is $c_1$. After noticing that $c_1$ is down, the server begins executing the ARIES algorithm on behalf of $c_1$.

In the analysis pass, the active-transaction and modified-page tables of $c_1$ are reconstructed, observing only log records written at $c_1$. At the end of the pass, the active-transaction table contains the forward-rolling transaction $T_1$ with UNDO-NEXT-LSN $= n_2$, and the modified-page table contains $p$ with REC-LSN $= n_2$.

The redo pass thus begins at REDO-ADDR $=$ REC-ADDR$(p) =$ the log-record address corresponding to REC-LSN$(p) = n_2$. In the redo pass, since REC-LSN$(p)$ is less than or equal to $n_2$, the current version of page $p$ must be fixed and write-latched.

From the cached-page table, the server finds out that page $p$ is currently cached with write privilege at client $c_2$. Thus the server sends the request *callback-page*$(p, \text{"write"})$ to client $c_2$, which responds to the request by fixing and write-latching $p$, making a list $L'$ of log records not yet shipped to the server (thus including that with LSN $m_2$), shipping a copy of $p$, the list $L'$ and REC-LSN($p$) $= m_2$ to the server, waiting for an acknowledgement, and purging page $p$ from the cache.

The server then replaces the server version of $p$ by the received current version of $p$, appends the log records in $L'$ to the log, updating the LSN-to-address mapping, marks $p$ as a modified page in the buffer control block, updates the modified-page table, and acknowledges the receipt of the page and the log records, records $p$ as no longer cached at $c_2$, examines PAGE-LSN($p$) only to find out that PAGE-LSN($p$) $=$ $m_2 > n_2$, so that the action $W[x_1, u_1, v_1]$ must not be redone. This completes the redo pass.

In the undo pass, the active transaction $T_1$ is aborted and rolled back at the server, resulting in performing the undo action $W^{-1}[x_1, u_1, v_1]$ on the page $p$ buffered at the server and in writing the following log records:

$a_3 : n_3 : \langle c_1, T_1, A \rangle$

$a_4 : n_4 : \langle c_1, T_1, W^{-1}, p, x_1, u_1, v_1, n_1 \rangle$

$a_5 : n_5 : \langle c_1, T_1, C \rangle$

where $n_3 > n_2$ and $n_5 > n_4 > \max\{n_3, m_2\}$ and $a_3$, $a_4$, and $a_5$ are the addresses of the log records in the server's log.

Then consider the case that the crashed client is $c_2$ instead of $c_1$. After noticing that $c_2$ is down, the server discards from the cached-page table the information about page $p$ being cached at $c_2$, thus making the server version of $p$ as the current version, and begins executing the ARIES algorithm on behalf of $c_2$.

In the analysis pass, the active-transaction and modified-page tables of $c_2$ are reconstructed, observing only log records written at $c_2$. As no log records from $c_2$ have not yet been shipped to the server, the reconstructed tables are those written at the last checkpoint, if any. In the case that the reconstructed tables are empty, nothing is done in the redo and undo passes.                                      □

When a crashed client is up and running again, no actions need be done for database recovery. After initializing its main-memory structures and starting a client database process, it is ready to process new transactions.

## 14.8   Recovery from a Server Failure

For recovery from its own failures, the server takes periodically checkpoints of its own. At such a checkpoint, the contents of the server's modified-page table are written to the server's log. The modified-page table contains an entry $(p, \text{REC-ADDR}(p))$ for a page with page-id $p$ if the server version of page $p$ is different from the disk version of $p$. The entry is inserted when an updated copy of $p$ is shipped from a client for the first time since fetching $p$ from disk (or when the server itself is the first updater).

According to the general logic of the ARIES algorithm, the recovery from a server failure is based on the last completely taken checkpoint. In the analysis pass, the modified-page table is reconstructed by initializing it from the copy found logged at the checkpoint and inserting entries for pages that became modified after the checkpoint was taken. In the redo pass, the complete history is repeated from the log survived at the server's log disk, redoing all updates found logged at the server.

In the undo pass, nothing needs to be done, since we have assumed that the server runs no transactions of its own. On the other hand, if transactions are run at the server under an instance of the client database system, then that client has crashed along with the server, so that ARIES recovery must be performed for that client in the way explained in the previous section.

There is a pitfall that must be avoided in the analysis pass. A page updated by a client and shipped to the server and marked there as modified before the server's last checkpoint appears in the reconstructed modified-page table, because it appears in the logged checkpoint. Similarly, a page that turned modified after the last checkpoint and whose log records were shipped to the server and taken to the log disk before the server failure also appears in the reconstructed modified-page table, because during the log scan in the analysis pass all pages for which log records are found in the log are registered in the modified-page table if not found there already.

But the reconstructed modified-page table does not necessarily contain an unmodified page that before the checkpoint was shipped to some client for updating for the first time. The log records for updates on such a page—but not the page itself—may have arrived to the server and taken to the log disk before the checkpoint.

*Example 14.4* The following scenario is possible:

1. Page $p$ is fetched from disk and shipped to client $c$ for updating.
2. Transaction $T$ at client $c$ updates page $p$ and commits. The log records written for $T$ are shipped to the server and appended to the server's log:

$a_2 : n_2 : \langle c, T, W, p, x, u, v, n_1 \rangle,$

$a_3 : n_3 : \langle c, T, C \rangle.$

Page $p$ is not shipped; it remains at $c$ cached for writing.
3. The server takes a checkpoint:

$a_4 : \langle begin\text{-}checkpoint \rangle.$

$a_5 : \langle modified\text{-}page\text{-}table\{\ldots\} \rangle$ — does not contain an entry for $p$.

$a_6 : \langle end\text{-}checkpoint \rangle.$

4. The updated page $p$ is called back from client $c$ to be cached at another client.
5. Client $c$ responds with a copy of page $p$ and log records.
6. The server appends the log records to the log buffer, installs $p$ into its buffer, inserts an entry for $p$ in the modified-page table, acknowledges the receipt of the page and log records, and ships a copy of $p$ to the requesting client.
7. Client $c$ purges $p$ from its cache.
8. The server crashes due to a failure.

Among the log records shipped in Step 5, there are no log records for updates on page $p$ if the page was not updated after the commit of $T$. In this case, the modified-page table reconstructed in the analysis pass of server recovery does not contain an entry for $p$. Thus the REDO-ADDR value determined may be greater than $a_2$, the log-record address for the update by $T$ on $p$. This in turn means that in the redo pass, the update logged at $a_2$ is not redone but is lost. ☐

A simple solution to the pitfall above would be to record an unmodified page as a modified page in the server's modified-page table immediately after shipping the page to a client for updating. A downside of this solution is that the REC-ADDR value for the page may then be left too conservative, leading to unnecessary work to be done in the redo pass.

A better solution is to include into a server's checkpoint copies of the local modified-page tables of all clients. When starting a checkpoint, the server requests all clients to send a copy of their modified-page tables. After receiving these copies, the server merges them with its own table and writes the merged table to its log, mapping REC-LSN values to REC-ADDR values as explained above. In this way the analysis pass can determine a sufficiently early REDO-ADDR value so as to get all logged updates to be redone on the pages buffered at the server.

*Example 14.5* The pitfall shown in Example 14.4 is avoided, if the server, when taking the checkpoint, requests client $c$ to ship a copy of its modified-page table, because that table must contain an entry for $p$ with REC-LSN $= n_2$. An entry for $p$ will thus also appear in the merged modified-page table included in the checkpoint. ☐

After the server has recovered from the failure, it must perform recovery on behalf of those clients that have crashed during the time the server was down. Moreover, the server must request each client that is up and running to send information about pages they currently cache, for the reconstruction of the cached-page table. The clients must resend to the server those log records whose receipt was not acknowledged by the server due to the failure.

## 14.9 Concurrency Control

The concurrency-control protocols presented earlier can all be adapted to controlling the concurrency of transactions in a page-server system. For the management of locks held by its transactions, each client $c$ maintains a *local lock table*. The lock-table entries are, as before, tuples $(T, x, m, d)$, where $x$ is the lock name, $m$ is the lock mode, $d$ is the duration, and $T$ is the identifier of the transaction of client $c$ that is the owner of the lock.

For detecting incompatibility of locks across different clients, a *global lock table* is maintained at the server. In this table, locks are recorded in the name of the client whose transaction holds the lock. Thus, the lock-table entries in the global lock table are tuples $(c, x, m, d)$, where $c$ is the identifier of the client whose transaction $T$ holds the lock $(T, x, m, d)$ recorded in the local lock table at $c$. If $m$ is a shared mode, so that several transactions can hold an $m$ lock on $x$, the lock is recorded in the global lock table at most once per client.

A client can grant a shared lock to any of its transactions when some of its transactions currently holds the lock, while an exclusive lock can only be granted by communicating with the server.

*Example 14.6* Transaction $T$ at client $c$ starts and requests a commit-duration S lock on key $x$, in order to perform a read action $R[x]$. Client $c$ examines its local lock table to see if some other transaction at $c$ holds a lock on $x$ that is incompatible with mode $S$. If so, $T$ must wait. Otherwise, if some other transaction at $c$ holds an S lock on $x$, then client $c$ can grant the requested lock to $T$ without communicating with the server. The lock $(T, x, S, \text{commit duration})$ is recorded in the local lock table of $c$.

If, on the other hand, no transaction currently holds a lock on $x$, client $c$ requests from the server an S lock on $x$ in the name of $c$. The server examines the global lock table to see if some other client holds an incompatible lock on $x$. If so, the client request must wait. When there are no longer any incompatible locks on $x$, the server grants the lock $(c, x, S, \text{commit-duration})$ to client $c$ and records it in the global lock table. Client $c$ records the lock $(T, x, S, \text{commit duration})$ in its local lock table, after which $T$ can read $x$.

Next, transaction $T$ requests an X lock on $x$ in order to perform a write action $W[x]$. Client $c$ examines its local lock table to see if some other transaction at $c$ holds a lock on $x$. If so, $T$ must wait. When there are no longer any local locks on $x$, client $c$ requests from the server an X lock on $x$ in the name of $c$. The server examines the global lock table to see if some other client holds a lock on $x$. If so, the client request must wait. When there are no longer any lock on $x$, the server grants the lock $(c, x, X, \text{commit duration})$ to client $c$ and records it in the global lock table. Client $c$ records the lock $(T, x, X, \text{commit duration})$ in its local lock table. Now $T$ can write $x$. □

## 14.10 Shared-Disks Systems

Many of the techniques needed in the management of a page-server system apply as such or as somewhat modified to the management of a *shared-disks system*. Differences are due to the characteristic that in a shared-disks system, there is no distinguished server node, but all nodes in the system are *peers*: every node runs an instance of exactly the same database management system.

Every node has a direct access to the shared database via its local buffer: the node can fetch a page from the disk into its buffer and take an updated page from the buffer back to its location on disk. A page can also be transferred from the buffer of a node via an interconnection network to the buffer of another node. This is usually faster than transferring the page via disk.

As in a page-server system, every transaction in a shared-disks system runs from start to end at one node. Such a transaction can naturally be a subtransaction of some distributed transaction in a distributed database system having the shared-disks system as one of its sites.

Every node maintains its own log, to which the log records written for the node's transactions are first appended. Every node determines the LSNs of log records in the same way as the clients in a page-server system. One of the nodes has access to the log files of all the other nodes, and a *log-merge process* merges the log files into a single file.

The buffering (or caching) of pages in the buffers of the nodes is controlled by the same privileges as in a page-server system: a page can reside for reading in the buffers of many nodes at the same time, but buffering a page for updating at one nodes is possible only if the page is not buffered at any other node.

The write privilege held by node $s_1$ on page $p$ can be transferred to another node $s_2$ in several ways. In the simplest (and also the slowest) way, page $p$ is transferred by the disk: node $s_1$ flushes $p$ from its buffer onto the disk, applying the WAL protocol, purges $p$ from its buffer, and gives up its caching privilege on $p$, after which node $s_2$ gets its write privilege on $p$ and fetches $p$ from the disk to its buffer. In a slightly faster way, node $s_1$ flushes $p$ onto disk and at the same time ships $p$ to node $s_2$ via a communication link, thus saving a disk access at $s_2$.

In the fastest ways, node $s_1$, the current holder of the write privilege on page $p$, does not flush $p$ onto disk but only takes the log records up to PAGE-LSN($p$) to the log disk and then ships $p$ to node $s_2$ via a communication link. In this way a page can hold updates performed at more than one node until it is flushed onto disk. This adds complication to failure recovery. When the node that last updated $p$ crashes, the current version of $p$ is lost; in redoing the lost updates on the disk version of $p$ the merged log must be scanned from a certain REC-ADDR.

In the very fastest way of transferring, node $s_1$ does not even take the log records onto disk before shipping page $p$ to node $s_2$. To make this possible, the WAL protocol must be changed so that a modified page can be flushed onto disk only after all the nodes that have updated the page have taken their logs onto the log disk.

## Problems

**14.1** Consider a transaction generated by the following application-program fragment running at client $c$:

**exec sql select sum($V$) into :$v$ from $r$;**

**exec sql update** $r$ **set** $V = :v$ **where** $X = 0$;

**exec sql commit**.

There exists a sparse (primary) B-tree index to $r(X, V)$ on attribute $X$ (unique key).

(a) Assuming that the database system is a conventional transaction-server system, what data are transferred between the disk and the server buffer and between the server and the client at different phases of the execution of the transaction? We assume that at the start of the transaction, there are no pages of relation $r$ in the server buffer.

(b) Repeat (a) in the case that the system is a page server. We assume that updates are propagated lazily. We also assume that at the start of the transaction, there are no pages of $r$ in the server buffer nor in any client cache and that there are no other transactions in progress simultaneously at other clients.

**14.2** Compare an object server and a page server by considering the number of messages exchanged between the server and a client and the amount of actual data (objects or data pages) transferred between the server and a client. In this respect, when is an object server more efficient than a page server (and vice versa)? You may assume that objects (or tuples in a relation) are considerably smaller than a page.

**14.3** Show that the following history of two transactions running at different clients is possible in a page-server system, when the tuples with keys $x$, $y$, and $z$ all reside in data page $p$ and the transactions are run at the serializable isolation level.

$$
\begin{array}{llllll}
T_1 \text{ at } c_1: & BR[x] & W[x] & & R[z]C & \\
T_2 \text{ at } c_2: & BR[y] & & W[y] & R[z]A & W^{-1}[y]C
\end{array}
$$

Give all transfers of page $p$ between the server and the clients, the log records generated and shipped, and the locks acquired and released, at each phase of the execution of the history. Assuming that the server buffer and the client caches are empty at the beginning, how do the disk version, the server version, and the current version of the database differ from each other at the end?

**14.4** It is possible that a client $c$ ships to the server log records of updates on a page $p$ while not shipping the updated page $p$ to the server. In other words, the server version of page $p$ may not contain all updates that are found logged at the server. Explain why. What happens when some transaction running at an other client wants to read or update page $p$? Is the transaction guaranteed to see the current version of page $p$? Also consider the case that client $c$ crashes at that point.

**14.5** Consider using the Commit-LSN technique of Sect. 9.6 for controlling the concurrency of transactions run at the read-committed isolation level. Give cheap ways of maintaining a global Commit-LSN value in a page-server system and of communicating its current value to the clients.

**14.6** To ease the acquisition of S locks on keys of tuples in a data page cached at a client, the server could, before shipping a data page to the client *mark unavailable* those keys on the page that are currently X-locked, so that the client can grant S locks on the unmarked keys without communicating with the server. With this policy, the callback protocol is made *adaptive* so that the server can call back single keys, besides whole pages. When a key is called back, the client just marks the key as unavailable if it is currently unlocked at the client and ships the S locks on the key to the server (to be installed into the global lock table) otherwise. Work out this extension to the caching and locking protocols.

**14.7** With the extended caching and locking protocols discussed in the previous exercise, we can allow a previous version of a page that is currently cached for updating at a client to be cached for reading at other clients. Work out this relaxed caching protocol.

**14.8** A shared-disks system must keep track, for each node, of the pages buffered and the modes of buffering and of the locks held at that node. In other words, a shared-disks system must maintain the equivalents of the cached-page table and the global lock table of a page-server system. Where are these tables kept and how are they managed during normal processing and after recovery from failure at one node?

## Bibliographical Notes

DeWitt et al. [1990] analyze the three alternative designs for a data-shipping client-server architecture: the object server, page server, and file server. Of these, the page server has been found to be the simplest to implement, and it can utilize the clustering of objects in pages [Franklin et al., 1996]. Different methods such as callbacks for ensuring cache consistency in page-server and shared-disks systems are reviewed and analyzed by Franklin et al. [1997].

Most of the material in this chapter comes from the article by Mohan and Narang [1994] that describes ARIES/CSA, the ARIES algorithm for client-server architectures. A different design of an ARIES-compatible page-server system has been implemented in the client-server version of the EXODUS storage manager [Franklin et al. 1992]. The page-transfer methods for shared-disks systems mentioned in Sect. 14.10 are discussed in detail by Mohan and Narang [1991].

Carey et al. [1994] and Zaharioudakis et al. [1997] analyze different caching protocols for page-server environments with fine-grained concurrency control, involving techniques such as adaptive callbacks (discussed in Problems 14.6 and 14.7). Jaluta [2002] and Jaluta et al. [2006] present ARIES-compatible algorithms for the management of transactions in a page-server system, including B-tree management with redo-only structure modifications, key-range locking, and adaptive callbacks; a solution to Problem 14.7 can be found here.

# Chapter 15
# Processing of Write-Intensive Transactions

Many traditional transaction-processing applications such as banking and stock trading are *write intensive* in nature: they involve a great number of concurrent, relatively short updating transactions that have stringent response-time requirements, besides strict consistency requirements. The shift from reads to writes in modern web applications has also been observed in recent years. This trend poses challenges to the performance of a traditional transaction-processing system based on write-ahead logging and random writes of B-tree pages.

With the traditional commit protocol for transactions doing write-ahead logging, when a large number of transactions want to commit at about the same time, it takes as many rotational delays as there are committing transactions until the last of these succeeds to commit, because the log is forced to disk after each append of a commit log record to the log buffer. When waiting for being committed, these transactions hold their commit-duration locks. To alleviate this logging bottleneck, a commit protocol called *group commit* can be applied with which a group of transactions requesting to be committed can release their locks and be committed by a single force write of the log containing all the commit log records for the group.

The fact that sequential disk accesses are much faster than random ones offers opportunities to important optimizations. In a *write-optimized B-tree*, a set of pages randomly updated in the buffer are flushed onto new contiguous locations on disk using a single large sequential disk write. A *merge tree* consists of a small main-memory B-tree (or some other kind of a balanced tree structure) that receives the updates from transactions and of a series of one or more disk-based B-trees of growing size each receiving updates through occasional merges with a smaller tree.

The separation of the log file that records the update actions and the data pages that contain the current data still incurs write overheads observed in write-intensive environments. This has given rise to recent proposals of *log-structured databases*, in which the log serves as the unique data repository in the system.

These new access structures designed for write-intensive transaction processing are used in several key-value stores acting as database servers for popular Internet

© Springer International Publishing Switzerland 2014
S. Sippu, E. Soisalon-Soininen, *Transaction Processing*, Data-Centric Systems
and Applications, DOI 10.1007/978-3-319-12292-2_15

applications. The structures also lend themselves to implementations on new hardware such as flash-memory-based solid-disk drives (SSDs), where reads and sequential append-only writes are faster than on hard disk, but random in-place writes are expensive.

## 15.1  Group Commit

According to the basic commit protocol presented in Sect. 3.8, the log is flushed onto the log disk every time a transaction commits (Algorithm 3.3). Assume that there are many short transactions, each updating only a single tuple in the database, and that a log-file page can hold the log records (begin-transaction, update log record, commit transaction) for $n$ such transactions. Then the same log-file page is written from the log buffer onto disk $n$ times, and these writes are all random disk writes, with at least one rotational delay in between. Moreover, there is a fixed overhead associated with initializing and completing any disk write.

The log file being a sequential file, a natural optimization is to flush only full pages and to flush as many pages as possible using a single sequential disk write. If a group of $k$ full log-file pages are written by a single sequential disk write, the amortized disk-write cost of committing a short transaction may be as low as $1/(kn)$ times the cost of flushing a single page. The $kn$ transactions are all considered committed once their log records are on the log disk.

The above procedure, in which the log is not flushed at every commit action, is called *group commit*. The application process on whose behalf a transaction is executed by some server-process thread is notified of the commit only after the log has been flushed next time, that is, when the transaction has committed in the original sense of the concept.

Obviously, a group commit delays the commit of individual transactions and hence decreases the durability of successfully terminated transactions under system crashes: a transaction that has appended its commit log record into the log buffer may still abort and roll back if a system crash occurs before the next log flush. Moreover, a large set of transactions waiting for their group to be committed by the next log flush may all hold commit-duration locks on data items they have read or written, thus possibly blocking other transactions that want to access those data items. This raises the question of whether the transactions that wait to be committed can release their locks before the log has been flushed. We will show in the following that this indeed is possible.

In the context of write-ahead logging and lock-based concurrency control, we say that a transaction is *precommitted* if it has appended its commit log record into the log buffer and released its locks and is waiting for the log to be flushed. A precommitted transaction $T$ will commit, and the application will be notified of the commit, if no system crash occurs before the next log flush; otherwise, $T$ will roll back. If the server-process thread executing $T$ fails after $T$ has precommitted, $T$

will still commit if no system crash occurs before the next log flush; however, in this case the application will not be notified of the commit.

The commit action $C$ of a transaction $T$ is now executed using the call *precommit(T)* (Algorithm 15.1), instead of using the standard call *commit(T)* (Algorithm 3.3). The difference is that the call *precommit(T)* only appends the commit log record of $T$ to the log buffer, without flushing the log, releases all locks held by $T$, and, instead of notifying immediately of the commit, waits for the next group commit to occur. The group commit is executed using the call *group-commit()* (Algorithm 15.2), which just flushes the log and wakes up the server-process threads waiting for the group commit.

---

**Algorithm 15.1** Procedure *precommit(T)*

---

$log(n, \langle T, C \rangle)$
Release all locks held by $T$
Delete the transaction record of $T$ from the active-transaction table
Wait for the event "group commit done"
Return a notification of the commit of $T$ to the application process

---

**Algorithm 15.2** Procedure *group-commit()*

---

*flush-the-log()*
Raise the event "group commit done"

---

With group commit, *early lock release* refers to the feature that a transaction releases its commit-duration locks after appending the commit log record to the log buffer but before the log records of the transaction are flushed onto the log disk. Because a transaction actually commits only after the log has been flushed, early lock release permits the appearance of any of the isolation anomalies dirty writes, dirty reads, and unrepeatable reads, in the strict meaning of these concepts. We have to show that an anomaly that appears only if a $C$ action is taken to mean precommit rather than commit cannot cause any harm during normal transaction processing or in recovering from a failure.

During normal transaction processing, because the commit-duration locks held by a transaction are released in performing the commit action, a transaction history that is possible under the group commit protocol (with the $C$ action meaning precommit) is always possible under the standard commit protocol (with $C$ meaning standard commit). This implies further that for any committed transaction $T$ (i.e., one that has its commit log record on the log disk), the transaction history including the commit action of $T$ does not contain any isolation anomalies for $T$ that are not permitted by the locking protocol. This is because all the transactions that have their commit action before that of $T$ are then also committed.

We then show that recoverability is maintained, so that a transaction that wants to perform a partial or total rollback during normal processing or that is active at the time of a system crash or process failure will be correctly rolled back. Naturally, we assume that the locking protocol used prevents dirty writes. During normal processing, in the absence of system crashes and process failures, it follows immediately from Theorem 5.24 that any active, not-precommitted transaction can do (or complete) a partial or total rollback, because then, for the purpose of the proof, we can assume that each $C$ action in the history means true commit rather than precommit.

Assume then that a transaction is active or precommitted, but not committed at the time of a system crash. Such a transaction does not have its commit log record on the log disk. It may have or may not have had its commit log record in the log buffer at the time of the failure. In any case, the application has not been notified of the commit of the transaction. Thus it is correct to roll back such a transaction. The sequence of log records found on the log disk at the time of restart recovery is one that can also be produced in the case that transactions are committed individually rather than in groups, up to the last log record flushed in the last group commit. Thus, the transaction is correctly rolled back in the undo pass of ARIES recovery.

We have yet to consider what must be done in the event of a failure of a single server-process thread that executes a transaction. Actually, nothing special needs to be done besides what is explained in Sect. 4.10. If the failed server-process thread was waiting for the next group commit, it does not hold any page latched; so the failure cannot have left any page corrupted. Thus nothing whatsoever needs to be done. The precommitted transaction either will be committed in the next group commit or aborted and rolled back if a system crash occurs before the commit log record goes to disk. In the former case the application is just left missing of the commit notification.

## 15.2  Online Page Relocation

In write-intensive transaction-processing environments where tuples are frequently inserted, updated, or deleted, the buffer soon fills up with modified pages, so that many pages must be flushed onto disk during normal processing and when taking checkpoints. Performance of flushing can be improved by using large *sequential disk writes* in which a sequence of pages is taken to a number of consecutive disk blocks by a single operation.

A large disk write necessarily *relocates* the flushed pages to newly allocated consecutive disk addresses, implying that the flushed pages must change their page identifiers accordingly. Page relocation naturally occurs in a reorganization of a database structure, with the overhead of adjusting pointers (tuple identifiers) in indexes pointing to tuples in relocated data pages. In offline reorganization of a sparse primary index, the secondary indexes are usually dropped and then rebuilt on the relocated data, because the index records in a secondary index usually carry tuple identifiers besides the primary keys. When done online during normal inserts, updates, and deletes on an indexed relation, page relocation also introduces the overhead of maintaining references within the index structure, such as parent-to-child links pointing to relocated child pages in a B-tree.

A straightforward solution to the problem of maintaining references to relocated pages is to use *logical page identifiers* that do not change in page relocation. A logical page-id can be regarded as identifying a *logical page* on a virtual device. A logical page that is currently in use (i.e., as a part of some physical database structure) is mapped to a *physical page* on a concrete device, that is, a database disk. Relocating a page then involves an update on this logical-to-physical page mapping, but the page references within a database structure need not be changed.

Naturally, the logical-to-physical page mapping must be part of the permanent database, and hence reads and updates on it must be protected by latching. When a set of modified pages is to be flushed from the buffer onto disk, the pages are fixed and write-latched, the pages covering the relevant parts of the logical-to-physical page mapping are fixed and write-latched, a consecutive area on disk is allocated, the modified pages are taken to that area, and the page mapping is updated so as to reflect the new locations of the flushed pages.

The previous disk versions of the relocated pages remain in their locations, but they can no longer be located through the current logical-to-physical page mapping. Those pages are called *shadow pages* of the now current pages; together with appropriate checkpointing, the shadow pages can be used to recover an earlier version of the database, and, if at least logical log records have been written for updates on the data tuples, the current version of the logical database that existed at about the time of the crash can be reconstructed. Indeed, some early database management systems relied heavily on shadow paging, rather than write-ahead logging, in implementing restart recovery. The logical-to-physical mapping is an essential part of *log-structured file systems*, which improve write performance by replacing a set of small writes by a single large write.

Logical-to-physical mappings are also part of some modern devices such as *solid-state drives* or SSDs, which are based on *flash memory*. Flash memory does not tolerate in-place updates on pages. The space occupied by a page on flash memory can be rewritten only after an entire block of many pages containing that page is first erased, involving an expensive operation. Thus, modified pages are always flushed onto free page addresses in previously erased blocks, and only when no such address is available, an erase operation is invoked to free a block of previously written pages that are no longer used (i.e., are not among the currently mapped physical pages). The logical-to-physical mapping is maintained as part of a structure called the *flash-translation layer* or FTL.

## 15.3  Write-Optimized B-Trees

In this section we consider a B-tree variant called the *write-optimized B-tree*, which incorporates online page relocation without the overhead of maintaining a logical-to-physical page mapping. When a modified B-tree page is to be flushed from the buffer onto disk as part of a large sequential disk write, the page goes to a new disk location and, accordingly, gets a new page-id. The changed page-id of the page replaces the old page-id in the index record stored in the parent page, thus avoiding the overhead of indirect addressing involved in logical-to-physical mappings.

Conventional in-place updates remain to be possible on the write-optimized B-tree. Of the B-tree pages, at least the root page should retain its original page-id and not be relocated. Note that relocating the root page would involve updating some system-catalog pages and would also violate our assumption made previously in Sect. 7.1 that the page-id of the root page never changes during the lifetime of the index.

Because of the online maintenance of changed page-ids within the index, the write-optimized B-tree makes sense only if the parent-to-child links are the only links that appear between the B-tree pages. Observe that if, for example, the leaf pages were linked from left to right, as is usual in many B-tree implementations, an update on the rightmost leaf page would cause sequences of updates that eventually propagate through the entire tree. Moreover, if two-way linking were used at the leaf level, which is also common, then a single update on any leaf page would cause infinite sequences of updates through the tree. This is our main motivation for not using sideways linking in the B-tree.

When a modified child page $q$ of a page $p$ is to be relocated, the parent page $p$ must also be buffered and write-latched, so that the page-id $q$ in the index record $(x, q)$ in the parent page pointing to the child page can be replaced by the new page-id $q'$ of the child page. The problem here is how to keep track of the parent pages.

The write-optimized B-tree employs an inexpensive method to maintain approximate information about the page-ids of parents of modified B-tree pages. We assume each page $q$ carries a *parent-link* field that stores the page-id of the parent of $q$ as it was observed when the page was last updated. The parent information is not kept strictly up to date, and hence it cannot be trusted. The parent-link field is updated only when the page itself is updated in an ordinary update action by a transaction or in a structure modification. No separate log record is written for the parent-link update.

The parent links can turn invalid in several ways. The parent-links of all child pages of a relocated page are left to point to the old location of the parent. Whenever a structure modification occurs on a non-leaf page, the parent links of at least a half of the children remain to point to an incorrect parent. For example, when a page $q$ is split, the parent links in the child pages of $q$ whose index records are moved from $q$ to its new sibling $q'$ continue to point to $q$.

The parent-link field of a B-tree leaf page is updated when the page is held write-latched for updating some tuple in it, by one of the single-tuple-update Algorithms 3.1, 3.2, 4.8, or 4.10, or by one of the bulk-update Algorithms 10.4 or 10.5. The page-id of the parent page of the updated page is taken from the saved path, if one is available; otherwise, the parent-link field remains unchanged. A B-tree page $q$ that is involved as a modified child page in a structure modification gets its parent-link updated as a part of the modification. For example, when performing a page split logged with $\langle page\text{-}split, p, q, x', q', \ldots \rangle$, the page-id $p$ of the parent page is assigned to the parent-link fields of the child pages $q$ and $q'$, and the page-id of the parent page $p'$ of page $p$, taken from the saved path, is assigned to the parent-link field of $p$. Initially, when a page is allocated for the B-tree, the parent-link field is set to null.

To aid the reconstruction of the parent information in the redo pass of restart recovery, when performing a forward-rolling single-tuple or bulk insert or delete on a leaf page, the value assigned to the parent-link field can be included into the log record written for the update, such as in $\langle T, I, p, x, v, p', n' \rangle$, where $p'$ is the page-id of the parent of the updated page $p$ (as of the saved path that existed during the update). Now when the update logged with such a log record is redone, also the update on the parent-link field can be redone. Similarly, the page-id of grandparent $p'$ of the split page $q$ could be included in the log record: $\langle page\text{-}split, p, q, x', q', p', \ldots \rangle$.

When a modified B-tree page $q$ is to be relocated and flushed, the page $p$ whose page-id appears in the parent-link field is fixed and write-latched. The contents of pages $p$ and $q$ are inspected so as to check if both pages still belong to the same B-tree and that $q$ still is a child page of $p$. If so, page $q$ can be relocated and flushed. Otherwise, this cannot be done for now. The correctness of this decision heavily relies on correct maintenance (in the page header) of the page type and internal identifier of the structure the page belongs to, as discussed in Sect. 2.2 and shown in detail in the structure-modification algorithms (Algorithms 8.1–8.4) that allocate or deallocate B-tree pages.

The procedure call *relocate-and-flush-pages($M$)* (Algorithm 15.3) is used by the buffer manager when it wants to relocate and flush $M$ least-recently used modified B-tree pages using a single disk write of $M$ pages. If the parent of one of these pages is not buffered or cannot be resolved using the information stored in the parent-link field, then the next page in the LRU chain is tried, and so on. The changing in the parent page $p$ of the page-id $q$ to $q'$ is logged with the redo-only log record

$$\langle page\text{-}id\text{-}change, p, x, q, q' \rangle. \tag{15.1}$$

Naturally, we apply the write-ahead-logging protocol, so that first are flushed onto the log disk all log records up to and including the one with the greatest LSN in the PAGE-LSNs of the pages to be relocated.

Observe that the page-id change on the parent page must be logged after, not before, the flushing of the page. If a crash occurs after the flushing and before the first *page-id-change* log record goes to the log disk, all the flushed pages just remain as garbage. The previous disk versions of the flushed pages, the shadow pages, remain in their locations in the physical database, and they are used to recover the pages in the redo pass of ARIES. If some but not all of the *page-id-change* log records happen to reach the log disk before the failure, the redo pass sees the relocated page for some of the pages and the shadow page for the rest.

---

**Algorithm 15.3** Procedure *relocate-and-flush-pages*($M$)

---

$P \leftarrow \emptyset$
**for** all buffer control blocks $b$ in the LRU chain **do**
    $q \leftarrow page\text{-}id(b)$
    **if** *modified-bit*($b$) $= 1$ **and** page $q$ is unlatched **then**
        *fix-and-write-latch*($q$)
        $p \leftarrow parent\text{-}link(q)$
        **if** $q$ is a B-tree page **and** $p \neq$ null **and** page $p$ is buffered and unlatched **then**
            *fix-and-write-latch*($p$)
            **if** $p$ and $q$ are pages of the same B-tree **and** $p$ contains an index record $(x, q)$ for child
            $q$ **then**
                $q' \leftarrow$ new page-id
                $P \leftarrow P \cup \{(p, x, q, q')\}$
                **if** $|P| \geq M$ **then**
                    exit from the **for** loop
                **end if**
            **else**
                *parent-link*($q$) $\leftarrow$ null
                *unlatch-and-unfix*($p$)
                *unlatch-and-unfix*($q$)
            **end if**
        **else**
            *unlatch-and-unfix*($q$)
         **end if**
    **end if**
**end for**
**for** all $(p, x, q, q') \in P$ **do**
    stamp $q'$ in the header of page $q$
    *parent-link*($q'$) $\leftarrow p$
    change the buffer control block of page $q$ to that of $q'$
    change the write latch on page $q$ to that on $q'$
**end for**
$n = \max\{\text{PAGE-LSN}(q') \mid (p, x, q, q') \in P\}$
flush all the not-yet-flushed log records up to and including that with LSN $n$
flush the pages $q'$ with $(p, x, q, q') \in P$ using a single disk write
**if** the disk write succeeded **then**
    **for** all $(p, x, q, q') \in P$ **do**
        $b' \leftarrow$ the buffer control block of page $q'$
        *modified-bit*($b'$) $\leftarrow 0$
        delete the entry for $q$ from the modified-page table
        replace the index record $(x, q)$ in page $p$ by $(x, q')$
        $log(n, \langle page\text{-}id\text{-}change, p, x, q, q'\rangle)$
        PAGE-LSN($p$) $\leftarrow n$
        *unlatch-and-unfix*($q'$)
    **end for**
    *unlatch-and-unfix*($p$) for all $p$ with some $(p, x, q, q') \in P$
**else**
    invalidate the pages $q$ with $(p, x, q, q') \in P$ and initialize redo recovery for them
**end if**

---

When a checkpoint is to be taken, the call *relocate-and-flush-pages*($M$) is first executed one or more times, maybe with a slightly modified selection logic applied for the pages to be flushed, one that ensures that the REDO-LSN, that is, the minimum of the REC-LSNs of the modified pages, is advanced. This is then followed by writing to the log the checkpoint log records, with copies of the modified-page and active-transaction tables.

The redo pass of ARIES must not do any page relocations, besides redoing logged page-id changes that are missing from parent pages. The redo pass is supposed to just repeat previously done logged updates, not to generate any new log records. If some modified pages must be evicted during the redo pass, which is unlikely, then those pages go to the locations defined by the page-ids current at the time of the redone updates.

In the redo pass, a page-id change on page $p$ logged with a log record of the form $n$ : ⟨*page-id-change, p, x, q, q'*⟩ is processed exactly as expected: if REC-LSN($p$) $\leq$ $n$, page $p$ is fixed and write-latched, and its PAGE-LSN is examined; if less than $n$, the index record $(x, q)$ in page $p$ is replaced by $(x, q')$, PAGE-LSN($p$) is advanced to $n$, page $p$ is unlatched, and so on.

Missing parent information for a modified B-tree page $q$ can be added using the call *find-parent*($q$) (Algorithm 15.4), which latches page $q$, takes some key $x$ from the page, unlatches $q$, and performs a traversal from the root page along the search path for $x$, correcting parent links on all modified pages encountered until $q$ is entered. However, we are never actually forced to use this call, because we always have the option of not relocating the modified page.

To make it possible to redo the parent-link additions performed by the call *find-parent*($q$), the additions could be logged with redo-only log records of the form:

$$\langle add\text{-}parent\text{-}link, p', p \rangle. \tag{15.2}$$

In the redo pass, such a log record is processed as expected: if REC-LSN($p'$) is less than or equal to the LSN $n$ of the log record, page $p'$ is fixed and write-latched, and its PAGE-LSN is examined; if less than $n$, the parent-link of $p'$ is set to $p$, PAGE-LSN($p'$) is advanced to $n$, page $p'$ is unlatched, and so on.

---

**Algorithm 15.4** Procedure *find-parent(q)*

---

*fix-and-read-latch(q)*
**while** *q* is a modified B-tree page **do**
    *x* ← some key in page *q*
    *unlatch-and-unfix(q)*
    *p* ← the page-id of the root page
    *fix-and-read-latch(p)*
    **while** *height(p)* > 1 **and** *p* ≠ *q* **do**
        *p'* ← the page-id of the child of *p* that covers *x*
        *fix-and-read-latch(p')*
        **if** page *p'* is modified **and** *parent-link(p')* ≠ *p* **then**
            *unlatch-and-unfix(p')*
            *fix-and-write-latch(p')*
            *parent-link(p')* ← *p*
        **end if**
        *unlatch-and-unfix(p)*
        *p* ← *p'*
    **end while**
    **if** *p* = *q* **then**
        exit from the **while** loop
    **end if**
    *fix-and-read-latch(q)*
**end while**
*unlatch-and-unfix(q)*

---

We observe that the only additional restriction that must be imposed besides the basic rules of buffering, namely, write-ahead logging and steal-and-no-force, is that if a non-root modified B-tree page is to be relocated, then its parent page must also be buffered and write-latched and its child link pointing to the relocated page must be updated to reflect the new location of the child page. Most notably, we do not enforce the requirement that all the modified child pages that are to be relocated should be flushed before flushing the parent.

As lower-level pages of a B-tree tend to be less frequently accessed than upper-level pages, it is likely that in the LRU chain, child pages appear mostly before their parent pages, and hence children are more likely subjects of being relocated or flushed than their parents.

## 15.4  Merge Trees

A *merge tree* is composed of a sequence $C_0, C_1, \ldots, C_n$ of B-trees (called *components*) of growing size with the smallest one, $C_0$, receiving all updates from transactions and kept small enough to fit in main memory in its entirety. When a component $C_i$ reaches a preset size limit, it is merged with the next larger component $C_{i+1}$, resulting in a B-tree $C'_{i+1}$ which is then substituted for $C_{i+1}$

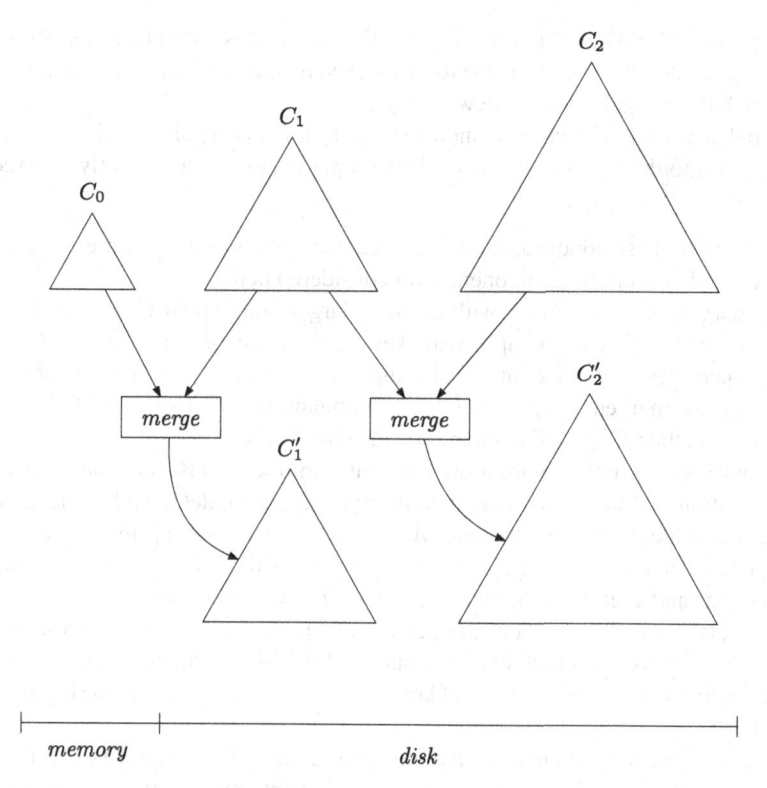

**Fig. 15.1** Merging entire trees in a three-level merge tree

(Fig. 15.1). The idea is that the new tree $C_{i+1}$ resulting from the merge is built into a new disk space using large sequential writes.

In merging components $C_i$ and $C_{i+1}$, the resulting component $C'_{i+1}$ can be built in a manner similar to that presented for B-tree loading in Sect. 8.8. Because the components $C_i$ with $i > 0$ are only updated via merging and never receive any normal updates, we use a fill-factor near 100%, so that when a page is split, only two records go from a full page to the new sibling page. A modification of the B-tree loading algorithm uses sequential writes (Problem 15.1).

The following conditions must be satisfied by a merge tree on relation $r(\underline{X}, V)$:

1. Each component $C_i$ indexes a set of tuples $(x, v)$ that either currently belong to $r$ or used to belong to $r$. In addition, a component $C_i$ with $i < n$ may index tuples $(x, \perp)$ where $\perp$ is a special *delete marker* indicating that a tuple with key $x$ used to belong to $r$ but was deleted.
2. The keys $x$ are unique within each $C_i$.
3. If $(x, v)$ currently belongs to $r$, then it is indexed by the component $C_i$ with the least $i$ such that $C_i$ indexes a tuple with key $x$.

4. If $(x, v)$ currently does not belong to $r$, then either it is not indexed by any of the components $C_i$, or $(x, \perp)$ is the tuple indexed by the component $C_i$ with the least $i$ such that $C_i$ indexes a tuple with key $x$.
5. A delete action $D[x, v]$ or an undo-insert action $I^{-1}[x, v]$ on $r$ produces the tuple $(x, \perp)$ into the set indexed by $C_0$ if the tuple $(x, v)$ is not currently indexed by $C_0$.

In addition to these conditions, a merge tree may have to satisfy some relative size constraints between the components, not considered here.

In merging component $C_i$ with the next larger component $C_{i+1}$ (Fig. 15.1), if both components index a tuple with key $x$, then the one indexed by $C_i$ takes precedence. Thus, if $C_i$ contains the tuple $(x, v_i)$ and $C_{i+1}$ contains the tuple $(x, v_{i+1})$, the merged component $C'_{i+1}$ will contain $(x, v_i)$. However, if $i + 1 = n$ and $v_i = \perp$, then $C'_{i+1}$ will contain no tuple with key $x$.

That the tuples of the database are distributed into several B-trees adds substantial complication to the implementation of read, insert, delete, and write actions, especially under key-range locking. A read action $R[x, > z, v]$ must access every component $C_i$ to determine the tuple $(x_i, v_i)$ with the least key $x_i > z$ in that component and then to return the tuple with the least such key $x_i$. Although all insert, delete, and write actions are performed on the $C_0$ component, these actions must also access the other components to check the correctness of the action, including maintaining uniqueness of keys, and to find the key next to the key inserted or deleted.

The smallest component $C_0$ of a merge tree actually does not need to be a B-tree. Taking advantage of the requirement that $C_0$ is kept small so that it can reside in main memory all the time, it can be any balanced search-tree structure with nodes of any size. In the following we assume that a genuine main-memory structure, such as an AVL-tree or a $(2, 3)$-tree, is used as the physical structure for $C_0$.

The updates by transactions on $C_0$ are *logged logically*, so that, for example, an insert action $I[x, v]$ by transaction $T$ is logged with $\langle T, I, x, v, n' \rangle$ and its undo action $I^{-1}[x, v]$ with $\langle T, I^{-1}, x, n' \rangle$, where $n'$ is the UNDO-NEXT-LSN of $T$ (recall the discussion of logical logging in Sect. 3.2). Structure modifications (balancing operations) triggered by updates on $C_0$ are not logged. In the event of a failure, $C_0$ is rebuilt from the logical update log records written since the last merge of $C_0$ with $C_1$.

As transactions perform only reads on components $C_i$ with $i > 0$, the merge of component $C_i$ with component $C_{i+1}$, $i > 0$, is straightforward: $C_i$ and $C_{i+1}$ are scanned in parallel in key order, generating a single sequence of tuples in ascending key order and building the new B-tree $C'_{i+1}$. During the scan, pages of $C_i$ and $C_{i+1}$ are read-latched; no locks are acquired. The new tree $C'_{i+1}$ can be constructed using the B-tree loading algorithm outlined in Sect. 8.8. When $C'_{i+1}$ is ready, it is atomically substituted for $C_{i+1}$: the system-catalog page holding the identifiers of the components is write-latched, the identifier of $C'_{i+1}$ is substituted for that of $C_{i+1}$, the substitution is redo-only logged, and the write latch is released.

The complicated part of merge-tree maintenance is the merge of $C_0$ with $C_1$. We must allow transactions to perform updates on $C_0$, while the merge process is in progress. The problem here is that transactions may perform updates on the key range that the merge process has already scanned. Those updates are missed by the merge process. The solution is to maintain a *side file* to capture the updates missed by the merge process, as done in online index construction in Chap. 11. When the merge process has finished the construction of $C_1'$, the side file is processed, and a new main-memory component $C_0'$ is built from the updates recorded in the side file. Finally, $C_0'$ is atomically substituted for $C_0$.

## 15.5 Log-Structured Databases

The conventional database organization separates the log from the application data stored in the data pages, allowing the application data to be organized efficiently for key-based clustered access, while the sequential log, which necessarily duplicates the application data and even retains previous versions of data items, is only used for recovery purposes. Write-ahead logging must be enforced so as to order properly the log flushings and the database-page flushings. In the case of a write-intensive application in which new data is often inserted once but rarely updated, and recent data is read often, the database buffer may be full of modified pages all the time, and a new page access may often cause some page from the buffer to be flushed onto the database disk. In the presence of many short updating transactions, also the log is flushed constantly.

A database organization called the *log-structured database*, or *log-only database*, tries to alleviate the write bottleneck of write-intensive applications as follows. The sequence of log records written by transactions for their update actions serves as the sole repository of the tuples of the database. Such a log is necessarily a logical log, because there are no data pages on which the log records should be applied. The log is flushed from the log buffer onto disk, as before, when a transaction commits or completes its rollback or when the log buffer becomes full or when a checkpoint is taken, but otherwise the write-ahead logging protocol is not observed.

The logical log for transactions on relation $r(\underline{X}, V)$ is interpreted as a single-version database as follows. Given key $x$, the current version of the tuple with key $x$ is $(x, v)$, if the most recent log record for an update on key $x$ is $\langle T, I, x, v, n' \rangle$, $\langle T, D^{-1}, x, v, n' \rangle$, $\langle T, W, x, u, v, n' \rangle$, or $\langle T, W^{-1}, x, v, u, n' \rangle$. The database does not contain a tuple with key $x$ if the log does not contain any log record for an update on $x$ or if the most recent log record for an update on $x$ is $\langle T, D, x, v, n' \rangle$ or $\langle T, I^{-1}, x, n' \rangle$.

For efficient access to single tuples, a log-structured database must come with a dense index. The index record for a tuple $(x, v)$ in $r$ is $(x, n)$, if the most recent log record for an update on $x$ is $\langle T, I, x, v, n' \rangle$, $\langle T, D^{-1}, x, v, n' \rangle$, $\langle T, W, x, u, v, n' \rangle$, or $\langle T, W^{-1}, x, v, u, n' \rangle$, and $n$ is the LSN of the log record. We assume here that the LSNs are physical addresses (byte offsets from the beginning of the log file).

The index is assumed to fit in main memory, and modifications on it are not logged. To accelerate the redo pass of restart recovery, the index is checkpointed periodically or after the number of updates reaches a preset threshold: the relation is S-locked, the binary image of the index is flushed to a disk file, the identifier of the file is logged, and the S lock is released. The redo pass only involves downloading the index from the file written at the last checkpoint and bringing the index up to date by scanning and applying the log records written after the checkpoint.

Since the log contains a complete history of all update actions performed by transactions, a slight modification to the above setting allows a log-structured database to function as a transaction-time versioned database (Sect. 12.1). Index records for newly created versions initially take the form $(x, T, n)$, where $T$ is the identifier of a transaction that has updated (i.e., created a new version of) the tuple with key $x$ and $n$ is the LSN of the log record for the most recent update by $T$ on $x$.

When transaction $T$ has committed, that is, written the commit log record and flushed the log, the UNDO-NEXT-LSN chain of $T$ is traversed in order to find out the updated keys, and the commit timestamp of $T$ is substituted for the transaction identifier $T$ in each index record $(x, T, n)$. Under snapshot isolation, transaction $T$ holds its X locks on the keys $x$ during the substitution of the commit timestamps, after which the locks can be released. To make possible the substitution of the timestamps during index recreation after a failure, the commit timestamps must be logged.

When a log-structured database organization is applied to a key-value store $r(\underline{X}, V)$ that partitions the key space logically into many disjoint key groups, the log can be split into several logs, one for each key group, thus reducing contention in log writes. Recall that each transaction on such a key-value store can only update tuples with keys in a single key group. If there are too many key groups, a single log can serve a subset of key groups, where the partitioning into the subsets is balanced according to the expected frequency of updates. Each log is indexed by its own index.

A log-structured database is especially amenable to the remote backup technique (Sect. 13.8) for disaster recovery. Log records written at the primary site are shipped to the remote site where they are appended to the log buffer, from where they go to the log disk of the remote site. The index to the log at the remote site is built as the log records arrive. As indexing is light-weight and the entire index resides in main memory, we probably do not need to divide the work into many work queues. If there are several logs, each serving a set of key groups, then the remote site has as many work queues, one for each log.

## 15.6   Databases on Flash Memory

New storage technologies such as *flash-memory*-based *solid-state drives* (SSDs) offer the promise of overcoming the performance limitations of hard-disk drives. The absence of mechanical parts implies that flash-memory-based SSDs are not limited by any seek or rotational delays, do not suffer mechanical failures, and consume less power than hard disks.

Reading or writing a random page on a flash-memory-based SSD may be tens of times faster than reading or writing a random page on hard disk, and the bandwidth (megabytes per second) of sequential reads or writes may also be distinctly (two to five times) better on SSDs than on hard disk. The performances of different SSDs vary greatly, but in many cases, reads on SSDs may be ten times faster than writes, whereas there is no such performance difference between hard-disk reads and writes. Contrary to what is commonly believed, sequential reads on SSDs are more efficient than random reads, as they are on hard disk, so that existing optimizations for hard disks such as data clustering and readahead are still valid.

A flash-memory-based SSD is normally built on an array of flash-memory packages. A typical *flash-memory package* is composed of one or more dies, where each *die* (chip) is segmented into multiple *planes*, each containing a few thousand (such as 2048) blocks and one or two registers of the page size as an input/output buffer. A *block* usually contains 64 to 128 pages, where each *page* has a 2 or 4 kB data part and a smaller metadata area.

Read operations on flash-memory-based SSDs are performed in units of pages. Also writes are normally performed in page granularity, but with some flash memories, sub-page operations are possible. The pages in a block must be written sequentially, from the least to the greatest page address. We may view a page identifier as a pair $(b, p)$, where $b$ is a block number and $p$ is the number of the page within the block.

A unique requirement of flash memory is that a block must be *erased* before its pages can be written. An erase operation can take three to four times the time taken by a write operation and must be conducted in block granularity. Therefore, a block is also called an *erase block*. Only whole blocks, not individual pages, can be erased. Flash-memory blocks have limited erase cycles. After wearing out, say after 100,000 erase cycles, a flash-memory cell can no longer store data.

The controller of a flash-memory-based SSD contains a mapping mechanism called a *flash translation layer* (FTL) that makes it possible to emulate a hard disk with in-place writes. Pages to be read or written are identified by *logical page identifiers* that remain unchanged in page-write operations. The address of the current physical page corresponding to a logical page identifier is given by the FTL mapping. When a modified page $p$ is to be flushed onto flash memory, a clean (i.e., erased) page location is allocated to hold the contents of the page, the page is written to the allocated location, the FTL mapping is adjusted to point to the allocated location, and the old page location is marked as invalid by updating its metadata.

The mapping granularity used in the FTL can be as large as a whole block or as small as a single page. With *page-level mapping*, updating a valid page $(b, p)$ (i.e., one belonging to the current database) involves fixing the page into the buffer, modifying the page, and, when the page is flushed, locating a block $b'$ that has at least one clean page left, writing the modified page to the first clean page location $p'$ in the block, invalidating the old page location, and adjusting the FTL mapping so that the logical page identifier of the updated page points to $(b', p')$ instead of

$(b, p)$. Obviously, page-level mapping is flexible but requires a large amount of memory space to store the mapping table.

A *block-level mapping*, on the contrary, needs less space to store the mapping table, because only logical blocks are mapped. But, because the page number components of page identifiers cannot be changed, an expensive *copy-erase-write* operation sequence is needed when updating a valid page in a block: all the valid pages in the block are fixed in the buffer, a clean block is allocated, the updated page with all the other valid pages from the block are written to the erased block into their (unchanged) page addresses, and the logical block is mapped to the new block in the FTL.

Assume that there are $n$ pages in a block. Writing pages sequentially into clean locations in the blocks in page-number order incurs at most one page-write operation and $1/n$ block-erase operations per page on the average. Updating a valid page incurs one page-read operation, one block-erase operation, and one page-write operation in the worst case, when a page-level mapping is used, but with a block-level mapping, an update of a valid page may incur up to $n - 1$ page reads, one block erase, and $n$ page writes.

Many FTLs adopt a hybrid approach by using a block-level mapping to manage most blocks as "data blocks" and using a page-level mapping to manage a small set of "log blocks," which works as a buffer to accept incoming write requests efficiently. The FTL mapping table is maintained in persistent flash memory and rebuilt in volatile buffer memory at startup time.

To ensure that clean blocks are available when updated pages are to be flushed, a *garbage collector* is run in the background, scanning flash-memory blocks and recycling invalidated pages. If a page-level mapping is used, the valid pages in the scanned block are copied out and condensed into a new block, the mapping is adjusted, and the old locations are marked as invalid, thus obtaining a whole block of invalidated pages that can be erased.

With the behavior described above, a flash-memory-based SSD is truly efficient for permanent database storage only if most of the write operations are sequential or if they can be turned into sequential writes using a large buffer from which modified pages are flushed in large groups. Due to constant page relocation, physical fragmentation of data is also a major problem with flash memory.

The transaction log, being append-only, is most suitable for implementation on flash memory, because the pages in each block are written one by one, in the sequential order of their addresses, and a page once written is never rewritten and hence never relocated. At the other extreme, there are the conventional index structures such as the B-tree with their frequent random in-place writes, which, if implemented as such on flash memory, would incur constant page relocation, at every update on a page, and thus also a big overhead in the maintenance of the FTL mapping.

The database structures considered in the previous sections of this chapter have features that make them more suitable than the conventional database structures for implementation on flash memory. The distinctive property of the write-optimized B-tree, that is, turning a set of random page writes into a single large sequential write onto a new disk location, readily contributes to a feasible implementation on a flash-memory-based SSD. Although some updates still remain to be in-place, so that an updated page goes into its old logical page address and hence must be mapped by the FTL to its new physical address, it is expected that the FTL remains moderate in size, so that a page-level mapping can be used. Naturally, garbage collection of relocated pages whose logical page identifiers change remains as the responsibility of the database system software.

Similar arguments of suitability for implementation on flash memory can be stated about the other structures discussed in this chapter. In a merge tree of Fig. 15.1, the merge process that merges the B-trees $C_i$ and $C_{i+1}$ to produce the B-tree $C'_{i+1}$ uses sequential reads to scan $C_i$ and $C_{i+1}$ and sequential writes to create $C'_{i+1}$. In a log-structured database, the log part of the database fits perfectly to flash-memory implementation.

## Problems

**15.1** Modify the B-tree loading algorithm suggested in Sect. 8.8 so as to apply sequential writes. Recall that the group of pages to be written sequentially must reside at consecutive addresses.

**15.2** Consider the structure modifications on the B-tree of Fig. 7.1 needed to accommodate an insertion, resulting in the B-tree of Fig. 8.8, or a deletion, resulting in the B-tree of Fig. 8.9. Assume that the B-tree is maintained as a write-optimized B-tree. Assuming that all parent links are correct initially (which is rare in practice), which of the parent links turn incorrect due to the structure modifications and which are updated to correct values?

**15.3** In the original design of the write-optimized B-tree, the parent links are only maintained as main-memory addresses between the buffer-control blocks of buffered B-tree pages. Work out the maintenance of such linking.

**15.4** Explain how read actions $R[x, > z, v]$, insert actions $I[x, v]$, delete actions $D[x, v]$, undo-insert actions $I^{-1}[x, v]$, and undo-delete actions $D^{-1}[x, v]$ are performed on a relation $r(\underline{X}, V)$ implemented as a merge tree, when key-range locking is applied.

**15.5** Elaborate the algorithm that merges disk-resident components $C_i$ and $C_{i+1}$ ($i > 0$) and creates the component $C'_{i+1}$ using the method of Problem 15.1.

**15.6** Explain how a read action $R[x, > z, v]$ is performed on a committed version $\tau$ of a transaction-time log-structured database. Also explain how the update actions $I[x, v]$ and $D[x, v]$ and their undo actions $I^{-1}[x, v]$ and $D^{-1}[x, v]$ are performed, under snapshot isolation.

**15.7** Assume a database that consists of several relations, some of which have a log-structured organization, possibly with a main-memory index, and the others are conventional disk-based B-trees. There is a single log to which all updates on all the relations are logged, either logically or physiologically, depending on the organization of the relation. Any transaction can operate on any number of the relations. Modifications on the disk-based index structures are logged, while those on the main-memory structures are not. Explain how ARIES-based recovery works in this setting: what needs to be done in the analysis, redo, and undo passes and what needs to be checkpointed?

**15.8** Assume that a flash-memory-based SSD applies a block-level FTL mapping. What actions are involved in garbage collection? Give a worst-case scenario in which garbage collection involves much reading and writing.

# Bibliographical Notes

DeWitt et al. [1984] observe that a transaction can release its commit-duration locks before the commit log record is forced onto disk, as long as it does not return results to the application and declare as having committed until the commit log record has reached the log disk. The issue of early lock release is formalized with the notion of partial strictness by Soisalon-Soininen and Ylönen [1995].

Johnson et al. [2010, 2012] identify four logging-related impediments to database system scalability: (a) the high volume of small-sized write requests may saturate the disk, (b) transactions hold locks while waiting for the log flush, (c) excessive process-context switching overwhelms the operating-system scheduler with threads executing log writes, and (d) contention appears as transactions serialize accesses to main-memory log data structures.

The write-optimized B-tree is from Graefe [2004], who suggests maintaining the parent links only between buffer control blocks of buffered pages rather than linking the pages themselves (as we do in Sect. 15.3). Graefe [2004] also suggests enforcing a stricter buffering policy for modified pages, so that a modified child page must always be relocated and flushed before the parent page.

The idea of turning costly random disk writes to append-only ones to be merged later as larger bulks into the main file dates back to differential files [Severance and Lohman, 1976] and log-structured file systems [Rosenblum and Ousterhout, 1992]. The log-structured merge tree or LSM tree is from O'Neil et al. [1996]. Jermaine et al. [2007] introduce a structure called the partitioned exponential file that consists of a set of merge trees each covering a subrange of the entire key space. Sears et al. [2008] describe a database storage engine that uses LSM trees to create full database replicas. The bLSM tree of Sears and Ramakrishnan [2012] is an LSM tree with near-optimal read and scan performance and with an improved merge scheduler. The bLSM tree is used in Walnut, a cloud object store [Chen et al. 2012].

The buffer tree of Arge [1995, 2003] is another index structure that incorporates the idea of collecting updates into larger bulks before posting them to their final destinations. The buffer tree has a buffer attached to each node of the B-tree; updates go to the buffer of the node nearest to the root until the buffer fills up, after which the buffer is emptied and the tuples from it are distributed to the buffers of the children, eventually reaching the leaves, if not deleted earlier.

The description of flash-memory-based solid-disk drives (SSDs) in Sect. 15.6 is mainly from an article by Chen et al. [2009], who have conducted intensive experiments and measurements on different types of state-of-the-art SSDs. Athanassoulis et al. [2010] describe techniques for making effective use of flash memory in three contexts: (1) as a log device for transaction processing on memory-resident data, (2) as the main data store for transaction processing, and (3) as an update cache for hard-disk-resident data warehouses.

B-tree variants designed to work efficiently on flash-memory-based SSDs include the B-tree flash-translation layer or BFTL of Wu et al. [2007] and the Bw-tree of Levandoski et al. [2013]. The performance of the write-optimized B-tree of Graefe [2004] on flash-memory-based SSDs is analyzed by Jørgensen et al. [2011]. Buffer-tree-like index structures optimized for SSDs include the FD-tree of Li et al. [2009, 2010] and the lazy-adaptive tree (LA-tree) of Agrawal et al. [2009].

Recent log-structured database systems that use the log as the sole data repository include LogBase [Vo et al. 2012] and Hyder [Bernstein et al. 2011]. LogBase aims to provide scalable distributed storage service for database-centric applications in the cloud. Main-memory multiversion indexes are maintained for efficient access to the relational data in the log. Hyder is designed for a shared-flash system in which each transaction executes at one server of the system accessing the shared database (cf. Sect. 14.10), logs its updates in one log record, and broadcasts the log record to all servers. Each server rolls forward the log against its locally cached partial copy of the last committed state. The database in Hyder is stored in the log as a multiversion binary-search-tree index in which updates on the leaves propagate up to the root (copy-on-write).

# References

Robert K. Abbott and Hector Garcia-Molina. Scheduling real-time transactions. *SIGMOD Record*, 17(1):71–81, 1988a.

Robert K. Abbott and Hector Garcia-Molina. Scheduling real-time transactions: a performance evaluation. In *VLDB 1988, Proc. of the 14th Internat. Conf. on Very Large Data Bases*, pages 1–12. Morgan Kaufmann, 1988b.

Robert K. Abbott and Hector Garcia-Molina. Scheduling real-time transactions with disk resident data. In *VLDB 1989, Proc. of the 15th Internat. Conf. on Very Large Data Bases*, pages 385–396. Morgan Kaufmann, 1989.

Devesh Agrawal, Deepak Ganesan, Ramesh K. Sitaraman, Yanlei Diao, and Shashi Singh. Lazy-adaptive tree: An optimized index structure for flash devices. *Proc. of the VLDB Endowment*, 2(1):361–372, 2009.

Divyakant Agrawal, Amr El Abbadi, Richard Jeffers, and Lijing Lin. Ordered shared locks for real-time databases. *VLDB J.*, 4(1):87–126, 1995.

Gustavo Alonso, Radek Vingralek, Divyakant Agrawal, Yuri Breitbart, Amr El Abbadi, Hans-Jörg Schek, and Gerhard Weikum. A unified approach to concurrency control and transaction recovery (extended abstract). In *Advances in Database Technology—EDBT'94, Proc. of the 4th Internat. Conf. on Extending Database Technology*, volume 779 of *Lecture Notes in Computer Science*, pages 123–130. Springer, 1994.

Lars Arge. The buffer tree: A new technique for optimal I/O-algorithms (extended abstract). In *WADS 1995, Proc. of the 4th Internat. Workshop on Algorithms and Data Structures*, volume 955 of *Lecture Notes in Computer Science*, pages 334–345. Springer, 1995.

Lars Arge. The buffer tree: A technique for designing batched external data structures. *Algorithmica*, 37(1):1–24, 2003.

Morton M. Astrahan, Mike W. Blasgen, Donald D. Chamberlin, Kapali P. Eswaran, Jim N. Gray, Patricia P. Griffiths, W. Frank King III, Raymond A. Lorie, Paul R. McJones, James W. Mehl, Gianfranco R. Putzolu, Irving L. Traiger, Bradford W. Wade, and Vera Watson. System R: Relational approach to database management. *ACM Trans. Database Syst.*, 1(2):97–137, 1976.

Manos Athanassoulis, Anastasia Ailamaki, Shimin Chen, Phillip B. Gibbons, and Radu Stoica. Flash in a DBMS: Where and how? *IEEE Data Eng. Bull.*, 33(4):28–34, 2010.

Jason Baker, Chris Bond, James Corbett, J. J. Furman, Andrey Khorlin, James Larson, Jean-Michel Leon, Yawei Li, Alexander Lloyd, and Vadim Yushprakh. Megastore: Providing scalable, highly available storage for interactive services. In *CIDR 2011, Proc. of the Fifth Biennial Conf. on Innovative Data Systems Research*, pages 223–234, 2011.

Rudolf Bayer and Edward M. McCreight. Organization and maintenance of large ordered indices. *Acta Inf.*, 1:173–189, 1972.

© Springer International Publishing Switzerland 2014

S. Sippu, E. Soisalon-Soininen, *Transaction Processing*, Data-Centric Systems and Applications, DOI 10.1007/978-3-319-12292-2

Rudolf Bayer and Mario Schkolnick. Concurrency of operations on b-trees. *Acta Inf.*, 9:1–21, 1977.

Rudolf Bayer and Karl Unterauer. Prefix b-trees. *ACM Trans. Database Syst.*, 2(1):11–26, 1977.

Rudolf Bayer, Hans Heller, and Angelika Reiser. Parallelism and recovery in database systems. *ACM Trans. Database Syst.*, 5(2):139–156, 1980.

Bruno Becker, Stephan Gschwind, Thomas Ohler, Bernhard Seeger, and Peter Widmayer. An asymptotically optimal multiversion B-tree. *VLDB J.*, 5(4):264–275, 1996.

Hal Berenson, Philip A. Bernstein, Jim Gray, Jim Melton, Elizabeth J. O'Neil, and Patrick E. O'Neil. A critique of ANSI SQL isolation levels. In *Proc. of the 1995 ACM SIGMOD Internat. Conf. on Management of Data*, pages 1–10. ACM Press, 1995.

Philip A. Bernstein. Transaction processing monitors. *Commun. ACM*, 33(11):75–86, 1990.

Philip A. Bernstein and Nathan Goodman. Multiversion concurrency control—theory and algorithms. *ACM Trans. Database Syst.*, 8(4):465–483, 1983.

Philip A. Bernstein and Eric Newcomer. *Principles of Transaction Processing for Systems Professionals*. Morgan Kaufmann, 1996.

Philip A. Bernstein and Eric Newcomer. *Transaction Processing. Second Edition*. Morgan Kaufmann, 2009.

Philip A. Bernstein, David W. Shipman, and James B. Rothnie. Concurrency control in a system for distributed databases (SDD-1). *ACM Trans. Database Syst.*, 5(1):18–51, 1980.

Philip A. Bernstein, Vassco Hadzilacos, and Nathan Goodman. *Concurrency Control and Recovery in Database Systems*. Addison Wesley, 1987.

Philip A. Bernstein, Colin W. Reid, and Sudipto Das. Hyder—a transactional record manager for shared flash. In *CIDR 2011, Proc. of the Fifth Biennial Conf. on Innovative Data Systems Research*, pages 9–20, 2011.

Alexandros Biliris. Operation specific locking in b-trees. In *Proc. of the Sixth ACM SIGACT-SIGMOD-SIGART Symp. on Principles of Database Systems*, pages 159–169. ACM Press, 1987.

Mike W. Blasgen, Morton M. Astrahan, Donald D. Chamberlin, Jim Gray, W. Frank King III, Bruce G. Lindsay, Raymond A. Lorie, James W. Mehl, Thomas G. Price, Gianfranco R. Putzolu, Mario Schkolnick, Patricia G. Selinger, Donald R. Slutz, H. Raymond Strong, Irving L. Traiger, Bradford W. Wade, and Robert A. Yost. System R: An architectural overview. *IBM Systems Journal*, 20(1):41–62, 1981.

Mike W. Blasgen, Morton M. Astrahan, Donald D. Chamberlin, Jim Gray, W. Frank King III, Bruce G. Lindsay, Raymond A. Lorie, James W. Mehl, Thomas G. Price, Gianfranco R. Putzolu, Mario Schkolnick, Patricia G. Selinger, Donald R. Slutz, H. Raymond Strong, Irving L. Traiger, Bradford W. Wade, and Robert A. Yost. System R: An architectural overview. *IBM Systems Journal*, 38(2/3):375–396, 1999.

Mark R. Brown and Robert Endre Tarjan. Design and analysis of a data structure for representing sorted lists. *SIAM J. Comput.*, 9(3):594–614, 1980.

Michael J. Cahill, Uwe Röhm, and Alan David Fekete. Serializable isolation for snapshot databases. *ACM Trans. Database Syst.*, 34(4), 2009.

Michael J. Carey, Michael J. Franklin, and Markos Zaharioudakis. Fine-grained sharing in a page server OODBMS. In *Proc. of the 1994 ACM SIGMOD Internat. Conf. on Management of Data*, pages 359–370. ACM Press, 1994.

Rick Cattell. Scalable SQL and NoSQL data stores. *SIGMOD Record*, 39(4):12–27, 2010.

Donald D. Chamberlin, Morton M. Astrahan, Mike W. Blasgen, Jim Gray, W. Frank King III, Bruce G. Lindsay, Raymond A. Lorie, James W. Mehl, Thomas G. Price, Gianfranco R. Putzolu, Patricia G. Selinger, Mario Schkolnick, Donald R. Slutz, Irving L. Traiger, Bradford W. Wade, and Robert A. Yost. A history and evaluation of System R. *Commun. ACM*, 24(10):632–646, 1981.

Fay Chang, Jeffrey Dean, Sanjay Ghemawat, Wilson C. Hsieh, Deborah A. Wallach, Mike Burrows, Tushar Chandra, Andrew Fikes, and Robert E. Gruber. Bigtable: A distributed storage system for structured data. *ACM Trans. Comput. Syst.*, 26(2):4:1–4:26, 2008.

Feng Chen, David A. Koufaty, and Xiaodong Zhang. Understanding intrinsic characteristics and system implications of flash memory based solid state drives. In *Proc. of the Eleventh Internat. Joint Conf. on Measurement and Modeling of Computer Systems, SIGMETRICS/Performance*, pages 181–192. ACM Press, 2009.

Jianjun Chen, Chris Douglas, Michi Mutsuzaki, Patrick Quaid, Raghu Ramakrishnan, Sriram Rao, and Russell Sears. Walnut: a unified cloud object store. In *Proc. of the 2012 ACM SIGMOD Internat. Conf. on Management of Data*, pages 743–754. ACM Press, 2012.

Douglas Comer. The ubiquitous b-tree. *ACM Comput. Surv.*, 11(2):121–137, 1979.

Sudipto Das, Divyakant Agrawal, and Amr El Abbadi. G-store: a scalable data store for transactional multi key access in the cloud. In *SoCC'10, Proc. of the 1st ACM Symp. on Cloud Computing*, pages 163–174, 2010.

David J. DeWitt, Randy H. Katz, Frank Olken, Leonard D. Shapiro, Michael Stonebraker, and David A. Wood. Implementation techniques for main memory database systems. In *Proc. of the 1984 ACM SIGMOD Internat. Conf. on Management of data*, pages 1–8. ACM Press, 1984.

David J. DeWitt, Philippe Futtersack, David Maier, and Fernando Vélez. A study of three alternative workstation-server architectures for object oriented database systems. In *VLDB 1990, Proc. of the 16th Internat. Conf. on Very Large Data Bases*, pages 107–121, 1990.

Kapali P. Eswaran, Jim Gray, Raymond A. Lorie, and Irving L. Traiger. On the notions of consistency and predicate locks. Technical report, IBM Research Laboratory, San Jose, Nov. 1974.

Kapali P. Eswaran, Jim Gray, Raymond A. Lorie, and Irving L. Traiger. The notions of consistency and predicate locks in a database system. *Commun. ACM*, 19(11):624–633, 1976.

Georgios Evangelidis, David B. Lomet, and Betty Salzberg. The hB-Pi-tree: A multi-attribute index supporting concurrency, recovery and node consolidation. *VLDB J.*, 6(1):1–25, 1997.

Alan Fekete, Dimitrios Liarokapis, Elizabeth J. O'Neil, Patrick E. O'Neil, and Dennis Shasha. Making snapshot isolation serializable. *ACM Trans. Database Syst.*, 30(2):492–528, 2005.

Michael J. Franklin, Michael J. Zwilling, C. K. Tan, Michael J. Carey, and David J. DeWitt. Crash recovery in client-server EXODUS. In *Proc. of the 1992 ACM SIGMOD Internat. Conf. on Management of Data*, pages 165–174. ACM Press, 1992.

Michael J. Franklin, Björn Thór Jónsson, and Donald Kossmann. Performance tradeoffs for client-server query processing. In *Proc. of the 1996 ACM SIGMOD Internat. Conf. on Management of Data*, pages 149–160. ACM Press, 1996.

Michael J. Franklin, Michael J. Carey, and Miron Livny. Transactional client-server cache consistency: Alternatives and performance. *ACM Trans. Database Syst.*, 22(3):315–363, 1997.

Ada Wai-Chee Fu and Tiko Kameda. Concurrency control of nested transactions accessing B-trees. In *Proc. of the Eighth ACM SIGACT-SIGMOD-SIGART Symp. on Principles of Database Systems*, pages 270–285. ACM Press, 1989.

Seth Gilbert and Nancy A. Lynch. Brewer's conjecture and the feasibility of consistent, available, partition-tolerant web services. *SIGACT News*, 33(2), 2002.

Goetz Graefe. Sorting and indexing with partitioned B-trees. In *CIDR 2003, Proc. of the First Biennial Conf. on Innovative Data Systems Research*, 2003a.

Goetz Graefe. Partitioned B-trees—a user's guide. In *BTW 2003, Datenbanksysteme für Business, Technologie und Web, Tagungsband der 10. BTW-Konferenz*, pages 668–671, 2003b.

Goetz Graefe. Write-optimized B-trees. In *VLDB 2004, Proc. of the 30th Internat. Conf. on Very Large Data Bases*, pages 672–683. Morgan Kaufmann, 2004.

Goetz Graefe. A survey of B-tree locking techniques. *ACM Trans. Database Syst.*, 35(3):16:1–26, 2010.

Goetz Graefe. Modern B-tree techniques. *Foundations and Trends in Databases*, 3(4):203–402, 2011.

Goetz Graefe. A survey of B-tree logging and recovery techniques. *ACM Trans. Database Syst.*, 37(1):1:1–35, 2012.

Goetz Graefe, Hideaki Kimura, and Harumi A. Kuno. Foster B-trees. *ACM Trans. Database Syst.*, 37(3):17:1–29, 2012.

Jim Gray. Notes on database operating systems. In *Operating Systems: An Advanced Course*, volume 60 of *Lecture Notes in Computer Science*, pages 393–481. Springer, 1978.

Jim Gray. A transaction model. In *Automata, Languages and Programming, Proc. of the 7th Colloquium*, volume 85 of *Lecture Notes in Computer Science*, pages 282–298. Springer, 1980.

Jim Gray. The transaction concept: Virtues and limitations. In *VLDB 1981, Proc. of the 7th Internat. Conf. on Very Large Data Bases*, pages 144–154. IEEE Computer Society, 1981.

Jim Gray and Leslie Lamport. Consensus on transaction commit. *ACM Trans. Database Syst.*, 31 (1):133–160, 2006.

Jim Gray and Andreas Reuter. *Transaction Processing: Concepts and Techniques*. Morgan Kaufmann, 1993.

Jim Gray, Raymond A. Lorie, Gianfranco R. Putzolu, and Irving L. Traiger. Granularity of locks in a large shared data base. In *VLDB 1975, Proc. of the Internat. Conf. on Very Large Data Bases*, pages 428–451. ACM, 1975.

Jim Gray, Raymond A. Lorie, Gianfranco R. Putzolu, and Irving L. Traiger. Granularity of locks and degrees of consistency in a shared data base. In *IFIP Working Conf. on Modelling in Data Base Management Systems*, pages 365–394, 1976.

Jim Gray, Paul R. McJones, Mike W. Blasgen, Bruce G. Lindsay, Raymond A. Lorie, Thomas G. Price, Gianfranco R. Putzolu, and Irving L. Traiger. The recovery manager of the System R database manager. *ACM Comput. Surv.*, 13(2):223–243, 1981.

Jim Gray, Pat Helland, Patrick E. O'Neil, and Dennis Shasha. The dangers of replication and a solution. In *Proc. of the 1996 ACM SIGMOD Internat. Conf. on Management of Data*, pages 173–182. ACM Press, 1996.

Antonin Guttman. R-trees: a dynamic index structure for spatial searching. In *Proc. of the 1984 ACM SIGMOD Internat. Conf. on Management of data*, pages 47–57. ACM Press, 1984.

Tuukka Haapasalo. *Accessing Multiversion Data in Database Transactions*. PhD thesis, Department of Computer Science and Engineering, Aalto University School of Science and Technology, Espoo, Finland, 2010. http://lib.tkk.fi/Diss/2010/isbn9789526033600/.

Tuukka Haapasalo, Ibrahim Jaluta, Bernhard Seeger, Seppo Sippu, and Eljas Soisalon-Soininen. Transactions on the multiversion B$^+$-tree. In *EDBT 2009, Proc. of the 12th Internat. Conf. on Extending Database Technology*, volume 360 of *ACM International Conference Proceeding Series*, pages 1064–1075. ACM Press, 2009a.

Tuukka Haapasalo, Ibrahim Jaluta, Seppo Sippu, and Eljas Soisalon-Soininen. Concurrent updating transactions on versioned data. In *IDEAS 2009, Proc. of the Internat. Database Engineering and Applications Symp.*, pages 77–87. ACM Press, 2009b.

Tuukka Haapasalo, Ibrahim Jaluta, Seppo Sippu, and Eljas Soisalon-Soininen. On the recovery of R-trees. *IEEE Trans. Knowl. Data Eng.*, 25(1):145–157, 2013.

Ibrahim Jaluta. *B-Tree Concurrency Control and Recovery in a Client-Server Database Management System*. PhD thesis, Department of Computer Science and Engineering, Helsinki University of Technology, Espoo, Finland, 2002. http://lib.tkk.fi/Diss/2002/isbn9512257068/.

Ibrahim Jaluta and Dibyendu Majumda. Efficient space management for B-tree structure-modification operations. In *ICTTA'06, Proc. of the 2nd Internat. Conf. on Information and Communication Technologies*, pages 2909–2912, 2006.

Ibrahim Jaluta, Seppo Sippu, and Eljas Soisalon-Soininen. Recoverable B$^+$-trees in centralized database management systems. *Internat. J. of Applied Science & Computations*, 10(3):160–181, 2003.

Ibrahim Jaluta, Seppo Sippu, and Eljas Soisalon-Soininen. Concurrency control and recovery for balanced B-link trees. *VLDB J.*, 14(2):257–277, 2005.

Ibrahim Jaluta, Seppo Sippu, and Eljas Soisalon-Soininen. B-tree concurrency control and recovery in page-server database systems. *ACM Trans. Database Syst.*, 31(1):82–132, 2006.

Christian S. Jensen and David B. Lomet. Transaction timestamping in (temporal) databases. In *VLDB 2001, Proc. of 27th Internat. Conf. on Very Large Data Bases*, pages 441–450. Morgan Kaufmann, 2001.

Christopher M. Jermaine, Edward Omiecinski, and Wai Gen Yee. The partitioned exponential file for database storage management. *VLDB J.*, 16(4):417–437, 2007.

Ricardo Jiménez-Peris, Marta Patiño-Martínez, Gustavo Alonso, and Bettina Kemme. Are quorums an alternative for data replication? *ACM Trans. Database Syst.*, 28(3):257–294, 2003.

Ryan Johnson, Ippokratis Pandis, Radu Stoica, Manos Athanassoulis, and Anastasia Ailamaki. Aether: A scalable approach to logging. *Proc. of the VLDB Endowment*, 3(1):681–692, 2010.

Ryan Johnson, Ippokratis Pandis, Radu Stoica, Manos Athanassoulis, and Anastasia Ailamaki. Scalability of write-ahead logging on multicore and multisocket hardware. *VLDB J.*, 21(2): 239–263, 2012.

Theodore Johnson and Dennis Shasha. B-trees with inserts and deletes: Why free-at-empty is better than merge-at-half. *J. Comput. Syst. Sci.*, 47(1):45–76, 1993.

Martin V. Jørgensen, René B. Rasmussen, Simonas Šaltenis, and Carsten Schjønning. Fb-tree: a B$^+$-tree for flash-based SSDs. In *IDEAS 2011, Proc. of the 15th Internat. Database Engineering and Applications Symp.*, pages 34–42. ACM, 2011.

Michael Kifer, Arthur Bernstein, and Philip M. Lewis. *Database Systems: An Application-Oriented Approach. Second Edition*. Pearson Addison-Wesley, 2006.

Donald E. Knuth. *The Art of Computer Programming, Volume III: Sorting and Searching*. Addison-Wesley, 1973.

Marcel Kornacker, C. Mohan, and Joseph M. Hellerstein. Concurrency and recovery in generalized search trees. In *Proc. of the 1997 ACM SIGMOD Internat. Conf. on Management of Data*, pages 62–72. ACM Press, 1997.

Yat-Sang Kwong and Derick Wood. A new method for concurrency in B-trees. *IEEE Trans. Software Eng.*, 8(3):211–222, 1982.

Leslie Lamport. The part-time parliament. *ACM Trans. Database Syst.*, 16(2):133–169, 1998.

Butler W. Lampson and Howard E. Sturgis. Crash recovery in a distributed data storage system. Technical report, Computer Science Laboratory, Xerox, Palo Alto Research Center, 1976.

Vladimir Lanin and Dennis Shasha. A symmetric concurrent b-tree algorithm. In *Proc. of the Fall Joint Computer Conference*, pages 380–389. IEEE Computer Society, 1986.

Philip L. Lehman and S. Bing Yao. Efficient locking for concurrent operations on b-trees. *ACM Trans. Database Syst.*, 6(4):650–670, 1981.

Justin J. Levandoski, David B. Lomet, and Sudipta Sengupta. The Bw-tree: A B-tree for new hardware platforms. In *ICDE 2013, Proc. of the 29th IEEE Internat. Conf. on Data Engineering*, pages 302–313. IEEE Computer Society, 2013.

Yinan Li, Bingsheng He, Qiong Luo, and Ke Yi. Tree indexing on flash disks. In *ICDE 2009, Proc. of the 25th Internat. Conf. on Data Engineering*, pages 1303–1306. IEEE Computer Society, 2009.

Yinan Li, Bingsheng He, Jun Yang, Qiong Luo, and Ke Yi. Tree indexing on solid state drives. *Proc. of the VLDB Endowment*, 3(1):1195–1206, 2010.

Timo Lilja, Riku Saikkonen, Seppo Sippu, and Eljas Soisalon-Soininen. Online bulk deletion. In *ICDE 2007, Proc. of the 23rd IEEE Internat. Conf. on Data Engineering*, pages 956–965. IEEE Computer Society, 2007.

Yi Lin, Bettina Kemme, Ricardo Jiménez-Peris, Marta Patiño-Martínez, and José Enrique Armendáriz-Iñigo. Snapshot isolation and integrity constraints in replicated databases. *ACM Trans. Database Syst.*, 34(2), 2009.

David Lomet. Advanced recovery techniques in practice. In Vijay Kumar and Meichun Hsu, editors, *Recovery Mechanisms in Database Systems*, pages 697–710. Prentice Hall, 1998.

David B. Lomet. MLR: A recovery method for multi-level systems. In *Proc. of the 1992 ACM SIGMOD Internat. Conf. on Management of Data*, pages 185–194. ACM Press, 1992.

David B. Lomet. Key range locking strategies for improved concurrency. In *VLDB 1993, Proc. of the 19th Internat. Conf. on Very Large Data Bases*, pages 655–664. Morgan Kaufmann, 1993.

David B. Lomet. The evolution of effective B-tree: Page organization and techniques: A personal account. *SIGMOD Record*, 30(3):64–69, 2001.

David B. Lomet. Simple, robust and highly concurrent B-trees with node deletion. In *ICDE 2004, Proc. of the 20th Internat. Conf. on Data Engineering*, pages 18–28. IEEE Computer Society, 2004.

David B. Lomet and Feifei Li. Improving transaction-time DBMS performance and functionality. In *ICDE 2009, Proc. of the 25th Internat. Conf. on Data Engineering*, pages 581–591. IEEE Computer Society, 2009.

David B. Lomet and Betty Salzberg. Access methods for multiversion data. In *Proc. of the 1989 ACM SIGMOD Internat. Conf. on Management of Data*, pages 315–324. ACM Press, 1989.

David B. Lomet and Betty Salzberg. Access method concurrency with recovery. In *Proc. of the 1992 ACM SIGMOD Internat. Conf. on Management of Data*, pages 351–360. ACM Press, 1992.

David B. Lomet and Betty Salzberg. Exploiting a history database for backup. In *VLDB 1993, Proc. of the 19th Internat. Conf. on Very Large Data Bases*, pages 380–390. Morgan Kaufmann, 1993.

David B. Lomet and Betty Salzberg. Concurrency and recovery for index trees. *VLDB J.*, 6(3): 224–240, 1997.

David B. Lomet, Roger S. Barga, Mohamed F. Mokbel, German Shegalov, Rui Wang, and Yunyue Zhu. Immortal DB: Transaction time support for SQL server. In *Proc. of the 2005 ACM SIGMOD Internat. Conf. on Management of Data*, pages 939–941. ACM Press, 2005a.

David B. Lomet, Richard T. Snodgrass, and Christian S. Jensen. Using the lock manager to choose timestamps. In *IDEAS 2005, Proc. of the 9th Internat. Database Engineering and Applications Symp.*, pages 357–368. IEEE Computer Society, 2005b.

David B. Lomet, Roger S. Barga, Mohamed F. Mokbel, German Shegalov, Rui Wang, and Yunyue Zhu. Transaction time support inside a database engine. In *ICDE 2006, Proc. of the 22nd Internat. Conf. on Data Engineering*, page 35. IEEE Computer Society, 2006.

David B. Lomet, Mingsheng Hong, Rimma V. Nehme, and Rui Zhang. Transaction time indexing with version compression. *Proc. of the VLDB Endowment*, 1(1), 2008.

Raymond A. Lorie. Physical integrity in a large segmented database. *ACM Trans. Database Syst.*, 2(1):91–104, 1977.

C. Mohan. ARIES/KVL: A key-value locking method for concurrency control of multiaction transactions operating on B-Tree indexes. In *VLDB 1990, Proc. of the 16th Internat. Conf. on Very Large Data Bases*, pages 392–405. Morgan Kaufmann, 1990a.

C. Mohan. Commit_LSN: A novel and simple method for reducing locking and latching in transaction processing systems. In *VLDB 1990, Proc. of the 16th Internat. Conf. on Very Large Data Bases*, pages 406–418. Morgan Kaufmann, 1990b.

C. Mohan. ARIES/LHS: A concurrency control and recovery method using write-ahead logging for linear hashing with separators. In *ICDE 1993, Proc. of the 9th Internat. IEEE Conf. on Data Engineering*, pages 243–252. IEEE Computer Society, 1993a.

C. Mohan. A cost-effective method for providing improved data availability during DBMS restart recovery after a failure. In *VLDB 1993, Proc. of the 19th Internat. Conf. on Very Large Data Bases*, pages 368–379. Morgan Kaufmann, 1993b.

C. Mohan. Concurrency control and recovery methods for $B^+$-Tree indexes: ARIES/KVL and ARIES/IM. In Vijay Kumar, editor, *Performance of Concurrency Control Mechanisms in Centralized Database Systems*, pages 248–306. Prentice Hall, 1996a.

C. Mohan. Commit_LSN: A novel and simple method for reducing locking and latching in transaction processing systems. In Vijay Kumar, editor, *Performance of Concurrency Control Mechanisms in Centralized Database Systems*, pages 307–335. Prentice Hall, 1996b.

C. Mohan. Repeating history beyond ARIES. In *VLDB 1999, Proc. of the 25th Internat. Conf. on Very Large Data Bases*, pages 1–17. Morgan Kaufmann, 1999.

C. Mohan. An efficient method for performing record deletions and updates using index scans. In *VLDB 2002, Proc. of the 28th Internat. Conf. on Very Large Data Bases*, pages 940–949. Morgan Kaufmann, 2002.

C. Mohan. History repeats itself: Sensible and NonsenSQL aspects of the NoSQL hoopla. In *EDBT 2013, Proc. of the 16th Internat. Conf. on Database Technology*, pages 11–16. ACM Press, 2013.

C. Mohan and Frank E. Levine. ARIES/IM: An efficient and high concurrency index management method using write-ahead logging. In *Proc. of the 1992 ACM SIGMOD Internat. Conf. on Management of Data*, pages 371–380. ACM Press, 1992.

C. Mohan and Inderpal Narang. Recovery and coherency-control protocols for fast intersystem page transfer and fine-granularity locking in a shared disks transaction environment. In *VLDB 1991, Proc. of the 17th Internat. Conf. on Very Large Data Bases*, pages 193–207. Morgan Kaufmann, 1991.

C. Mohan and Inderpal Narang. Algorithms for creating indexes for very large tables without quiescing updates. In *Proc. of the 1992 ACM SIGMOD Internat. Conf. on Management of Data*, pages 361–370. ACM Press, 1992.

C. Mohan and Inderpal Narang. ARIES/CSA: A method for database recovery in client-server architectures. In *Proc. of the 1994 ACM SIGMOD Internat. Conf. on Management of Data*, pages 55–66. ACM Press, 1994.

C. Mohan and Hamid Pirahesh. ARIES-RRH: Restricted repeating of history in the ARIES transaction recovery method. In *ICDE 1991, Proc. of the 7th Internat. IEEE Conf. on Data Engineering*, pages 718–727. IEEE Computer Society, 1991.

C. Mohan, Bruce G. Lindsay, and Ron Obermarck. Transaction management in the R* distributed database management system. *ACM Trans. Database Syst.*, 11(4):378–396, 1986.

C. Mohan, Donald J. Haderle, Yun Wang, and Josephine M. Cheng. Single table access using multiple indexes: Optimization, execution, and concurrency control techniques. In *Advances in Database Technology—EDBT'90, Proc. of the Internat. Conf. on Extending Database Technology*, volume 416 of *Lecture Notes in Computer Science*, pages 29–43. Springer, 1990.

C. Mohan, Donald J. Haderle, Bruce G. Lindsay, Hamid Pirahesh, and Peter M. Schwarz. ARIES: A transaction recovery method supporting fine-granularity locking and partial rollbacks using write-ahead logging. *ACM Trans. Database Syst.*, 17(1):94–162, 1992a.

C. Mohan, Hamid Pirahesh, and Raymond A. Lorie. Efficient and flexible methods for transient versioning of records to avoid locking by read-only transactions. In *Proc. of the 1992 ACM SIGMOD Internat. Conf. on Management of Data*, pages 124–133. ACM Press, 1992b.

C. Mohan, Kent Treiber, and Ron Obermarck. Algorithms for the management of remote backup data bases for disaster recovery. In *ICDE 1993, Proc. of the 9th Internat. IEEE Conf. on Data Engineering*, pages 511–518. IEEE Computer Society, 1993.

Yehudit Mond and Yoav Raz. Concurrency control in $B^+$-trees databases using preparatory operations. In *VLDB 1985, Proc. of the 11th Internat. Conf. on Very Large Data Bases*, pages 331–334. Morgan Kaufmann, 1985.

Patrick E. O'Neil, Edward Cheng, Dieter Gawlick, and Elizabeth J. O'Neil. The log-structured merge-tree (LSM-tree). *Acta Inf.*, 33(4):351–385, 1996.

M. Tamer Özsu and Patrick Valduriez. *Principles of Distributed Database Systems. Third Edition.* Springer, 2011.

Christos H. Papadimitriou. *The Theory of Database Concurrency Control.* Computer Science Press, 1986.

Stacy Patterson, Aaron J. Elmore, Faisal Nawab, Divyakant Agrawal, and Amr El Abbadi. Serializability, not serial: Concurrency control and availability in multi-datacenter datastores. *Proc. of the VLDB Endowment*, 5(11):1459–1470, 2012.

Kerttu Pollari-Malmi, Eljas Soisalon-Soininen, and Tatu Ylönen. Concurrency control in B-trees with batch updates. *IEEE Trans. Knowl. Data Eng.*, 8(6):975–984, 1996.

Dan R. K. Ports and Kevin Grittner. Serializable snapshot isolation in PostgreSQL. *Proc. of the VLDB Endowment*, 5(12):1850–1861, 2012.

Saeed K. Rahimi and Frank S. Haug. *Distributed Database Management Systems. A Practical Approach.* Wiley, 2010.

Mendel Rosenblum and John K. Ousterhout. The design and implementation of a log-structured file system. *ACM Trans. Comput. Syst.*, 10(1):26–52, 1992.

Kurt Rothermel and C. Mohan. ARIES/NT: A recovery method based on write-ahead logging for nested transactions. In *VLDB 1989, Proc. of the 15th Internat. Conf. on Very Large Data Bases*, pages 337–346. Morgan Kaufmann, 1989.

James B. Rothnie, Philip A. Bernstein, Stephen Fox, Nathan Goodman, Michael Hammer, Terry A. Landers, Christopher L. Reeve, David W. Shipman, and Eugene Wong. Introduction to a system for distributed databases (SDD-1). *ACM Trans. Database Syst.*, 5(1):1–17, 1980.

Yehoshua Sagiv. Concurrent operations on B*-trees with overtaking. *J. Comput. Syst. Sci.*, 33(2): 275–296, 1986.

Betty Salzberg. Timestamping after commit. In *PDIS 1994, Proc. of the Third Internat. Conf. on Parallel and Distributed Information Systems*, pages 160–167. IEEE Computer Society, 1994.

Betty Salzberg and Allyn Dimock. Principles of transaction-based on-line reorganization. In *VLDB 1992, Proc. of the 18th Internat. Conf. on Very Large Data Bases*, pages 511–520. Morgan Kaufmann, 1992.

Behrokh Samadi. B-trees in a system with multiple users. *Inf. Process. Lett.*, 5(4):107–112, 1976.

Hans-Jörg Schek, Gerhard Weikum, and Haiyan Ye. Towards a unified theory of concurrency control and recovery. In *Proc. of the 12th ACM SIGACT-SIGMOD-SIGART Symp. on Principles of Database Systems*, pages 300–311. ACM Press, 1993.

Russell Sears and Raghu Ramakrishnan. bLSM: a general purpose log structured merge tree. In *Proc. of the 2012 ACM SIGMOD Internat. Conf. on Management of Data*, pages 217–228. ACM Press, 2012.

Russell Sears, Mark Callaghan, and Eric A. Brewer. *Rose*: Compressed, log-structured replication. *Proc. of the VLDB Endowment*, 1(1):526–537, 2008.

Dennis G. Severance and Guy M. Lohman. Differential files: Their application to the maintenance of large databases. *ACM Trans. Database Syst.*, 1(3):256–267, 1976.

Avi Silberschatz, Henry F. Korth, and S. Sudarshan. *Database System Concepts. Fifth Edition.* McGraw-Hill, 2006.

Seppo Sippu and Eljas Soisalon-Soininen. A theory of transactions on recoverable search trees. In *Database Theory—ICDT 2001, Proc. of the 8th Internat. Conf.*, volume 1973 of *Lecture Notes in Computer Science*, pages 83–98. Springer, 2001.

Gary H. Sockut and Balakrishna R. Iyer. A survey on online reorganization in IBM products and research. *IEEE Data Eng. Bull.*, 19(2):4–11, 1996.

Gary H. Sockut and Balakrishna R. Iyer. Online reorganization of databases. *ACM Comput. Surv.*, 41(3):14:1–136, 2009.

Gary H. Sockut, Thomas A. Beavin, and Chung-C. Chang. A method for on-line reorganization of a database. *IBM Systems Journal*, 36(3):411–436, 1997.

Eljas Soisalon-Soininen and Tatu Ylönen. Partial strictness in two-phase locking. In *Database Theory—ICDT 1995, Proc. of the 5th Internat. Conf.*, volume 893 of *Lecture Notes in Computer Science*, pages 139–147. Springer, 1995.

Jagannathan Srinivasan, Souripriya Das, Chuck Freiwald, Eugene Inseok Chong, Mahesh Jagannath, Aravind Yalamanchi, Ramkumar Krishnan, Anh-Tuan Tran, Samuel DeFazio, and Jayanta Banerjee. Oracle8i index-organized table and its application to new domains. In *VLDB 2000, Proc. of the 26th Internat. Conf. on Very Large Data Bases*, pages 285–296. Morgan Kaufmann, 2000.

V. Srinivasan and Michael J. Carey. On-line index construction algorithms. In *Proc. of the High Performance Transaction Systems Workshop*, 1991.

V. Srinivasan and Michael J. Carey. Performance of on-line index construction algorithms. In *Advances in Database Technology—EDBT'92, Proc. of the 3rd Internat. Conf. on Extending Database Technology*, volume 580 of *Lecture Notes in Computer Science*, pages 293–309. Springer, 1992.

V. Srinivasan and Michael J. Carey. Performance of B+ tree concurrency control algorithms. *VLDB J.*, 2(4):361–406, 1993.

Michael Stonebraker. Operating system support for database management. *Commun. ACM*, 24(7): 412–418, 1981.

Michael Stonebraker, editor. *The INGRES Papers: Anatomy of a Relational Database System.* Addison-Wesley, 1986.

Michael Stonebraker. The design of the POSTGRES storage system. In *VLDB 1987, Proc. of the 13th Internat. Conf. on Very Large Data Bases*, pages 289–300. Morgan Kaufmann, 1987.

Michael Stonebraker. The case for partial indexes. *SIGMOD Record*, 18(4):4–11, 1989.

Michael Stonebraker. In search of database consistency. *Commun. ACM*, 53(10):8–9, 2010a.

Michael Stonebraker. Sql databases v. NoSQL databases. *Commun. ACM*, 53(4):10–11, 2010b.

Michael Stonebraker and Rick Cattell. 10 rules for scalable performance in 'simple operation' datastores. *Commun. ACM*, 54(6):72–80, 2011.

Michael Stonebraker and Lawrence A. Rowe. The design of Postgres. In *Proc. of the 1986 ACM SIGMOD Internat. Conf. on Management of Data*, pages 340–355. ACM Press, 1986.

Michael Stonebraker, Eugene Wong, Peter Kreps, and Gerald Held. The design and implementation of INGRES. *ACM Trans. Database Syst.*, 1(3):189–222, 1976.

Michael Stonebraker, Lawrence A. Rowe, and Michael Hirohama. The implementation of Postgres. *IEEE Trans. Knowl. Data Eng.*, 2(1):125–142, 1990.

Michael Stonebraker, Daniel J. Abadi, David J. DeWitt, Samuel Madden, Erik Paulson, Andrew Pavlo, and Alexander Rasin. MapReduce and parallel DBMSs: Friends or foes? *Commun. ACM*, 53(1):64–71, 2010.

Xiaowei Sun, Rui Wang, Betty Salzberg, and Chendong Zou. Online B-tree merging. In *Proc. of the 205 ACM SIGMOD Internat. Conf. on Management of Data*, pages 335–346. ACM Press, 2005.

Irving L. Traiger. Trends in system aspects of database management. In *Proc. of the 2nd Internat. Conf. on Databases*, pages 1–21, 1983.

Hoang Tam Vo, Sheng Wang, Divyakant Agrawal, Gang Chen, and Beng Chin Ooi. Logbase: A scalable log-structured database system in the cloud. *Proc. of the VLDB Endowment*, 5(10): 1004–1015, 2012.

Gerhard Weikum. Principles and realization strategies of multilevel transaction management. *ACM Trans. Database Syst.*, 16(1):132–180, 1991.

Gerhard Weikum and Gottfried Vossen. *Transactional Information Systems. Theory, Algorithms, and the Practice of Concurrency Control and Recovery*. Morgan Kaufmann, 2002.

Chin-Hsien Wu, Tei-Wei Kuo, and Li-Ping Chang. An efficient b-tree layer implementation for flash-memory storage systems. *ACM Trans. Embedded Comput. Syst.*, 6(3), 2007.

Markos Zaharioudakis, Michael J. Carey, and Michael J. Franklin. Adaptive, fine-grained sharing in a client-server OODBMS: A callback-based approach. *ACM Trans. Database Syst.*, 22(4): 570–627, 1997.

Chendong Zou and Betty Salzberg. On-line reorganization of sparsely-populated B$^+$-trees. In *Proc. of the 1996 ACM SIGMOD Internat. Conf. on Management of Data*, pages 115–124. ACM Press, 1996a.

Chendong Zou and Betty Salzberg. Towards efficient online database reorganization. *IEEE Data Eng. Bull.*, 19(2):33–40, 1996b.

# Index

© Springer International Publishing Switzerland 2014
S. Sippu, E. Soisalon-Soininen, *Transaction Processing*, Data-Centric Systems and Applications, DOI 10.1007/978-3-319-12292-2

Printed in the United States
By Bookmasters